Edible Seaweeds of the World

Edible Seaweeds of the World

Leonel Pereira

Department of Life Sciences
MARE – Marine and Environmental
Sciences Centre/IMAR-Institute of Marine Research
University of Coimbra
Coimbra
Portugal

CRC Press
Taylor & Francis Group
Boca Raton London New York

CRC Press is an imprint of the
Taylor & Francis Group, an **informa** business

A SCIENCE PUBLISHERS BOOK

Cover illustrations provided by the author of the book, Leonel Pereira

CRC Press
Taylor & Francis Group
6000 Broken Sound Parkway NW, Suite 300
Boca Raton, FL 33487-2742

First issued in paperback 2021

Version Date: 20151110

ISBN-13: 978-0-367-78323-5 (pbk)
ISBN-13: 978-1-4987-3047-1 (hbk)

Library of Congress Cataloging-in-Publication Data

Names: Pereira, Leonel.
Title: Edible seaweeds of the world / Leonel Pereira.
Description: Boca Raton : Taylor & Francis, 2016. | "A CRC title." | Includes bibliographical references and species index.
Identifiers: LCCN 2015042674 | ISBN 9781498730471 (hardcover : alk. paper)
Subjects: LCSH: Marine algae as food. | Marine plants--Nutrition. | Plants, Edible.
Classification: LCC TX402 .P47 2016 | DDC 641.3/03--dc23
LC record available at http://lccn.loc.gov/2015042674

Visit the Taylor & Francis Web site at
http://www.taylorandfrancis.com

and the CRC Press Web site at
http://www.crcpress.com

Foreword

If you were not (fully) aware of the culinary and functional food value of seaweeds, then this book is for you. If you are not aware … then where have you been? This is the time to get some (more) seaweed in your diet! Start now and use this book as one of your guides.

Leonel's book is much needed for the interested general public, phycologist (those people that study seaweeds for a living or out of interest), marine resource scientist and practitioners, students and academics.

Professor Pereira has taken on a grand task—to compile a useful repository of information of all edible seaweeds on a global basis.

The text is clear and informative. The tome is very well illustrated using some older drawings as well as photographs where appropriate. It is such a pleasure to get a title like this in the hands and not on a computer screen. One can attack the book by reading it cover to cover, or one can open up at any given page and find something interesting to dip in and out of. Either way it is a real smorgasbord of (edible) phycological information.

All of the "common gang" are in there... wakame, konbu (kombu) and the now ubiquitous nori. There must be few people consuming sushi and miso soup who are not aware of their seaweed content? However, this book contains so much more! Leonel has gone far and wide in his collation of information—the book has a very rich bibliography as a source of further and more detailed information.

Hardly a day goes by without some new, mainstream culinary revelation about how good seaweeds are for health, we even now know that pan or flash fried dulse (*Palmaria* spp., e.g., *P. palmata* or *P. mollis*) tastes "just like bacon"—a vegan delight!

The time for increasing the amount of seaweed we consume daily is now! There is more than enough evidence to suggest that not only does it taste good, but by golly it does you good.

Professor Pereira has provided a considerable service is bringing this information into sharp focus. He caters for everyone with illustration, description, distribution and common uses. There is an excellent glossary of terms and also suggestions for recipes. This book has it all—it is bound to please. Get one for yourself and buy a copy for that friend of yours that was always waiting to eat seaweed but had not yet tried. The "proof is in the pudding"... enjoy!

Alan T. Critchley
Phycologist

Preface

Seaweeds have inhabited the oceans for over 2 billion; they are used as food by Asian people since the seventeenth century. Today, seaweeds are used in many countries for various purposes: directly as food, especially in sushi and other traditional recipes, as a source of phycocolloids, for extraction of compounds with antiviral, antibacterial or antitumor properties and as biofertilizers. About twenty-four million tons of seaweeds are harvested annually worldwide. Of the various species known, less than 20 account for 90% of the biomass exploited commercially. In addition to the multiple applications mentioned above, which have greatly expanded over the last 50 years, largely on account of the phycocolloids (Agar, Carrageenans and Alginates) which are used as thickeners in the food industry, in soups, preserved meat, dairy products and pastries. We notice an upward trend in the consumption including in North America and in Europe.

The criteria for the search and selection of edible species with commercial value are based primarily on the texture and taste of each alga (rather than nutritional value) and, secondly on, the creation of new dictary habits in the West, i.e., the calorific value or health benefits.

The question that arises is simple—what benefits does seaweed bring to the human diet, in terms of food, culinary or dietary? The answer is that—they are exactly opposite to the concept of fast food: seaweeds are a natural food which grows in the wild and is abundant (and with a growth rate capable of sustaining intensive cropping), and capable of providing a high nutritional value and, at the same time, is a low calorie food. Blood glucose levels are not affected by seaweed polysaccharides because of the high fiber content. Algae appear to be the best way to fix not only the lack of food for consumption, but to counter nutritional deficiencies worldwide (in developed, emerging and/or developing countries) since it contains a wide range of essential constituents—minerals (iron and calcium), proteins (with all essential amino acids), vitamins and fibers (Pereira 2011) which are absolutely necessary nutrients for primary metabolism.

In fact algae constitute a huge potential source of food and thus, the importance of this book which provides the most complete listing ever made of edible species worldwide. This book describes about 345 red algae (Rhodophyta), 125 green algae (Chlorophyta) and 195 brown algae

(Phaeophyceae). The book contains 34 color illustrations of the main edible species as well as a thorough review of the published literature from the earliest records to the present.

Leonel Pereira

Contents

CHAPTER 1

Introduction

1.1 Role of Algae in Nature

Algae are uni- or multi-cellular organisms which live in water or in wet locations and have chlorophyll (organic pigment capable of absorbing and transforming energy from sunlight) and is therefore capable of performing photosynthesis, namely transforming light energy into chemical, capturing carbon dioxide (CO_2) to form complex organic compounds (along with water and mineral salts) and release oxygen (O_2); this organic synthesis process, proposes that the algae are the true "lungs" of the planet Earth, that cover extensive woodlands, such as the Amazon rainforest. So, algae are primary producers, i.e., they are able to produce oxygen and organic compounds that serve as food for other living beings, fundamental to food chains of all aquatic ecosystems (Pereira 2015a, Pereira and Correia 2015).

1.2 Main Taxonomic Groups of Marine Algae

Microalgae constitute the phytoplankton, thus living in surface water and suspended in the water column and restricted to the photic zone and eventually forming the basis of the marine food chain—are invertebrates such as crustaceans (copepods, which are part of zooplankton), mollusks (mussels), and vertebrates (such as the famous juvenile sardines, among others).

Macroalgae are macroscopic marine algae, some of which can reach up to several meters long (some stems of these algae can reach up to 65 m long) and whose stems exhibit a high degree of complexity and tissue organization. They too, as primary producers, are at the base of the marine food chain sustaining several benthic animal communities (invertebrates such as some sea urchins and/or gastropods, and vertebrates such as herbivorous fish) and which also eventually find refuge from predators.

The color of algae is no more than the visible expression of the combination of the different photosynthetic pigments in cells; for this reason, for over a century the distinction of the different phyla and seaweeds classes is made with the help of their coloring. The macroalgae have extremely varied colors, however but they all have chlorophyll. These pigments are located in small organelles, the chloroplasts, and are responsible for the green color of most plants (vascular and non-vascular).

Macroalgae (or Seaweeds) are aquatic photosynthetic organisms belonging to Eukarya (or Eukaryota) Domain, Plantae (green and red algae) and Chromista (brown algae) Kingdoms, respectively. Although the classification systems have evolved over the centuries, it is generally agreed that:

A. Green seaweeds are included in the Phylum Chlorophyta, and their pigments are chlorophylls a, b and carotenoids;
B. The red seaweeds belong to the Phylum Rhodophyta, and have photosynthetic pigments such as chlorophyll, carotenoids and some phycobilins;
C. The brown seaweeds belong to the phylum Heterokontophyta (or Ochrophyta) and all of them are grouped in Class Phaeophyceae; their pigments are chlorophyll a, c and carotenoids (dominated by fucoxanthin, responsible for their brownish color).

1.3 Importance of Algae for Mankind

Today, seaweed is used in many countries for very different purposes: directly as food, phycocolloids extraction, extraction of compounds with antiviral, antibacterial, or antitumor activity and as biofertilizers. More than 20 tons of seaweeds are harvested annually worldwide. The major producers are China and Japan, followed by America and Norway. France used to import Japanese seaweed in the 1970s but 10 years later it went on to produce algae for food and biological products users. Contrary to what happens in East Asia, the West is more interested in thickeners and gelling properties of hydrocolloids extracted from seaweeds: carrageenan, agar, and alginate (E407, E406, and E400, respectively).

1.3.1 The Seaweed and the Food Industry

Microalgae and macroalgae have been utilized by man for hundreds of years as food, fodder, remedies, and fertilizers. Ancient records show that people collected macroalgae for food as long as 500 B.C. in China and one thousand of years later in Europe. Microalgae such as *Arthrospira* (formerly

Spirulina, Cyanobacteria) have a history of human consumption in Mexico and Africa. In the 14th century the Aztecs harvested *Arthrospira* from Lake Texcoco (Farrar 1966) and used to make a sort of dry cake called "Tecuitlatl", and very likely the use of this cyanobacterium as food in Chad dates back to the same period, or even earlier, to the Kanem Empire (9th century A.D.).

People migrating from countries such as China, Japan, and Korea, but also from Indonesia and Malaysia, where algae have always been used as food, have brought this custom with them, so that today there are many more countries all over the world where the consumption of algae is not unusual, including Europe (Barsanti and Paolo 2014).

On the east and west coasts of the US and Canada, around Maine, New Brunswick, Nova Scotia and British Columbia, some companies have begun cultivating macroalgae onshore, in tanks, specifically for human consumption, and their markets are growing, both in those two countries and with exports to Japan. Ireland and northern Ireland are showing a renewed interest in macroalgae that were once a traditional part of the diet. In addition to direct consumption, agars and carrageenans extracted from red macroalgae and alginates from brown macroalgae and microalgae have been included in a remarkable array of prepared food products, serving mostly to modify viscosity or texture. Global utilization of macroalgae is on increase, and in terms of harvested biomass per year, macroalgae are among the most important cultivated marine organisms (Barsanti and Paolo 2014).

1.3.2 Phycocolloids

Colloids are extracted compounds that form colloidal solutions, an intermediate state between a solution and a suspension, and are used as thickeners, gelling agents, and stabilizers for suspensions and emulsions (see Table 1). Hydrocolloids are carbohydrates that when dissolved in water form viscous solutions. The phycocolloids are hydrocolloids extracted from algae and represent a growing industry, with more than one million tons of seaweeds extracted annually for hydrocolloid production (Ioannou and Roussis 2009, Pereira et al. 2009c, 2010b, Pereira 2011).

Many seaweeds produce hydrocolloids, associated with the cell wall and intercellular spaces. Members of the red algae (Rhodophyta) produce galactans (e.g., carrageenans and agars) and the brown algae (Ochrophyta, Phaeophyceae) produce uronates (alginates) (Pereira et al. 2009c, Bixler and Porse 2011, Pereira and van de Velde 2011, Pereira et al. 2013).

Sulfated galactans (e.g., agars and carrageenans) can be obtained from red algae, and alginates and other sulfated polysaccharides (e.g., laminaran and fucoidan) are obtained from brown algae. Phycocolloids are used in food industries as natural additives and have different European codes: E400 (alginic acid), E401 (sodium alginate), E402 (potassium alginate),

Table 1. Applications of seaweed phycocolloids (Adapted from van de Velde and De Ruiter 2002, Dhargalkar and Pereira 2005, Pereira 2004, 2008, Pereira and Ribeiro-Claro 2014).

Use	Phycocolloid	Function
Food additives		
Baked food	Agar Kappa, Iota, Lambda	Improving quality and controlling moisture
Beer and wine	Alginate Kappa	Promotes flocculation and sedimentation of suspended solids
Canned and processed meat	Alginate Kappa	Hold the liquid inside the meat and texturing
Cheese	Kappa	Texturing
Chocolate milk	Kappa, Lambda	Keep the cocoa in suspension
Cold preparation puddings	Kappa, Iota, Lambda	Thicken and gell
Condensed milk	Iota, Lambda	Emulsify
Dairy creams	Kappa, Iota	Stabilize the emulsion
Fillings for pies and cakes	Kappa	Give body and texture
Frozen fish	Alginate	Adhesion and moisture retention
Gelled water-based desserts	Kappa + Iota Kappa + Iota + CF	Gelling
Gums and sweets	Agar Iota	Gelling, texturing
Hot preparation flans	Kappa, Kappa + Iota	Gelling and improving taste
Jelly tarts	Kappa	Gelling
Juices	Agar Kappa, Lambda	Viscosity, emulsifier
Low calorie gelatins	Kappa + Iota	Gelling
Milk ice-cream	Kappa + GG, CF, X	Stabilize the emulsion and prevent ice crystals formation
Milkshakes	Lambda	Stabilize the emulsion
Salad dressings	Iota	Stabilize the suspension
Sauces and condiments	Agar Kappa	Thicken
Soymilk	Kappa + Iota	Stabilize the emulsion and improve taste
Cosmetics		
Shampoos	Alginate	Vitalization interface
Toothpaste	Carrageenan	Increase viscosity

Table 1. contd....

Table 1. contd.

Use	Phycocolloid	Function
Lotions	Alginate	Emulsification, elasticity and skin firmness
Lipstick	Alginate	Elasticity, viscosity
Medicinal and Pharmaceutical uses		
Dental mold	Alginate	Form retention
Laxatives	Alginate Carrageenan	Indigestibility and lubrication
Tablets	Alginate Carrageenan	Encapsulation
Metal poisoning	Carrageenan	Binds metal
HSV	Alginate	Inhibit virus
Industrial and Lab uses		
Paints	Alginate	Viscosity and suspension glazing
Textiles	Agar, Carrageenan	Sizing and glazing
Paper making	Alginate, Agar, Carrageenan	Viscosity and thickening
Analytical separation	Alginate Carrageenan	Gelling
Bacteriological media	Agar	Gelling
Electrophoresis gel	Agar, Carrageenan	Gelling

E403 (ammonium alginate), E404 (calcium alginate), E405 (propylene glycol alginate), E406 (agar), E407 (carrageenan), E407a (semi-refined carrageenan or "processed Eucheuma seaweed"), and E408 (Furcellaran) (Pereira et al. 2013). Agar, alginates and carrageenans are the ones with the highest economic and commercial significance, since these polysaccharides exhibit high molecular weights, high viscosity and excellent gelling, stabilizing and emulsifying properties. They are extracted in fairly large amounts from the algae. All these polysaccharides are water soluble and could be extracted with hot water or alkaline solution (Minghou 1990).

Agar

Agar is the Malay name for red algae, also called agar-agar and Kanten. It is an extract of red algae, sold in granular or powder form or as flakes or long strips. This colloid is widely used as a gelling agent and rarely eaten on its own (see recipe at the end of this book). In its pure form, agar is a tasteless and odorless polysaccharide; which normally contains proteins, vitamins and minerals, as the red algae from which it is derived (Pereira 2010b, Mouritsen 2013).

Kanten was discovered by accident. For more than a thousand years, the Japanese have eaten a dish called "Tokoroten", which is prepared for the boiling of red algae *Gelidium amansii* (in Japan called "Tengusa"), and then letting the mixture stiffen. At some point toward the end of the 17th century, leftover "tokoroten" was taken outside on a freezing cold day. When it was found later, it had become a dry, whitish solid. The result was giving the name Kanten (Shimamura 2010, Mouritsen 2013).

Agar is light in color, semi-transparent, and very brittle when dry. When soaked it absorbs water and, in contrast to gelatin (prepared from animal origin), it should be allowed to swell up completely before the water is warmed to a temperature above its melting point of 85°C. Agar can be used as a gelling agent as it cools (see Table 1) (Pereira 2011, Mouritsen 2013).

Alginate

"Alginate" is the term usually used for the salts of alginic acid, but it can also refer to all the derivatives of alginic acid and alginic acid itself; in some publications the term "algin" is used instead of alginate. Chemically, alginates are linear copolymers of β-D-mannuronic acid (M) and α-L-guluronic acid (G) (1-4)-linked residues, arranged either in heteropolymeric (MG) and/or homopolymeric (M or G) blocks (Larsen et al. 2003, Pereira et al. 2003, Leal et al. 2008).

Alginic acid was discovered in 1883 by E.C.C. Stanford, a British pharmacist who called it algin. Alginic acid is extracted as a mixed salt of sodium and/or potassium, calcium and magnesium. Since Stanford discovered algin, the name has been applied to a number of substances, e.g., alginic acid and all alginates, derived from alginic acid. The extraction process is based on the conversion of an insoluble mixture of alginic acid salts of the cell wall in a soluble salt (alginate) which is appropriate for the water extraction (Lobban et al. 1988, Lahaye 2001). Alginic acid is present in the cell walls of brown seaweeds, where it is partially responsible for their flexibility (McHugh 2003).

While any brown seaweed could be used as a source of alginate, the actual chemical structure of the alginate varies from one genus to another, and similar variability is found in the properties of the alginate that is extracted from the seaweed. Since the main applications of alginate are in thickening aqueous solutions and forming gels, its quality is judged on how well it performs in these uses (McHugh 2003).

Twenty five to 30 years ago almost all extraction of alginates took place in Europe, the USA, and Japan. The major change in the alginates industry over the last decade has been the emergence of producers in China in the 1980s. Initially, production was limited to low cost, low quality alginate for the internal, industrial markets produced from the locally cultivated *Saccharina japonica*. By the 1990s, Chinese producers were competing in western industrial markets to sell alginates, primarily based on low cost (Pereira 2011).

A high quality alginate forms strong gels and gives thick, aqueous solutions. A good raw material for alginate extraction should also give a high yield of alginate. Brown seaweeds that fulfill the above criteria are species of *Ascophyllum, Durvillaea, Ecklonia, Fucus, Laminaria, Lessonia, Macrocystis* and *Sargassum*. However, *Sargassum* is only used when nothing else is available: its alginate is usually borderline quality and the yield usually low (Pereira 2008).

The goal of the extraction process is to obtain dry, powdered, sodium alginate. The calcium and magnesium salts do not dissolve in water; the sodium salt does. The rationale behind the extraction of alginate from the seaweed is to convert all the alginate salts to the sodium salt, dissolve this in water, and remove the seaweed residue by filtration (McHugh 2003).

Water-in-oil emulsions such as mayonnaise and salad dressings are less likely to separate into their original oil and water phases if thickened with alginate. Sodium alginate is not useful when the emulsion is acidic, because insoluble alginic acid forms; for these applications Propylene Glycol Alginate (PGA) is used since this is stable in mild acid conditions. Alginate improves the texture, body and sheen of yoghurt, but PGA is also used in the stabilization of milk proteins under acidic conditions, as found in some yoghurts. Some fruit drinks have fruit pulp added and it is preferable to keep this in suspension; addition of sodium alginate, or PGA in acidic conditions, can prevent sedimentation of the pulp and to create foams. In chocolate milk, the cocoa can be kept in suspension by an alginate/phosphate mixture, although in this application it faces strong competition from carrageenan. Small amounts of alginate can thicken and stabilize whipped cream (Nussinovitch 1997, Onsoyen 1997).

Alginates have several commercial applications based on their thickening, gelling, emulsifier and stabilizing abilities (see Table 1). They are used in the food industry for improving the textural quality of numerous products such as salad dressing, ice-cream, beer, jelly and lactic drinks, but also in cosmetics, pharmaceuticals, textiles and painting industries (Murata et al. 2001, Kim and Lee 2008, Pereira 2011).

Carrageenan

For several hundred years, carrageenan has been used as a thickening and stabilizing agent in food in Europe and in Asia. In Europe, the use of carrageenan started more than 600 years ago in Ireland. In the village of Carraghen, on the south Irish coast, flans were made by cooking the so-called Irish moss (red seaweed species *Chondrus crispus*) in milk. The name carrageenin, as it was first called, was used for the first time in 1862 for the extract from *C. crispus* and was dedicated to this village (Tseng 1945). The name was later changed to carrageenan so as to comply with the "-an" suffix for the names of polysaccharides (Pereira et al. 2009c, Pereira 2011). The extraction method was first described by Smith in 1844 (van de Velde and de Ruiter 2002).

The Irish moss has been used in industry since the 19th century, in the clarification of beer (Therkelsen 1993). The industrial extraction of carrageenan started in 1930 in New-England, from *Chondrus crispus* and *Mastocarpus stellatus* thalli, for the preparation of chocolate milk. The interruption of agar imports during World War II, led to its replacement by carrageenan. This situation was the starting point of a booming industry (Ribier and Godineau 1984).

Fractionation of crude carrageenan extracts started in the early 1950s (Smith et al. 1955), resulting in the characterization of the different carrageenan types. A Greek prefix was introduced to identify the different carrageenans. In the same period, the molecular structure of carrageenans was determined (O'Neill 1955a,b). The structure of 3,6-anhydro-D-galactose in Kappa-carrageenan, as well as the type of linkages between galactose and anhydrogalactose rings, was determined.

Today, the industrial manufacture of carrageenan is no longer limited to extraction from *C. crispus*, and numerous red seaweed species (Gigartinales, Rhodophyta) are used. For a long period of time, these seaweeds have been harvested from naturally occurring populations. Seaweed farming started almost 200 years ago in Japan. Scientific information about the seaweed life cycles allowed artificial seeding in the 1950s. Today, lots of seaweed taxa are cultivated, lowering the pressure on naturally occurring populations (Critchley and Ohno 1998).

Carrageenans represent one of the major texturizing ingredients used by the food industry (see Table 1); they are natural ingredients, which have been used for decades in food applications and are generally regarded as safe (GRAS). Carrageenans are commercially important hydrophilic colloids which occur as a matrix in numerous species of red seaweed (Stanley 1987). They are the third most important hydrocolloid in the food industry, after gelatin (animal origin) and starch (plant origin) (van de Velde and de Ruiter 2002). This phycocolloid has annual sales of over $US 200 million, and represents 15% of the world use of food hydrocolloids. The market for carrageenan has consistently grown at 5% per year, from 5.500 tons in 1970 to up to 20.000 tons in 1995 (Bixler 1996). Shortages of carrageenan-producing seaweeds suddenly appeared in mid-2007, resulting in doubling of the price of carrageenan; some of this price increase was due to increased fuel costs and a weak $US (most seaweed polysaccharides are traded in USD). The reasons for shortages of the raw materials for processing are less certain:

perhaps it is a combination of environmental factors, sudden increases in demand, particularly from China, and some market manipulation by farmers and traders. Most hydrocolloids are experiencing severe price movements (Pereira 2013).

The modern industry of carrageenans dates from the 1940s where it was found to be the ideal stabilizer for the suspension of cocoa in chocolate milk. In the last decades, due to its physical functional properties, such as gelling, thickening, emulsifying and stabilizing abilities, carrageenans have been employed in food industry to improve the texture of cottage cheese, puddings and dairy desserts, and in the manufacture of sausages, patties and low-fat hamburgers (van de Velde and de Ruiter 2002, van de Velde et al. 2004, Pereira et al. 2009a, Pereira and van de Velde 2011b, Li et al. 2014).

The most commonly used commercial carrageenans are extracted from *Kappaphycus alvarezii* and *Eucheuma denticulatum* (McHugh 2003). Primarily wild-harvested genera such as *Chondrus, Furcellaria, Gigartina, Chondracanthus, Sarcothalia, Mazzaella, Iridaea, Mastocarpus,* and *Tichocarpus* are also mainly cultivated as carrageenan raw materials, and producing countries include Argentina, Canada, Chile, Denmark, France, Japan, Mexico, Morocco, Portugal, North Korea, South Korea, Spain, Russia, and the US (Pereira et al. 2009c, Bixler and Porse 2011, Pereira et al. 2013).

As mentioned before, the original source of carrageenans was from the red seaweed *Chondrus crispus*, which continues to be used, but in limited quantities. *Betaphycus gelatinus* is used for the extraction of beta (β) carrageenan. Some South American red algae used previously only in minor quantities have, more recently, received attention from carrageenan producers, as they seek to increase diversification of raw materials in order to provide for the extraction of new carrageenan types with different physical functionalities and, therefore, increase product development, which in turn stimulates demand (McHugh 2003, Pereira 2011). *Gigartina skottsbergii, Sarcothalia crispata,* and *Mazzaella laminarioides* are currently the most valuable species and all are harvested from natural populations in Chile and Peru. Large carrageenan processors have fuelled the development of *Kappaphycus alvarezii* (which goes by the name "cottonii" in the trade) and *Eucheuma denticulatum* (commonly referred to as "spinosum" in the trade) farming in several countries including the Philippines, Indonesia, Malaysia, Tanzania, Kiribati, Fiji, Kenya, and Madagascar (McHugh 2003, Pereira 2011). Indonesia has recently overtaken the Philippines as the world's largest producer of dried carrageenophytes biomass (Pereira 2011).

1.4 Seaweed as Food (Sea vegetables)

Seaweeds referred as Seavegetables are daily food items among the countries of the East Asia, especially Japan, Korea and China, the island nations of the Pacific, and of the Celtic cultures of Europe, where they form the basis for a multi-billion USD business employing sea farmers and processors (Johnston 1966, Rao et al. 2007).

Seaweeds have been consumed as a vegetable since the beginning of fourth century in Japan and sixth century in China. On average, the Japanese eat 1.4 Kg seaweed per person per year. This ancient tradition and everyday habit have made possible a large number of epidemiological researches showing the health benefits linked to seaweed consumption (Johnston 1966, MacArtain et al. 2007, Mouritsen 2013). The spread of Japanese and Chinese cuisine and of health foods throughout the world has brought new attention to seavegetables. There is immense potential for increasing the consumption of seavegetables as food items, dietary supplements, food texturisers, flavoring, coloring and condiments especially among mainstream markets. France was one of first European countries to establish a specific regulation concerning the use of seaweeds for human consumption as non-traditional food substances. According to this regulation, 21 macroalgae and three microalgae are authorized as vegetables and condiments (Burtin 2003, CEVA 2014).

1.4.1 Nutritional Composition of Edible Seaweed

In general terms, it can be said that a varied diet includes a proportion of seaweed products, for example, up to 10% as in Japan, promotes wellness. This is due principally to the high concentration in marine algae of important minerals and vitamins. These minerals in seaweeds are in what are known as chelated and colloidal forms, which enhance their bioavailability in the body. Seaweeds are also a good source of proteins and essential amino acids. In addition, marine algae have much greater fiber content than traditional vegetables and fruits, as they are largely composed of both soluble and insoluble dietary fiber. Because dietary fiber is indigestible, it contributes to no calories, and enhances the digestive function by absorbing water and thereby easing the passage of food through the intestinal system. Soluble fiber, particularly, slows the rate at which polysaccharides are absorbed, which helps to lower blood sugar levels, a potential advantage for diabetics (Pereira 2011, Mouritsen 2013).

Carbohydrates

The chemical composition and the abundance of carbohydrates vary among seaweed species. Red seaweeds varieties consist of different typical carbohydrates types including: floridean starch (α-1,4-bindingglucan), cellulose, xylan, and mannan. Moreover, their water-soluble fiber fraction is formed by sulfur-containing galactans, e.g., agar and carrageenan (see Phycocolloids) (Jimenez-Escrig and Sanchez-Muniz 2000).

On the other hand, the typical carbohydrates in brown seaweeds varieties consist of fucoidan, laminaran (β-1,3-glucan), cellulose, alginates, and mannitol. In brown seaweeds, fibers are mainly cellulose and insoluble alginates (El-Said and El-Sikaily 2013). In contrast, the amorphous, slimy fraction of fibers consists mainly of water-soluble alginates and/or fucoidan. The main reserve polysaccharides of Phaeophyceae are laminaran (β-1,3-glucan) and mannitol (Kolb et al. 1999).

The typical seaweeds' carbohydrates are not digestible by the human gastrointestinal tract and, therefore, they are dietary fiber. The content of total dietary fiber ranges from 33–50 g/100 g DW (see Table 2) (Lahaye 1991, Jimenez-Escrig and Cambrodon 1999, Ruperez and Saura-Calixto 2001, Pereira 2011).

Sulfated polysaccharides

As previously described, seaweeds are rich sources of sulfated polysaccharides, including some that have become valuable additives in the food industry because of their rheological properties such as gelling and thickening agents. In addition, sulfated polysaccharides are recognized to possess a number of biological activities including anticoagulant, antiviral, and immuno-inflammatory activities that might find relevance in nutraceutical/functional food, cosmetic/cosmeceutical and pharmaceutical applications (Jiao et al. 2011).

Fucoidan

Fucans are sulfated polysaccharides that are composed of a fucose backbone. One of the best studied fucans from brown algae is fucoidan, which was first isolated by Kylin in 1913 (Kylin 1913). The fucoidan from *Fucus vesiculosus* has been available commercially for decades (Sigma-Aldrich Chemical Company, St. Louis, MO, US). Early work on its structure showed that it contained primarily (1→2) linked 4-O-sulfated fucopyranose residues. However, 3-linked fucose with 4-sulfated groups was subsequently reported to be present in some of the fucose residues. Subsequently, Chevolot and colleagues reported that the fucoidan from *F. vesiculosus* and *Ascophyllum nodosum* contains a predominant disaccharide motif containing sulfate at the 2-position of the 3-linked fucose and sulfate groups on the 2- and 3-positions of the 4-linked fucose (Chevolot et al. 2001, Pereira et al. 2013).

Table 2. Nutrient composition of selected edible seaweed (% dry weight).

Species	Protein	Ash	Dietary fiber	Carbohydrate	Lipid	Reference
Chlorophyta (Green seaweed)						
Caulerpa lentillifera	10–13	24–37	33	38–59	0.86–1.11	Pattama and Chirapart 2006 Matanjun et al. 2009
C. racemosa	17.8–18.4	7–19	64.9	33–41	9.8	El-Sarraf and El-Shaarawy 1994 Akhtar and Sultana 2002 Santoso et al. 2006, Kumar et al. 2011b
Codium fragile	8–11	21–39	5.1	39–67	0.5–1.5	Ortiz et al. 2009 Guerra-Rivas et al. 2011
Ulva compressa	21–27	18.6	33–45	48.2	0.3	Burtin 2003, Mamatha et al. 2007 Patarra et al. 2011
Ul. lactuca	10–25	12.9	29–38	36–43	0.6–1.6	Fleurence 1999 Morrissey et al. 2001 Manivannan 2008, Kumar et al. 2011b
Ul. pertusa	20–26	-	-	47.0	-	Fleurence 1999, Pengzhan et al. 2003
Ul. rigida	18–19	28.6	38–41	43–56	0.9–2.0	Santoso et al. 2006 Taboada et al. 2010, Kumar et al. 2011
Ul. reticulata	17–20	-	65.7	50–58	1.7–2.3	Shanmugam and Palpandi 2008 Kumar et al. 2011b
Phaeophyceae (Brown seaweed)						
Alaria esculenta	9–20	-	42.86	46–51	1–2	Applegate and Gray 1995 Morrissey et al. 2001
Eisenia bicyclis	7.5	9.72	10–12	60.6	0.1	Mitchell 2000, Mišurcová et al. 2010
Fucus spiralis	10.77	-	63.88	-	-	Patarra et al. 2011
F. vesiculosus	3–14	14–30	45–59	46.8	1.9	Applegate and Gray 1995 Fleurence 1999, Saá 2002 Truus et al. 2001, Ruperez 2002 Díaz-Rubio et al. 2008

Himanthalia elongata	5–15	30–36	33–37	44–61	0.5–1.1	Morrissey et al. 2001 Saá 2002, Burtin 2003 López-López et al. 2009 Gómez-Ordóñez et al. 2010
Laminaria digitata	8–15	37.59	37.3	48	1.0	Fleurence 1999, Morrissey et al. 2001 Ruperez 2002, Burtin 2003
Saccharina japonica	7.5	26.63	10–36	51.9	1.0	Mitchell 2000, Dawczynski 2007 Mišurcová et al. 2010
S. latissima	6–26	34.78	30	52–61	0.5–1.1	Morrissey et al. 2001, Saá 2002 Gómez-Ordóñez et al. 2010
Sargassum fusiforme	11.6	19.77	17–62	30.6	1.4	Mitchell 2000, Dawczynski 2007 Mišurcová et al. 2010
Undaria pinnatifida	12–23	26–39	16–46	45–51	1.5–4.5	Yamada et al. 1991 Mitchell 2000, Ruperez 2002 Saá 2002, Burtin 2003 Dawczynski 2007 López-López et al. 2009 Mišurcová et al. 2010
Rhodophyta (Red seaweed)						
Chondrus crispus	11–21	21.08	10–34	55–68	1.0–3.0	Indergaard and Minsaas 1991 Morrissey et al. 2001 Ruperez 2002, Saá 2002
Gracilaria chilensis	13.7	18.9	–	66.1	1.3	Ortiz et al. 2009
Palmaria palmata	8–35	15–30	28.57	46–56	0.7–3	Indergaard and Minsaas 1991 Fleurence 1999 Morrissey et al. 2001, Saá 2002
Porphyra tenera	33–47	20.5	12–35	44.3	0.7	Fleurence 1999, Mitchell 2000 Ruperez 2002, Burtin 2003 Mišurcová et al. 2010
P. umbilicalis	29–39	12	29–35	43	0.3	Saá 2002, López-López et al. 2009
P. yezoensis	31–44	7.8	48.6	44.4	2.1	Indergaard and Minsaas 1991 Noda 1993, Dawczynski 2007

Fucoidan can be easily cooked out of most edible brown algae by simmering 20–40 min in water. When consumed, it seems to reduce the intensity of the inflammatory response and promote more rapid tissue healing after wound trauma and surgical trauma. This means that brown seaweed broth is recommended after auto collision, sports injuries, bruising falls, muscle and joint damage, and deep tissue cuts, including surgery (Fitton 2011).

Laminaran

Laminaran is a small glucan present in either soluble or insoluble form. The first form is characterized by complete solubility in cold water, while the other is only soluble in hot water (Kylin 1913, Chevolot et al. 2001). This polysaccharide is composed of D-glucose with β-(1,3) linkages, with β-(1,6) intra-chain branching (Pereira and Ribeiro-Claro 2014a).

Laminaran and its derivatives are other sulfated polysaccharides that could become functional food. Extracted from some brown seaweed (Laminariales, Phaeophyceae), they have a lot of potential in the area of health, for example, it has known antitumor effects (Ibrahim et al. 2005).

Ulvan

Ulvan represents 8–29% of the algae dry weight and is produced by species belonging to the phylum Chlorophyta (green algae), mostly belonging to the class Ulvophyceae (Robic et al. 2009). It is mainly made up of disaccharide repeating sequences composed of sulfated rhamnose and glucuronic acid, iduronic acid, or xylose (Percival and McDowell 1967, Quemener et al. 1997).

The two major repeating disaccharides are aldobiuronic acids designated as: type A, ulvanobiuronic acid 3-sulfate (A3s) and type B, ulvanobiuronic acid 3-sulfate (B3s). Partially sulfated xylose residues at O-2 can also occur in place of uronic acids. Low proportions of galactose, glucose, mannose, and protein are also generally found in ulvan. Additionally, minor repeat units have been reported that contain sulfated xylose replacing the iduronic acid or glucuronic acid (Lahaye and Robic 2007, Jiao et al. 2011, Pereira and Ribeiro-Claro 2014a).

Marine sulfated polysaccharides other than fucans have also been shown to possess antioxidant, antilipidemic, and anticoagulant activities. Reports include sulfated galactan and ulvan-like sulfated polysaccharides obtained from green algae (Chlorophyta), in particular from *Codium* and *Ulva* (Jiao et al. 2011).

Lipids

Seaweeds are known as low-energy food (see Table 2). Despite low lipid content, ω-3 and ω-6 polyunsaturated fatty acids (PUFAs) introduce a significant part of seaweed lipids. PUFAs are the important components of all cell membranes and precursors of eicosanoids that are essential bio-regulators of many cellular processes. PUFAs effectively reduce the risk of cardiovascular diseases, cancer, osteoporosis, and diabetes. Because of the frequent use of seaweeds in Asia and their increasing utilization as food also in other parts of the world, seaweeds could contribute to the improvement of a low level of ω-3 PUFAs, especially in the western diet. The major commercial sources of ω-3 PUFAs are fish, but their wide use as food additives is limited for the typical fishy smell, unpleasant taste, and oxidative non-stability (Mišurcová et al. 2011).

Proteins

Seaweed protein is a source of all amino acids, especially glycine, alanine, arginine, proline, glutamic, and aspartic acids (see Table 2). In algae, Essential Amino Acids (EAAs) represent almost a half of total amino acids and their protein profile is close to the profile of egg protein. In case of non-EAAs, all three groups (green, brown, and red seaweeds) contain a similar amount. Red seaweed seems to be a good source of protein because its value reaches 47%. The issue of protein malnutrition supports the trend to find a new and cheap alternative source of protein. Algae could play an important role in the above-mentioned challenge because of its relatively high content of nitrogen compounds. Algae may be used in the industry as a source of ingredients with high nutritional quality (Černá 2011).

Vitamins

Vitamins can be divided into those that are either water or fat-soluble. Water-soluble vitamins include both B-complex vitamins and vitamin C. The B-complex vitamins are the largest group and have roles associated with metabolism, muscle tone, cell growth and the nervous system. For example, Nori (*Porphyra* spp.) and Sea Lettuce (*Ulva* spp.) are good sources of vitamin B_{12} which has an important role in DNA synthesis. Vitamin C is a water-soluble vitamin that is important for gum health; iron absorption and resistance to infection (see Table 3).

Fat-soluble vitamins include vitamin A, D, E and K. Vitamin A (retinol) plays an important role in bone growth, tooth development, reproduction and cell division. Vitamin D, another fat-soluble vitamin, is important for bone growth and maintenance. Vitamins E and K also have a number of biological functions including antioxidant activity and blood clotting. In addition to their biochemical functions and antioxidant activity, seaweed-derived vitamins have been demonstrated to have other health benefits such as reducing hypertension, preventing cardiovascular disease and reducing the risk of cancer (LTS 2005, Škrovánková 2011, Oceanharvest 2015).

Table 3. Vitamin content of some edible seaweed (mg/100 g edible portion).

Species	A	B₁ (Thiamin)	B₂ (Riboflavin)	B₃ (Niacin)	B₅ (Panththenic Acid)	B₆ (Pyridoxine)	B₈ (Biotin)	B₆ (Cobalamin)	C (Ascorbic Acid)	E	Folic acid	Reference
Chlorophyta (Green seaweed)												
Caulerpa lentillifera	-	0.05	0.02	1.09	-	-	-	-	1.00	2.22	-	Pattama et al. 2006
Codium fragile	0.527	0.223	0.559	-	-	-	-	-	<0.223	-	-	García et al. 1993
Ulva lactuca	0.017	<0.024	0.533	98*	-	-	-	6*	<0.242	-	-	García et al. 1993; Morrissey et al. 2001
Ulva pertusa	-	-	-	-	-	-	-	-	30–241**	-	-	Tsuchiya 1950
Ulva rigida	9581	0.47	0.199	< 0.5	1.70	<0.1	0.012	6	9.42	19.70	0.108	Taboada et al. 2009
Phaeophyceae (Brown seaweed)												
Alaria esculenta	-	-	0.3–1*	5*	-	0.1*	-	-	100–500*	-	-	Morrissey et al. 2001
Fucus vesiculosus	0.307	0.02	0.035	-	-	-	-	-	14.124	-	-	García et al. 1993 Saá 2002
Himanthalia elongata	0.079	0.020	0.020	-	-	-	-	-	28.56	-	0.176–0.258	García et al. 1993 Saá 2002 Quirós et al. 2004
Laminaria digitata	-	1.250	0.138	61.2	-	6.41	6.41	0.0005	35.5	3.43	-	MacArtain et al. 2007
Laminaria ochroleuca	0.041	0.058	0.212	-	-	-	-	-	0.353	-	0.479	García et al. 1993 Quirós et al. 2004
Saccharina japonica	0.481	0.2	0.85	1.58	-	0.09	-	-	-	-	-	Kolb et al. 2004

Saccharina latissima	0.04	0.05	0.21				0.0003	0.35	1.6		Saá 2002
Undaria pinnatifida	0.04–0.22	0.17–0.30	0.23–1.4	2.56	-	0.18	0.0036	5.29	1.4–2.5	0.479	García et al. 1993, Saá 2002, Kolb et al. 2004, Quirós et al. 2004
Rhodophyta (Red seaweed)											
Chondrus crispus	-	-	-	-	-	-	0.6–4*	10–13*	-	-	Morrissey et al. 2001, Rupérez 2002
Gracilaria spp.	-	-	-	-	-	-	-	16–149**	-	-	Tsuchiya 1950
Palmaria palmata	1.59	0.073–1.56	0.51–1.91	1.89	-	8.99	0.009	6.34–34.5	2.2–13.9	0.267	Morrissey et al. 2001, Saá 2002, Quirós et al. 2004
Porphyra umbilicalis	3.65	0.144	0.36	-	-	-	0.029	4.214	-	0.363	García et al. 1993, Saá 2002, Quirós et al. 2004
Porphyra yezoensis	16000***	0.129	0.382	11.0	-	-	0.052	-	-	-	Noda 1993, Watanabe et al. 2000

* expressed as *ppm*; ** expressed as *mg%*; *** expressed as *I.U.*

Although seaweed contains both water and fat-soluble vitamins, the vitamin composition of seaweed is variable and depends on a number of factors. For example, evidence exists of seasonal variation in the vitamin content of the seaweed *Eisenia arborea*, where fat-soluble vitamins follow a different pattern to those that are water-soluble. Another factor affecting seaweed vitamin content is light exposure, as plants growing in bright light can contain higher levels of some vitamins (Oceanharvest 2015).

Seaweed species is another critical factor which can affect vitamin composition. For example, the level of niacin (vitamin B$_3$) in some brown seaweed (e.g., *Laminaria* spp.) is approximately one tenth the level found in the red seaweed, *Porphyra tenera*. Other factors that can influence vitamin content include geographical location, salinity and sea temperature. Vitamin content can also be affected by processing as both heat and dehydration can have a significant effect on the vitamin levels (Škrovánková 2011, Oceanharvest 2015).

Minerals

All essential minerals are available in dietary seaweeds. No land plant even remotely approaches seaweeds as sources of metabolically-required minerals. Seaweeds can provide minerals often absent from freshwater and crops grown on mineral-depleted soils (Drum 2013).

For example, Japanese people consume more than 1.6 Kg seaweed DW per year per capita (Fleurence 1999). Moreover, because of their minerals presence (Na, K, Ca, Mg, Fe, Zn, Mn, etc.) they are needed for human nutrition. However, this wide range in mineral content (8–40%) is not found in edible land plants, due to many factors such as; seaweed phylum, geographical origin, and seasonal, environmental and physiological variations (Nisizawa et al. 1987, Ruperez 2002). Seaweeds are also one of the most important vegetable sources of calcium. Their calcium content may be as high as 7% of the dry weight and up to 25 to 34% in the chalky seaweed, *Lithothamnion* (El-Said and El-Sikaily 2013).

Seaweeds are 20–50% dry weight mineral (Kazutosi 2002). This figure is obtained by burning off the seaweed's organic material and weighing the remaining ash. The elements abundant in seaweeds include: potassium, sodium, calcium, magnesium, zinc, copper, chloride, sulfur, phosphorous, vanadium, cobalt, manganese, selenium, bromine, iodine, arsenic, iron, and fluorine (Drum 2013) (see Table 4).

Many human body substances require particular mineral elements as part(s) of their respective structure. Examples are iron for hemoglobin and iodine for thyroxin (Soetan et al. 2010).

Table 4. Mineral composition of some edible seaweeds (mg.100 g^{-1} DW).

Species	Na	K	P	Ca	Mg	Fe	Zn	Mn	Cu	I	Reference
Chlorophyta (Green seaweed)											
Caulerpa lentillifera	8917	970–1142	1030	780–1874	630–1028	9.3–21.4	2.6–3.5	7.9	0.11–2.2	-	Pattama and Chirapart 2006 Matanjun et al. 2009
C. racemosa	2574	318	29.71	1852	384–1610	30–81	1–7	4.91	0.6–0.8	-	Santoso et al. 2006 Kumar et al. 2011
Ulva lactuca		-	140	840	-	66	-	-	-	-	Castro-Gonzalez et al. 1996
U. rigida	1595	1561	210	524	2094	283	0.6	1.6	0.5	-	Taboada et al. 2009
Phaeophyceae (Brown seaweed)											
Fucus vesiculosus	2450–5469	2500–4322	315	725–938	670–994	4–11	3.71	5.50	<0.5	14.5	Ruperez 2002, Saá 2002
Himanthalia elongata	4100	8250	240	720	435	59	-	-	-	14.7	Saá 2002
Laminaria digitata	3818	11,579	-	1005	659	3.29	1.77	<0.5	<0.5	-	Rupérez 2002
Saccharina japonica	2532–3260	4350–5951	150–300	225–910	550–757	1.19–43	0.89–1.63	0.13–0.65	0.25–0.4	130–690	Kaas et al. 1992 Funaki et al. 2001 Kolb et al. 2004
S. latissima	2620	4330	165	810	715	-	-	-	-	15.9	Saá 2002
Sargassum fusiforme	-	-	-	1860	687	88.6	1.35	-	-	-	Sugawa-Katayama and Katayama 2009
Undaria pinnatifida	4880–6494	5691–6810	235–450	680–1380	405–680	1.54–30	0.944	0.332	0.185	22–30	Saá 2002 Kolb et al. 2004

Table 4. contd....

Table 4. contd.

Species	Na	K	P	Ca	Mg	Fe	Zn	Mn	Cu	I	Reference
Rhodophyta (Red seaweed)											
Chondrus crispus	1200–4270	1350–3184	135	420–1120	600–732	4–17	7.14	1.32	<0.5	24.5	Ruperez 2002 Saá 2002
Gracilaria spp.	5465	3417	-	402	565	3.65	4.35	-	-	-	Krishnaiah et al. 2008
Palmaria palmata	1595	7310	235	560	610	50	2.86	1.14	0.376	55	Saá 2002
Porphyra tenera	3627	3500	-	390	565	10.3	2.21	2.72	<0.5	-	Rupérez 2002
P. umbilicalis	940	2030	235	330	370	23	-	-	-	17.3	Saá 2002
P. yezoensis	570	2400	-	440	650	13	10	2	1.47	-	Noda 1993

Chronic dietary shortages or disease-related mineral depletions can produce both specific and general disease conditions: Iodine shortage results in varying degrees of thyroid dysfunction; poor absorption of dietary calcium can result in osteoporosis. Adequate residential body mineral supplies are critical for optimal body system functioning (Drum 2013, Karthick 2014).

Seaweed salt

In earlier times, salt was a precious commodity, which was sometimes referred to as white gold and was used by the Vikings as a medium of commercial exchange. As far back as 6050 BC, salt has been an important and integral part of the world's history, as it has been interwoven into the daily lives of countless historic civilizations. Used as a part of Egyptian religious offerings and valuable trade between the Phoenicians and their Mediterranean empire, salt and history have been inextricably intertwined for millennia, with great importance placed on salt by many different races and cultures of people. Even today, the history of salt touches our daily lives. The word "salary" was derived from the word "salt." Salt was highly valued and its production was legally restricted in ancient times, so it was historically used as a method of trade and currency. The word "salad" also originated from "salt," and began with the early Romans salting their leafy greens and vegetables. Undeniably, the history of salt is both broad ranging and unique, leaving its indelible mark in cultures across the globe (Mouritsen 2013, ASSC 2015).

Today, most ordinary white table salt is now derived from underground salt domes, formed millions of years ago. Called rock salt, it is almost pure sodium chloride (NaCl), about 99.3%, and has a smattering of other salts and minerals. Consequently, the major source of salt throughout the middle Ages was the sea. Coastal people were always able to obtain what is known as gray salt. In warm climates this was done by allowing seawater to evaporate in the sun and in colder areas by boiling the water over an open fire. The salt consists of 90 to 96% sodium chloride and ca. 4 to 7% water, as well as varying amounts of other salts and minerals, depending on how the salt was produced (Mouritsen 2013, Pereira and Correia 2015).

Sodium (Na) is vital for certain functions in one's body. One can't live without it. But it is very easy to consume far more than one actually needs as salt is often hidden in the food one eats. There is a large body of evidence to show that too much sodium causes high blood pressure which is one of the most significant contributors to both heart disease and stroke, and raises one's risk of developing eye problems and kidney disease. A natural alternative to salt is a great way to start cutting down sodium intake. Seaweed has, on average, between 9–12% sodium versus table and gourmet salts which have as much as 98% sodium. But what about all that salty flavour that tastes so good?

A healthy alternative to the consumption of sodium is the consumption of algae naturally rich in potassium. All living cells and that includes all of our cells, need potassium (K) all the time to function and stay alive. There are no exceptions. Our bodies have no innate potassium conservation mechanisms. The human evolutionary assumption seems to be that we will always have plenty of potassium available in our wild and live food diets, since all living cells require potassium. This is in contrast to sodium, also an essential element, for which we have a very rigorous sodium conservation mechanism.

The human tongue, just as the average beginning analytical chemistry students, seems to have difficulty distinguishing potassium from sodium: both taste salty. In equal amounts, potassium is up to eight times saltier than sodium. The use of edible seaweeds is a delicious high-potassium salt replacement in most foods and on popcorn, for example (Drum 2013).

1.5 Marine Algae and Health

Many people of the west are surrounded by fast food rich in calories and unsaturated fats, high powered advertising and over-consumption. The mass market has actually become accustomed to the expression of "junk food" to designate such offerings, but yet this highly processed "food" is consumed in large amounts. The consequences of consumption of these offerings for the mass (western) the lack of essential nutrients, obesity and diseases related to excessive intake of sugars (diabetes) and fat (arteriosclerosis), among others. It is worrying that the fast food trends of the west are being adopted seemingly without concern in developing countries as they become more prosperous, hence rates of associated disease are increasing.

What roles do seaweeds have in this picture?

They represent exactly the opposite: a natural food that gives us a highly nutritious but low in calories. Algae are therefore the best way to address the nutritional deficiencies of the current food, due to its wide range of constituents: minerals (iron and calcium), protein (with all essential amino acids), vitamins and fiber (Nisizawa 2006, Hotchkiss and Trius 2007, Pereira 2011).

It is thought that the overall content of certain traditional Asian diets contributes to the low incidence of cancer (Kanke et al. 1996), particularly breast cancer (Lawson et al. 2001). It is apparent that the unique levels of seaweed intake contribute to the variance in the levels of breast cancer (Teas 1983, Funahashi et al. 2001). There is a nine-fold lower incidence of breast cancer in the Japanese population and an even lower incidence in the Korean population compared to the incidence in the West (Teas 1983, Adami et al. 1998, Lawson et al. 2001).

The relative longevity and health of Okinawan Japanese populations has been attributed in part to dietary algae in studies (Yamori et al. 2001). These studies compared Okinawan descendants who were living in Brazil with Okinawans. The former have a higher risk of developing cardiovascular and other diseases. For a dietary intervention study, 3 g of DHA, 5 g of seaweed (Wakame) powder, and 50 mg of isoflavonoids from soybean (*Glycine max*) were given daily to immigrants, at high risk for developing diseases, in Brazil for 10 weeks. This combination reduced blood pressure and cholesterol levels, suppressed the urinary markers of bone re-absorption, and attenuated a tendency toward diabetes (Fitton 2003).

In recent years, there has been a growing interest in so-called functional food groups, amongst which seaweeds would seem to be able to play an important role since they can provide physiological benefits, in addition to nutritional as, for instance, anti-hypertensive, anti-oxidant or anti-inflammatory (Goldberg 1994, Madhusudan et al. 2011). A functional food can be defined as a food that produces a beneficial effect in one or more physiological functions, increases the welfare and or decreases the risk of suffering from the onset or development of a particular disease. The functionalities are far more preventative than curative. Furthermore, new types of products, derived from food, often referred to as nutraceuticals have recently been developed and marketed extensively. These products are usually used as food supplements, rather than whole foods and are marketed as tablets and pills and can provide important health benefits. Frequently, functional foods are obtained from traditional foods enriched with an ingredient which is able to provide or promote a beneficial action for human health. These are the so-called functional ingredients. According Madhusudan et al. (2011) and Pereira (2011), many biologically active compounds are present in seaweed, which can be used as therapeutic agents (see Table 5) in dietary supplements (Pereira 2011).

Table 5. Summary of nutraceutical significance of some seaweed compounds.

Category	Compounds	Seaweed source	Potential health benefit	References
Lipids and fatty acids	Omega 3 and omega 6 acids	*Porphyra* spp. Red algae *Undaria pinnatifida* and *Sargassum fusiforme* (formerly *Hizikia fusiformis*) Brown algae	Prevention of cardio-vascular diseases, osteoarthritis and diabetes	Saá 2002 Burtin 2003 Dawczynski et al. 2007 Venugopal 2009 Mišurcová et al. 2011
Carotenoids	β-carotene, lutein	*Chondrus crispus* *Porphyra yezoensis* Red algae	Antimutagenic; Protective against breast cancer	Okaih et al. 1996 Maruyama et al. 1991 Nam et al. 1998 Lohrmann et al. 2004 Yang et al. 2010
	β-carotene, lycopene	*Porphyra* spp. Red algae	Recent studies have shown the correlation between a diet rich in carotenoids and a diminishing risk of cardio-vascular disease, and cancers	Okuzumi et al. 1993 Yan et al. 1999 Dominguez 2013
	Lutein, zeaxanthin	Red algae	Diminishing risk of ophthalmological diseases	Okuzumi et al. 1993 Yan et al. 1999 Takaichi 2011
	Fucoxanthin	Brown algae	Antiangiogenic; Protective effects against retinol deficiency	Sugawara et al. 2006 Sangeetha et al. 2009 Peng et al. 2011

Minerals	Iodine	*Fucus vesiculosus* *Laminaria* spp. *Undaria pinnatifida*	The brown seaweeds have traditionally been used for treating thyroid goiter	Suzuki et al. 1965 Saá 2002
	Calcium	*Undaria pinnatifida* *Laminaria* spp. *Saccharina* spp.	Seaweed consumption may thus be useful in the case of expectant mothers, adolescents and elderly that all exposed to a risk of calcium deficiency	Saá 2002 Burtin 2003
Phycobilin pigments	Phycoerythrin, Phycocyanin	Red algae	Antioxidant properties, which could be beneficial in the prevention or treatment of neuro-degenerative diseases caused by oxidative stress (Alzheimer's and Parkinson's) as well as in the cases of gastric ulcers and cancers	Gonzales et al. 1999 Padula and Boiteux 1999 Remirez et al. 1999
	Phycoerythrin	Red algae	Amelioration of diabetic complications	Yabuta et al. 2010
Polyphenols	Flavonoids	*Palmaria palmata*	At high experimental concentrations that would not exist *in vivo*, the antioxidant abilities of flavonoids *in vitro* are stronger than those of vitamin C and E	Bagchi et al. 1999 Yuan et al. 2005
	Phlorotannins	Brown algae	Antioxidant activity of polyphenols extracted from brown and red seaweeds has already been demonstrated by *in vitro* assays	Nakamura et al. 1996 Shibata et al. 2008 Wijesekara et al. 2010
			Algicidal and bactericidal effect	Nagayama et al. 2002 Nagayama et al. 2003

Table 5. contd.....

Table 5. contd.

Category	Compounds	Seaweed source	Potential health benefit	References
Polysaccharides and dietary fibers	Agars, carrageenans, ulvans and fucoidans	Red, green and brown algae	These polysaccharides are not digested by humans and therefore can be regarded as dietary fibers	Lahaye et al. 1990 Lahaye et al. 1991
	Carrageenan, fucoidan	Red algae (carrageenophytes), brown algae	Antitumor	Haijin et al. 2003 Yuan and Song 2005 Choosawad et al. 2005 Yuan et al. 2011
	Carrageenan (lambda, iota and nu variants)	Red algae (carrageenophytes)	Antiviral, anti-HSV and anti-HIV	Spieler 2002 Smit 2004
	Fucoidan	Brown algae	Anticoagulant and antithrombotic activity	Li B et al. 2008
			Antitumor and immunomodulatory activity	Li B et al. 2008
			Antiviral and anti-HIV	Sugawara et al. 1989 Béress et al. 1993 Witvrouw and De Clercq 1997 Feldman et al. 1999 Smit 2004, Li B et al. 2008

Proteins and amino acids	Proteins	*Palmaria palmata* *Porphyra tenera*	Higher protein contents are recorded in green and red seaweeds (on average 10–30% of the dry weight). In some red seaweed, such as *Palmaria palmata* (dulse) and *Porphyra tenera* (nori), proteins can represent up to 35 and 47% of the dry matter, respectively	Burtin 2003
	Proteins, amino acids	*Undaria pinnatifida*	*Undaria pinnatifida* (wakame) has a high balance between the essential amino acids, which gives a high biological value to their proteins. Proteins, in addition, with a high bioavailability (85–90%)	Saá 2002
Vitamins	Vitamin B$_{12}$	*Porphyra* spp.	Is particularly recommended in the treatment of the effects of ageing, of CFS and anemia	Watanabe et al. 2000
	Vitamin C	*Himanthalia elongata* *Palmaria palmata*	Strengthens the immune defense system, activates the intestinal absorption of iron, controls the formation of conjunctive tissue and the protidic matrix of bony tissue, and also acts in trapping free radicals and regenerates Vitamin E	Saá 2002 Burtin 2003
	Vitamin E	*Fucus* spp.	Due to its antioxidant activity, vitamin E inhibits the oxidation of the low-density lipoproteins. It also plays an important part in the arachidonic acid chain by inhibiting the formation of prostaglandins and thromboxan	Burtin 2003

CHAPTER 2

Edible Seaweeds Listed by Geographic Region

2.1 North Atlantic and Mediterranean

On the European Atlantic coast, macroalgae have been harvested by coastal populations for a long time. Two main uses were identified: human consumption and agriculture (cattle food and soil enrichment). Archaeologists and ethnologists have mentioned other uses such as fuel, mattresses, etc.

The first recorded commercial use of seaweed was in the 17th century when it was used for the production of glass (France, Norway). Coastal populations, mainly farmers, gathered and burnt algae to produce ash, a source of potash for the production of glass and soap.

Some species are exploited and used for human consumption. In the countries studied, and particularly in France, Spain (Galicia) and Ireland, there are small and medium enterprises using edible seaweed. These new types of industry have been developed over the past few years following the increasing demand from European consumers. All the edible algae are harvested manually and dried in an artisanal way.

Macroalgae farming is not very well developed in Europe. Commercial aquaculture of seaweed is found in France (Brittany), Spain (Galicia) and on an experimental basis in Ireland, Asturias (Spain), Norway, Portugal (see www.algaplus.pt) and the United Kingdom. The main cultivated species are *Saccharina latissima* and *Undaria pinnatifida*. In Ireland, *Palmaria palmata* farming is being experimented with on the west coast but the results seem limited.

The main constraints on the development of seaweed farming have been the lack of markets and the high cost of European production compared to Asian production. Due to the lack of competitiveness of the European industry compared to its Asian counterpart and despite the rising world demand, the production of seaweed in Europe has decreased in the past decade. New markets may reverse this trend, such as the increase in edible seaweed production (this is currently a niche market), the growing demand of the biotech sector or the development of bio-fuel based on seaweed. The rise of conservation claims however may modify models of wild stock exploitation and increase the appeal of a sustainable seaweed farming sector (Netalgae 2012).

2.1.1 Chlorophyta

Bryopsis cupressina J.V. Lamouroux

Description: Thallus (conspicuous phase) usually of erect, feather-like uniaxial fronds from a rhizoidal holdfast; height about 2 cm; branching similar to a Cypress (*Cupressus* sp.).

Geographic distribution: Atlantic Is (Azores); Adriatic Sea; Mediterranean Sea; SW Asia (Sri Lanka).

Uses: It has been recommended for consumption in Israel (Lipkin and Friedlander 2006).

Bryopsis plumosa (Hudson) C. Agardh (Fig. 1)

Common name: English: Evenly branched mossy feather weed (Bunker et al. 2010), Hen pen, Sea-moss (Holland 2011), Stoneworts (Pereira 2015a).

Description: Thallus bushy, heteromorphic, richly branched, light to dark-green, single cell, tubular, erect fronds of; branched filaments 2–15 cm long; primary axis and main branches to 1.5 mm in diam.; lower part of axis bare, upper part pinnately branched with constrictions at base; branches may have thin filamentous (less 1 mm in diam.), pinnate branchlets with constrictions at base; length of branches decreases towards the apices making thallus pyramid-shaped; apical cells rounded at tip (Milchakova 2011).

Geographic distribution: N Europe; Arctic Canada, Adriatic Sea; Black Sea; Mediterranean; Atlantic Is (Azores, Bermuda, Canary Is, Cape Verde Is, Iceland, Madeira, Selvage Is); Tropical and subtropical Atlantic; Caribbean Sea; Indian Ocean; Asia (China, Japan, Korea, Russia, Taiwan); SE Asia (Indonesia, Malaysia, the Philippines, Singapore, Thailand, Vietnam); Australia; New Zealand; Papua New Guinea; Pacific Is (Micronesia, Polynesia, Mariana Is); NE Pacific (Alaska to British Columbia).

Uses: Considered an edible species (Holland 2011).

Figure 1. *Bryopsis plumosa* (Hudson) C. Agardh (Chlorophyta). Illustration from William H. Harvey (Plate III, Phycologia Britannica, 1846–1951).

B. plumosa is a potential source for antibacterial, antiviral, antifungal, antiprotozoal drugs and for extraction of lectin. The extracts have anti-fertility, hypoglycemic and diuretic properties and favors cardiovascular and nervous activity (Milchakova 2011, Pereira 2015a).

Cladophora prolifera (Roth) Kützing

Description: Plants tufted, to 20 cm tall, dark green, becoming blackish when dried, coarse and stiff; main filaments to 300–475 µ diam., the cells to

20 diam. long, copiously di-trichotomously branched, the branches rather erect, clustered towards the tips; branchlets 130–200 µ diam., cells 4–6 diam. long, the tip cells blunt.

Geographic distribution: NE Atlantic; Atlantic Is (Azores, Bermuda, Canary Is, Cape Verde Is, Madeira, Selvage Is); Adriatic Sea; Mediterranean; Caribbean Sea; Tropical and subtropical W and E Atlantic; Indian Ocean; Asia (China, Japan, Taiwan); SE Asia (Indonesia, Malaysia, the Philippines, Thailand); Australia; Pacific Is (Solomon Is).

Uses: Potential food use in Morocco (Zbakh et al. 2014).

Codium bursa (Olivi) C. Agardh (Fig. 2)

Common name: English: Green sponge ball (Bunker et al. 2010).

Description: Thallus a spongy sphere with a velvety soft, shiny surface; becoming flattened when increasing in size. The internal branched filamentous network becomes looser with increasing size of the sphere,

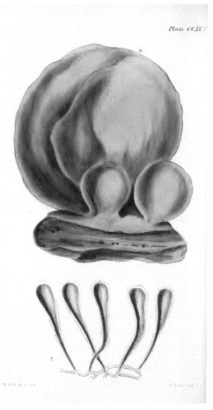

Figure 2. *Codium bursa* (Olivi) C. Agardh (Chlorophyta). Illustration from William H. Harvey (Plate CCXC, Phycologia Britannica, 1846–1951).

the space fills with water and the surface becomes indented; anchored to substratum by felted filaments (Braune and Guiry 2011).

Geographic distribution: NE Atlantic; Atlantic Is (Azores, Canary Is); Adriatic Sea; Mediterranean; New Zealand.

Uses: Considered an edible species in Turkey (Özvarol et al. 2010).

Monostroma grevillei (Thuret) Wittrock

Synonyms: *Enteromorpha grevillei, Ulva grevillei*

Common names: English: Green laver (Dickinson 1963); Korean: Hotparae (Sohn 1998); Swedish: Strutsallat (Tolstoy and Österlund 2003).

Description: Bright green seaweed that has blades with irregularly torn edges, originating from a small, basal disk; the torn blade edges are a result of the fact that juvenile *M. grevillei* are inflated sacs that burst open as they grow. The genus name *Monostroma* means "single layer", which is fitting due to the single layer of cells that makes up the blades of this alga. Blades can reach 5 to 10 cm in length.

Geographic distribution: NE Atlantic; Canadian Arctic; Atlantic Is (Azores, Greenland, Iceland); Asia (Japan, Korea, Russia); NE Pacific (Alaska, California); Antarctic Is.

Uses: *M. grevillei* is eaten in Scotland in the form of soup or boiled and spread on bread. In Japanese cuisine, *M. grevillei* is used in soups or in its dried and powdered form as a seasoning (Harrison 2013); in Korean cuisine is used in various seasonings with soy sauce (Sohn 1998, Zemke-White and Ohno 1999, Roo et al. 2007).

M. grevillei is used as a compress for wounds in traditional N European medicine (SIA 2014).

Monostroma latissimum Wittrock

Common names: English: Jade nori (Chapman and Chapman 1980); Chinese: Hai tsai (Madlener 1977), Hai-cai (Arasaki and Arasaki 1983); Japanese: Awo-Nori (Madlener 1977), Aonori, Hitoegusa (Arasaki and Arasaki 1983), Aonoriko (McConnaughey 1985).

Description: Green monostromatic foliose having a basic life history consisting of an alternation of heteromorphic generations; a haploid foliose thallus alternating with a diploid single cyst cell.

Geographic distribution: Adriatic Sea; NE Atlantic; SW Atlantic (Brazil); Asia (Japan, Korea, Taiwan, the Philippines); Australia and New Zealand.

Uses: An algae of this genus, is commercially cultivated in E Asia and S America for the edible product "Hitoegusa-nori" or "Hirohano-hitoegusa nori", popular sushi wraps. The highly valued food "Aonori" is produced from this species (Levring et al. 1969), and used in China, Taiwan and Japan, cooked as soup and as a seasoning (Chapman and Chapman 1980, Arasaki and Arasaki 1983, Simoons 1990).

It is cultivated in shallow, calm waters, such as are found in bays and estuaries, but, like *Porphyra*, it can also be grown in deeper waters using floating rafts. It is a flat, leafy plant and only one cell thick. It averages 20% protein and has a useful vitamin and mineral content. It has a life cycle involving an alternation of generations, one generation being the familiar leafy plant, the other microscopic and approximately spherical. It is this latter generation that releases spores that germinate into the leafy plant (McHugh 2003).

Ulva intestinalis Linnaeus (formerly *Enteromorpha intestinalis*) (Fig. 3)

Common names: English: Gut weed (Bunker et al. 2010, Holland 2011, Pereira 2015a), Grass kelp, Gut weed, Hollow green nori, Hollow green

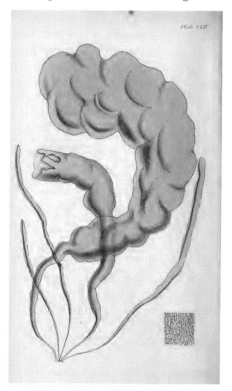

Figure 3. *Ulva intestinalis* Linnaeus (Chlorophyta). Illustration from William H. Harvey (Plate CLIV, Phycologia Britannica, 1846–1951).

wee, Tubular sea lettuce, Sea grass (Morrissey et al. 2001, MacArtain et al. 2007, Holland 2011); Filipino: Lumot (Zanevelt 1955); French: Ulve en tube (SIA 2014); German: Darmtang (Braune 2008); Hawaiian: Limu ele-ele (MacCaughey 1918); Japanese: Awo-nori (Smith 1904); Korean: Parae (Sohn 1998); Portuguese: Erva-patinha, Erva-patinha verde, Erva-do-calhau (Neto et al. 2005, Pereira 2009b, Pereira 2010a, Pereira and Correia 2015).

Description: Conspicuous bright grass-green seaweed, consisting of inflated irregularly constricted, tubular fronds that grow from a small discoid base. Fronds are typically unbranched. Fronds may be 10–30 cm or more in length and 6–18 mm in diam., the tips of which are usually rounded (Pereira 2010a).

Geographic distribution: *U. intestinalis* occurs worldwide, except in polar waters.

Uses: *U. intestinalis* is one of the main components of Japanese green Nori. In its dried form it is added to many Japanese soups and it is often crushed into a powder to be used on a variety of foods (Pereira 2011). *U. intestinalis* is also used in Welsh cooking to make a dish called Laverbread (SIA 2014). Used for preparation of Miso soup, Nori-jam, salads, and seaweed powder for various foods in the Japanese cuisine (Arasaki and Arasaki 1983, Ohno et al. 1998, Zemke-White and Ohno 1999, Lee 2008). In the coastal areas of Shandong (China), people collect *U. intestinalis* extensively, using it in bread by mixing with cornmeal; stuffing it and meat, after Chao frying, into bread rolls; and making it into a vegetable soup (Bangmei and Abbott 1987). In the Philippines the species is eaten raw as a salad (Zaneveld 1955, 1959), as well as in Malaysia (Sidik et al. 2012), Thailand (Lewmanomont 1978), Pakistan (Rahman 2002), Indonesia (Istini et al. 1998, Harrison 2013, Irianto and Syamdidi 2015), and Canada (Turner 1974). In Azores (Atlantic Is, Portugal), this species is used in cooking especially for the preparation of "tortas" (Neto et al. 2005); and in Korean cuisine is used in various seasonings with sesame oil, and sometimes vinegar (Kang 1968, Bonotto 1976, Madiener 1977, Sohn 1998, Roo et al. 2007); is also eaten in Hawaii (Reed 1906, MacCaughey 1918, Levring et al. 1969).

Used as food supplement or herbal medicine (Milchakova 2011), the extracts of this species have antimitotic and cytotoxic activity (Chenieux et al. 1980), are used for polysynaptic blocker (Baker 1984), and the treatment of aphthae, back pain, paronychia, lymphatic swellings, and goiter (Oh 1990).

See recipes made with *Ulva intestinalis* in Annex I.

Ulva lactuca Linnaeus (Fig. 4)

Common names: Chinese: Hai Tsai, Shih shun, Haisai Kun-po, Kwanpo (Madlener 1977); English: Salt seaweed, Sea lettuce, Lettuce laver, Green Laver, Sea Grass, Thin stone brick (Madlener 1977), Chicory sea lettuce

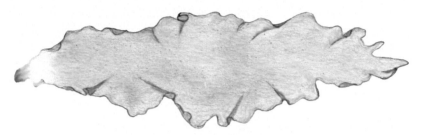

Figure 4. *Ulva lactuca* Linnaeus (Chlorophyta).

(FAO 1997); Filipino: Gamgamet (Agngarayngay et al. 2005); French: Laitue de mer (Boisvert 1984); German: Meersalat (Madlener 1977); Japanese: Aosa (Madlener 1977); Portuguese: Alface-do-mar, Limo (Oliveira 1990, Pereira 2009b, 2010a,b, Pereira and Correia 2015); Spanish: Luche, Luchi (Madlener 1977); Swedish: Havssallat (Tolstoy and Österlund 2003); Tahitian: Rimu miti (Payri et al. 2000); Tamil: Pattu pasi (Kaliaperumal et al. 1995).

Description: Light-green to deep grass-green, leaf-like flattened thallus rounded in shape, often apically widened, sometimes lobed; fairly soft, flaccid and smooth; undulated at the edges; margins smooth; short-stiped or almost sessile; attached by a discoidal holdfast (Braune and Guiry 2011).

Geographic distribution: *U. lactuca* is ubiquitous, common to most shorelines around the world.

Uses: Used as food in salads and vegetable soup in India, Pakistan, Korea and Indonesia (Johnston 1966, Bonotto 1976, Madlener 1977, Tseng 1983, Kaliaperumal et al. 1995, Istini et al. 1998, Rahman 2002, Agngarayngay et al. 2005, Nang 2006, Sidik et al. 2012, Irianto and Syamdidi 2015). *Ulva* has been and is still being used for human nutrition in E Asia, Ireland, Scotland and Pacific coast of N America; in more recent times also in France in specific preparations for seasoning and as an ingredient in pasta; significant Vitamin C, protein, iron and iodine content (Levring et al. 1969, Braune and Guiry 2011). *U. lactuca* known as "rimu miti" or "salt seaweed" is an edible species eaten by ancient Tahitians (Payri et al. 2000). In Caribbean Is and Asia is cooked fresh or dried, used as flour or meal, fried or baked (Radulovich et al. 2015).

See recipes of *Ulva* in Annex I.

Ulva rotundata Bliding

Description: Formed of a thin and flattened thallus, comprising two layers of cells with chloroplasts. This flexible blade can vary from light green to dark green and can reach 1 m in length in waters rich in organic matter.

Geographic distribution: NE Atlantic; Atlantic Is (Canary Is); Tropical and subtropical W Atlantic; Adriatic Sea.

Uses: Considered an edible species (Fleurence et al. 1995).

2.1.2 Ochrophyta – Phaeophyceae

Alaria esculenta (Linnaeus) Greville (Fig. 5)

Synonym: *Alaria macroptera*

Common names: English: Daberlocks, Bladderlocks (Ohmi 1968), Edible kelp, Honeyware, Wing Kelp, Bladderlochs, Tangle, Henware (Madlener 1977), Murlins, Dabberlocks, Stringy kelp (Arasaki and Arasaki 1983), Horsetail Kelp (McConnaughey 1985), Atlantic wakame; French: Wakamé (Boisvert 1984); German: Essebarer riementang (Braune 2008); Icelandic:

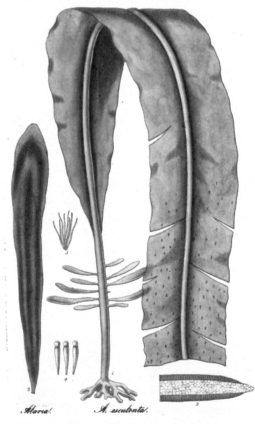

Figure 5. *Alaria esculenta* (Linnaeus) Greville (Ochrophyta, Phaeophyceae). Illustration from Robert K. Greville (Algae Britannicae, 1830).

Marinkjarni (Madlener 1977); Japanese: Chigaiso (Arasaki and Arasaki 1983).

Description: Short cylindrical stipe (exceptionally up to 75 cm) continuing as a distinct midrib throughout the length of the narrow, ribbon-like, slightly wavy blade; attached to substratum by claw-like holdfast termed haptera. The blade is yellowish, olive-green or rich brown in color, supple to the touch and very flexible. Blade length varies seasonally but is usually between 30 cm-1.5 m (exceptionally 4 m) in length. Blade may be tattered and torn by wave action sometimes leaving only the midrib at which point it may be confused with *Chorda filum*. Older plants may have flat, finger-like sporophylls, each up to 10 cm in length, growing from the stipe at the base of the blade. The sporophylls bear reproductive bodies called sori. When fertile the sori form a typical H-shaped figure on the sporophylls (Tyler-Walters 2008).

Geographic distribution: NE Atlantic (France, Scotland, Ireland, Greenland, Iceland, NE US, NE Canada); N Sea (England, Norway, the Netherlands); Pacific (Bering Sea and Sea of Japan).

Uses: The seaweed is edible and used to be commonly served either as a vegetable or a salad leaf in Ireland, Scotland and Iceland (Johnston 1970, Zemke-White and Ohno 1999); is also consumed in Siberia (Eidlitz 1969), Orkney Is and Norway (Kiple and Ornelas 2000); it is the only kelp-like seaweed in the British Is that has a distinct midrib and cannot be confused with anything else. It is also the only seaweed with sporophylls. In Canada and the US this seaweed is sometimes sold as "Atlantic wakame" and is presented as an alternative to Japanese wakame (*Undaria pinnatifida*). Indeed, though the colors differ (true dried wakame is green), "Dabberlocks" can be substituted for any Japanese recipe calling for "Wakame" (Tokida 1954, Hallsson 1961, Chapman and Chapman 1980, Arasaki and Arasaki 1983, Kuhnlein and Turner 1996, Zemke-White and Ohno 1999, Roo et al. 2007, Harrison 2013, Pereira 2015a).

See recipes made with *Alaria* in Annex I.

Alaria ochotensis Yendo

Common names: Japanese: Karafuto-wakame (Tokida 1954).

Description: The most striking character of this kelp is that the petioles of the dropped sporophylls remain persistently on the stipe. The blade, which is extremely thin and finely corrugated, is frosted with abundant cryptostomata on the surface and provided with numerous, peculiarly ramified, glandular cells in the superficial tissue (Tokida 1954).

Geographic distribution: Asia (Japan, Russia).

Uses: Used as food in Japan (Tokida 1954).

Alaria pylaiei (Bory de Saint-Vincent) Greville (Fig. 6)

Common names: Greenlandic: Suvdluistsit (Ostermann 1938); Inuktitut: Kjpilasat (Hoygaard 1937); Russian: Me'cgomei (Eidlitz 1969).

Description: The stipe is continued into the frond forming a long conspicuous midrib; the lamina is thin, membranous.

Geographic distribution: N Atlantic; Arctic Canada; Atlantic Is (Greenland, Iceland); NE Pacific (Alaska).

Uses: Eaten by Icelanders, Inuit (Canada), Ammassalik Is (Greenland), Alaska Indians, and Siberia (Hoygaard 1937, Ostermann 1938, Hallsson 1961, Eidlitz 1969, Kuhnlein and Turner 1996).

Figure 6. *Alaria pylaiei* (Bory de Saint-Vincent) Greville (Ochrophyta, Phaeophyceae).

Ascophyllum nodosum (Linnaeus) Le Jolis (Fig. 7)

Common names: English: Wrack (Madlener 1977), Yellow tang (Duddington 1966), Knotted wrack, Knobbed wrack (Arasaki and Arasaki 1983), Asco, Sea whistle, Egg wrack (Bunker et al. 2010); German: Knotentang, Sägetang (Braune 2008); Greenlandic: Miserarnat (Hoygaard 1937).

Description: *A. nodosum* is a large, common brown alga. The fronds are olive-brown. It is a species of the N Atlantic Ocean, also known as Norwegian kelp, knotted Kelp, or knotted wrack. It is common on the north-western coast of Europe (from Svalbard to Portugal) including east Greenland and the NE coast of N America. It has long fronds with large egg-shaped air-bladders. The fronds can reach 2 m in length. When ripe they are yellow, and as the tide goes out they can form huge piles of seaweed. It lives up to 15 years and is a dominant species of the middle shore (Pereira 2010a).

Figure 7. *Ascophyllum nodosum* (Linnaeus) Le Jolis (Ochrophyta, Phaeophyceae).

Geographic distribution: NE Atlantic (Arctic to Portugal, N Sea); NW Atlantic (Canadian Arctic to New Jersey).

Uses: *A. nodosum* is very effective at accumulating nutrients and minerals from the surrounding seawater, and this is what makes them a valuable resource for human enterprise, such as food (MacArtain et al. 2007, Brownlee et al. 2011, Abreu et al. 2014, Pereira 2015a); harvested in Ireland, Scotland and Norway for the manufacture of seaweed meal (Guiry and Garbary 1991); consumed also in Iceland and Greenland (Hoygaard 1937, Hallsson 1961); is used also to made a herbal tea (Urta Islandica 2015).

Source of alginates; manual harvest in Ireland; mechanical harvest in Norway; used for pinking selfish in some areas (Braune and Guiry 2011).

Bifurcaria bifurcata R. Ross (Fig. 8)

Synonym: *Bifurcaria tuberculata*

Common names: English: Brown tuning fork weed, Brown forking weed (Bunker et al. 2010); Portuguese: Frosque (INII 1966).

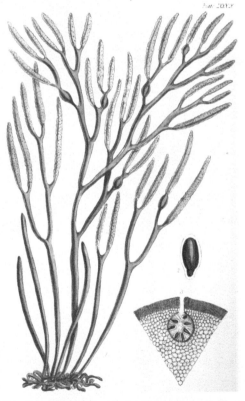

Figure 8. *Bifurcaria bifurcata* R. Ross (Ochrophyta, Phaeophyceae). Illustration from William H. Harvey (Plate LXXXLX, Phycologia Britannica, 1846–1951).

Description: Up to 50 cm in length; olive-yellow in color, but much darker when dry; holdfast expanded and knobby; frond cylindrical, unbranched near base then branching dichotomously. Elongate reproductive bodies present at ends of branches. Rounded air bladders sometimes present (Skewes 2008).

Geographic distribution: NE Atlantic.

Uses: Edible seaweed (Gómez-Ordóñez et al. 2010, Gómez-Ordóñez and Rupérez 2011, Kılınç et al. 2013), with antifouling activity (Hellio et al. 2001).

Chorda filum (Linnaeus) Stackhouse (Fig. 9)

Common names: English: Dead Men's ropes, Mermaid's tresses, Cat gut, Sea lace (Dickinson 1963), Mermaid's fishing line (Madlener 1977), Sea laces,

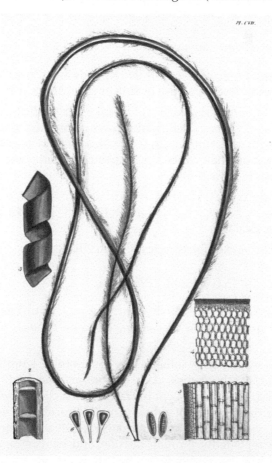

Figure 9. *Chorda filum* (Linnaeus) Stackhouse (Ochrophyta, Phaeophyceae). Illustration from William H. Harvey (Plate CVII, Phycologia Britannica, 1846–1951).

Mermaid's line (Chapman and Chapman 1980), Bootlace Weed (Bunker et al. 2010); German: Meersaite (Braune 2008); Japanese: Tsurumo (Madlener 1977); Swedish: Sudare (Tolstoy and Österlund 2003).

Description: *C. filum* is brown seaweed with long cord-like fronds, only 5 mm thick in diam. The fronds are hollow, slippery, unbranched and grow up to 8 m long. The species attaches to the substratum using a small discoid holdfast. It is an annual species, disappearing in winter (White 2006).

Geographic distribution: *C. filum* is widespread in the temperate waters of the Northern Hemisphere.

Uses: *C. filum* can be collected fresh in summer, when tender and used in soups or finely chopped in salads in Scotland (Launert 1981, Green 2014). Used also for foodstuff in Asia, namely in salads in China, Japan and Korea (Tokida 1954, Johnston 1966, Bonotto 1976, Madlener 1977, Chapman and Chapman 1980, Tseng 1983, Bangmei and Abbott 1987, Hotta et al. 1989).

Colpomenia peregrina Sauvageau

Common names: English: Oyster thief (Rhoads and Zunic 1978); French: Ballons (Dickinson 1963), Voleuse d'huitres (Duddington 1966).

Description: Sometimes regularly spherical (*C. peregrina*) or more or less irregular outline (*C. sinuosa*), yellowish-brown color, fixed to the substratum by filamentous rhizoids. Internally, the thallus is characterized by an outer cortex composed of small colored cells and inner pith composed of large pigmented cells (Pereira 2010a).

Geographic distribution: NE Atlantic (Norway to Canary Is); W Mediterranean; NW Pacific (Japan); NE Pacific (Alaska to California); Australia, New Zealand.

Uses: Considered and edible species (Green 2014).

Cystoseira baccata (S.G. Gmelin) P.C. Silva

Common names: English: Bushy berry wrack (Bunker et al. 2010); Portuguese: Pinheiro, Rabo-de-cavalo, Rabo-de-raposa, Rabo-de-zorro (INII 1966).

Description: Plants usually solitary, 1 m or more in length, attached by a thick, conical attaching disk. Axis simple or branched, up to 1 m in length, flattened, about 1 x 0.4 cm in transverse section; apex smooth and surrounded during periods of active growth by incurred young laterals. Lateral branch systems distichous, alternate, radially symmetrical, profusely branched in a repeatedly pinnate fashion and bearing sparse, filiform, occasionally bifurcate appendages on the branches of higher orders; deciduous, leaving decurrent bases which give an irregular, zigzag outline to the axis; cryptostomata lacking. Aerocysts present in axes of branches of higher order, sometimes in chains; seasonal, particularly numerous in autumn. Receptacles 1–5 cm long, formed from axes of ultimate ramuli, irregularly nodose and bearing simple, filiform appendages (Braune and Guiry 2011).

Geographic distribution: Widely distributed in the NE Atlantic (Baltic, the Netherlands, Belgium, Ireland, Britain, N France, N Spain, Portugal, Mauritania, and Canary Is).

Uses: Considered and edible species (Green 2014).

Cystoseira barbata (Stackhouse) C. Agardh

Description: Thallus large, bushy, richly branched, dark-brown to light-olive, 20–170 cm long, 4–12 mm wide, holdfast conical, unattached form is also known; stem very rough, bearing main and additional branches; main branches 10–40 cm long, cylindrical, arranged alternately or quasi-irregularly; branchlets often form panicles close to apices of main branches; additional branches 5–10 cm long; air vesicles present on branches during winter and spring, oblong, 7–15 mm long, 2–5 mm wide, solitary or moniliform, 2–10 per branch; receptacles develop on apices of branches, 2–20 mm long, 1–3 mm wide, oval or lanceolate, spinules absent, surface smooth and slightly sinuous; apices with a sterile spike; unattached plants prostrate, thin, 20–80 cm long; branches very thin and long, with sparse lateral branchlets; receptacles nearly always absent; air vesicles reduced (Milchakova 2011).

Geographic distribution: NE Atlantic; Adriatic Sea; Black Sea; Mediterranean; Atlantic Is (Canary Is, Selvage Is); Indian Ocean.

Uses: Used as a food additive in medical treatment of oncological patients in Ukraine and Bulgaria; a component in fragrance industry and cosmetology, also as a fodder for domestic animals and farmed fish, prawns and crabs (Milchakova 2011, Panayotova and Stancheva 2013).

C. barbata is a source of alginic acid salts, iodine-containing amino acids, PUFA, antimicrobial, antimitotic, antiviral, antibacterial and antitumor BASs; micro- and macroelements (e.g., iodine, bromine, potassium, calcium, magnesium, chlorine, sulfur, selenium) (Milchakova 2011, Panayotova and Stancheva 2013, Pereira 2015a).

Cystoseira crinita Duby

Description: Thallus large, bushy, 10–120 cm high, dark-brown; holdfast giving rise to as many as 20 shoots; richly branched; stem 5–80 cm long; 2–4 mm wide, resilient, flexible and smooth; main branches 6–18 cm long; additional branches few, 3–10 cm long; air vesicles large, 5–8 mm long,

4–5 mm width, triangular, inflated, solitary; receptacles forms on surface of air vesicles, cylindrical, abortive process absent; apices blunt (Milchakova 2011).

Geographic distribution: Adriatic Sea; Black Sea; Mediterranean; Atlantic Is (Canary Is).

Uses: Used as a food additive in medical treatment of oncological patients in Ukraine and Bulgaria; and as fodder for domestic animals and for farmed fish, prawn and crabs; also used in fragrance industry and cosmetology (Milchakova 2011, Ivanova et al. 2013, Panayotova and Stancheva 2013).

C. crinita is a source of alginic acid salts, PUFA, BASs with antimicrobial, antimitotic, antiviral, antibacterial and antitumor activity; as well as micro- and macroelements (e.g., iodine, bromine, potassium, calcium, magnesium, chlorine, sulfur, selenium) (Milchakova 2011, Panayotova and Stancheva 2013, Pereira 2015a).

Cystoseira nodicaulis (Withering) M. Roberts

Common name: English: Bushy noduled wrack (Bunker et al. 2010).

Description: Thallus to 1 m long, usually solitary, attached by an irregular conical disk. Axis cylindrical usually branched, with smooth, rounded apex immersed between bases or tophules of developing laterals. Lateral branch systems (below) radial or distichous, with greenish-blue iridescence when first formed, about 50 cm long, repeatedly branched in a pinnate manner, either regularly or irregularly, with infrequent cryptostomata and bearing spine-like appendages; deciduous in summer; first-formed laterals of the season with tophules, later without; tophules ovoid, to 15 mm long, smooth or covered with small tubercles, persistent on axis after rest of lateral has been shed. Receptacles formed in the ultimate branchlets, simple or branched, nodose, usually bearing spine-like appendages; air vesicles inconspicuous, dilations of ultimate branchlets, solitary, in series or confluent; sometimes absent (Braune and Guiry 2011).

Geographic distribution: NE Atlantic; Atlantic Is (Canary Is, Cape Verde Is, Madeira); Mediterranean.

Uses: Considered an edible species (Green 2014).

Cystoseira tamariscifolia (Hudson) Papenfuss (Fig. 10)

Common names: English: Rainbow bladderweed (Ager 2008), Bushy rainbow wrack (Bunker et al. 2010); Portuguese: Pinheiro, Rabo-de-cavalo, Rabo-de-raposa, Rabo-de-zorro (INII 1966).

Description: *C. tamariscifolia* is bushy seaweed, up to 60 cm in length but usually 30–45 cm. It has a cylindrical frond and branches irregularly. The

Plate CCLXV

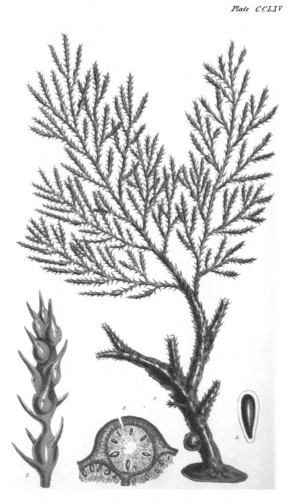

Figure 10. *Cystoseira tamariscifolia* (Hudson) Papenfuss (Ochrophyta, Phaeophyceae). Illustration from William H. Harvey (Plate CCLXV, Phycologia Britannica, 1846–1951).

reproductive bodies on the end of branches are long, oval and spiny. Small air bladders are usually found below the reproductive bodies. *C. tamariscifolia* is olive green in color, almost black when dry. When the plant is seen underwater it has a blue-green iridescence. Found in rockpools and on the lower shore; grows on both rocky shores and gravelly flats (Ager 2008).

Geographic distribution: NE Atlantic (United Kingdom to Mauritania); Mediterranean (France to Turkey).

Uses: Considered an edible species (Green 2014).

Dictyopteris polypodioides (A.P. De Candolle) J.V. Lamouroux (formerly *Dictyopteris membranacea*) (Fig. 11)

Common names: English: Netted wing weed (Bunker et al. 2010).

Description: Thallus flat and leaf-like, to 300 mm long and 20–30 mm broad; fronds olive to yellow-brown, translucent, and ± regularly dichotomously forked with a prominent midrib extending to the apices. Margins sometimes split to the midrib. Initially with an unpleasant smell shortly after collection, and degenerating quickly (Pereira 2010a).

Geographic distribution: Cold-temperate to warm NE Atlantic (Ireland and Britain to Canary Is); Mediterranean; Indian Ocean; W Pacific (the Philippines); NW Pacific (Japan).

Uses: Direct use as food (Levring et al. 1969).

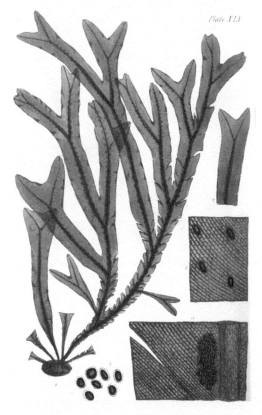

Figure 11. *Dictyopteris polypodioides* (A.P. De Candolle) J.V. Lamouroux (Ochrophyta, Phaeophyceae). Illustration from William H. Harvey (Plate XLX, Phycologia Britannica, 1846–1951).

Dictyota spiralis Montagne

Synonym: *Dilophus spiralis*

Description: Thallus leathery, 5–15 cm high, reddish-brown, attached by decumbent stolons; branching dichotomous, often irregular; some segments cease to ramify, others grow closely by 3–4 in tuft; segments near-linear, slightly wider at nodes, basal parts often narrowed; apices broad and blunt (Milchakova 2011).

Geographic distribution: NE Atlantic; Adriatic Sea; Black Sea; Mediterranean; Atlantic Is (Canary Is).

Uses: This species have nutritional and pharmaceutical applications (Pereira et al. 2012). The extracts of *D. spiralis* have antibacterial, antiviral and cytotoxic properties (Milchakova 2011, Pereira 2015a).

Fucus ceranoides Linnaeus (Fig. 12)

Common names: English: Horned wrack (Duddington 1966), Estuary wrack (Bunker et al. 2010); Portuguese: Bagão (INII 1966).

Figure 12. *Fucus ceranoides* Linnaeus (Ochrophyta, Phaeophyceae).

Description: Large brown intertidal seaweed, restricted to growing in estuaries or near freshwater streams on the shore. *F. ceranoides* does not have airbladders, but the side of the fronds are often inflated (White 2007).

Geographic distribution: *F. ceranoides* is widely distributed in NE Atlantic but is only common in brackish water; it is characteristic of estuaries and is often abundant where freshwater streams run onto the shore. The species is found on the middle part of the shore, where it attaches to stones, rocks or gravel (White 2007).

Uses: Considered an edible species (Green 2014).

Fucus distichus Linnaeus

Synonym: *Fucus gardneri*

Common names: English: Two-headed wrack (Bunker et al. 2010), Popweed (Garza 2005), Bladderwrack, Rockweed.

Description: Brown or yellow-brown in color, and is attached to rocks with a visible holdfast. There is no visible stipe. A thick stock branches into numerous little blades. *F. distichus* has a visible midline that runs down the center of the frond as well as down the center of the branching bladelets; mature fronds generally have bulbs or small sacs at branch tips. The surface of these small bulbs is textured or bumpy. *F. distichus* may grow up to 30 cm long in Alaska. The young specimens are shorter, 5–10 cm, and appear more yellowish and often do not have bulbs on branch tips. These younger ones are preferred for picking (Garza 2005).

Geographic distribution: NE Pacific (Bering Sea, Aleutian Is, and Alaska to California).

Uses: Humans can eat this species (young tips can be enjoyed fresh or blanched, and it can be dried); it can be purchased for food as tips, flakes, and powder (Zemke-White and Ohno 1999, Garza 2005, Harrison 2013, Kellogg and Lila 2013).

See recipes made with *Fucus* in Annex I.

Fucus serratus Linnaeus (Fig. 13)

Common names: English: Serrated wrack (Anonymous 1978), Saw wrack, Toothed wrack; Swedish: Sågtång (Tolstoy and Österlund 2003).

Description: *F. serratus* is flat, olive brown to golden brown algae with serrated edges. It grows about 150 cm long; recognized by the strap-like branching fronds, with jagged serrated edges. The frond, a divided leaf, can grow to be 2 cm wide. Serrated wrack is found on hard substances near the

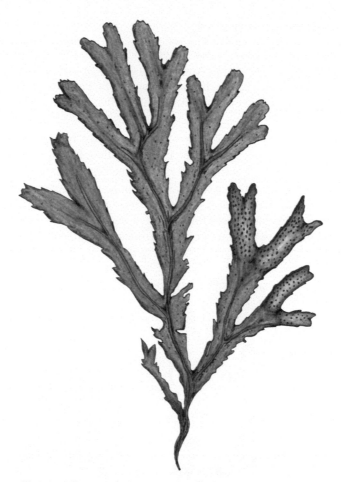

Figure 13. *Fucus serratus* Linnaeus (Ochrophyta, Phaeophyceae).

bottom of the ocean. Serrated wracks can be found in dense populations about 14 m down or in banks or low waters around a depth of 1.5 m. Most common, however, it is found on sheltered rocky substrata "wrack" seaweed (Pereira 2015a).

Geographic distribution: Arctic (White Sea); NE Atlantic; Atlantic Is (Canary Is, Iceland); Baltic Sea.

Uses: This seaweed is not commonly used as food; however, its edible properties are very similar to that of Bladderwrack. Stored dried it makes a very nutritious tea as well as being used in soups. Powdered down it makes a great stock for stews (Zemke-White and Ohno 1999, Roo et al. 2007, Green 2014, Martinez et al. 2014, Pereira 2015a).

Fucus spiralis Linnaeus (Fig. 14)

Common names: English: Jelly bags (Dickinson 1963), Spiral wrack (Duddington 1966), Flat wrack (Anonymous 1978), Spiraled wrack (Bunker et al. 2010); Portuguese: Bagão, Bodelha, Esgalhota, Fava-do-mar, Tremoço-do-mar (INII 1966, Neto et al. 2005, Pereira 2009b, Pereira 2010a, Pereira and Correia 2015).

Description: Well-grown fronds are usually easily recognizable by the flattened, twisted, dichotomously branched thallus, lacking bladders, and the large, oval receptacles at the frond tips, each receptacle being surrounded by a narrow rim of vegetative frond. Nevertheless, younger plants are not always so easy to identify, and even mature plants can be confused with *F. ceranoides* or with bladderless forms of *F. vesiculosus*. Both of these species, however, have narrower, more pointed, rimless receptacles (Pereira 2010a, Pereira 2014, SIA 2014).

Figure 14. *Fucus spiralis* Linnaeus (Ochrophyta, Phaeophyceae).

Geographic distribution: NE Atlantic; North Sea; Baltic; NW Atlantic; W Mediterranean; NE Pacific.

Uses: The brown seaweed *F. spiralis*, commonly known as "tremoço-do-mar" (in the Atlantic Is of Azores, Portugal), is considered a snack—the swollen reproductive parts of the frond (the receptacles) are picked and consumed fresh (Neto et al. 2005). Recently, students from the Professional School of Praia da Vitória (Azores) developed some innovative dishes with algae, including the *F. spiralis*, called "Pork Cheeks with Spiral Wrack Vinaigrette" (Matos 2013).

F. spiralis has been used historically for treatment of obesity, gout, goiter, and corns, and also in weight reducing and revitalizing bath treatments. It has been used for cattle feed, and as an organic manure. This alga, regularly exposed to sun radiation and its oxidative consequences, has developed optimal bioelectronics characters. It has a high concentration of phloroglucinol derivatives, including phenol acid, and in turn has been used in products from companies in France and the UK such as nutritional supplements, skin serum, body lotion, and compounds and extracts used as ingredients in other skin and hair products (SIA 2014).

Fucus vesiculosus Linnaeus (Fig. 15)

Common names: English: Paddy Tang (Dickinson 1963), Paddy-tang (Ohmi 1968), Red fucus, Dyers Fucus, Swine tang, Sea ware, Bladder, Rockweed, Bladderwrack, Popping wrack, Wrack (Madlener 1977), Bladder wrack (Anonymous 1978), Bladder-wrack (Rhoads and Zunic 1978); French: Fucus vésiculeux (Boisvert 1984); German: Blasentang (Braune 2008); Inuit (Greenland): Mikarkat (Hoygaard 1937); Portuguese: Trombolho, Estalos, Esgalhota, Bagão, Limo-bexiga (Oliveira 1990), Bodelha (Morton et al. 1998, Pereira 2010a), Fava-do-mar (Pereira and Correia 2015); Swedish: Blåstång (Tolstoy and Österlund 2003).

Description: The species of the genus *Fucus* are found in the intertidal zone on rocky shores in a wide range of exposures. It is common on the mid shore often with *Ascophyllum nodosum*. *F. vesiculosus* is generally larger and has a lighter color than *F. spiralis*, and has air bladders (aerocysts) arranged on both sides of the middle rib (Pereira 2010a). It can be found in high densities living for about four-five years. Under sheltered conditions, the fronds have been known to grow up to 2 m in Maine, America (Wippelhauser 1996).

Geographic distribution: N Atlantic; W Mediterranean.

Uses: *F. vesiculosus* is commonly used as a food in Japan, though less so in Europe and N America (Alaska); consumed in W Europe, Spain, Portugal, Scotland and Ireland (Hoygaard 1937, Madelener 1977, Indergaard and

Figure 15. *Fucus vesiculosus* Linnaeus (Ochrophyta, Phaeophyceae).

Minsaas 1991, Zemke-White and Ohno 1999, Roo et al. 2007, Lodeiro et al. 2012, Martinez et al. 2014). In Alaska is made into tea (Chapman and Chapman 1980). It can be stored dried, and added to soups and stews in flakes or powder form for flavor (Pereira 2015b).

Halopteris scoparia (Linnaeus) Sauvageau (Fig. 16)

Synonym: *Stypocaulon scoparium*

Common names: English: Sea flax weed (Bunker et al. 2010); Swedish: Taggtofs (Tolstoy and Österlund 2003); Japanese: Hake-kashirazaki, Yezo-kashirazaki (Tokida 1954).

Description: Dark brown algae that forms beautiful fluffy clumps in shallow rocky-bottomed water. Growing only up to 15 cm in length, *H. scoparia* has a main axis with alternate plumed branches which are more or less fan-shaped when flat, though when buoyed up by water they form inverted cone-shaped tufts with a very delicate appearance due to the many filamentous branches. Plants are usually attached to rocks with

Figure 16. *Halopteris scoparia* (Linnaeus) Sauvageau (Ochrophyta, Phaeophyceae). Illustration from William H. Harvey (Plate XXXVII, Phycologia Britannica, 1846–1951).

small to extensive disks often obscured by many matted rhizoids, though these are lacking in free-living plants. These plants are characterized by pure, sheltered waters with high light levels (Pereira 2008, Pereira 2010a, Pereira 2014f).

Geographic distribution: This species spans from Norway to Cape Verde (Atlantic) and the Mediterranean, and also in the Adriatic and Black Seas; it can also be found along the Atlantic provinces of Canada. There have been reports that *H. scoparia* may be in the waters of Japan as well.

Uses: *H. scoparia* could be likewise harvested on a smaller scale as a food supplement and for pharmaceutical purposes. They contain diverse antimicrobial and antifungal substances (Moreau et al. 1984, Munda 2006).

H. scoparia is known to be an ingredient in compounds used in personal care products; it contains growth substances (phytohormones) that include auxins, gibberellins, cytokinins, abscissic acid and betaines (Pereira 2008).

Himanthalia elongata (Linnaeus) S.F. Gray (Fig. 17)

Common names: English: Sea thong (Duddington 1966), Thongweed, Buttonweed, Sea haricots, Sea spaghetti (Anon 2000), Thong weed (Bunker

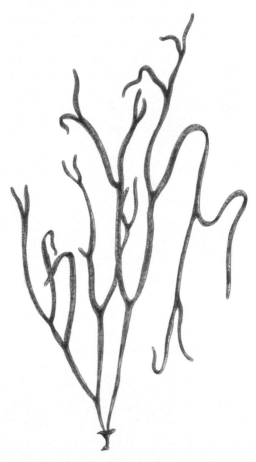

Figure 17. *Himanthalia elongata* (Linnaeus) S.F. Gray (Ochrophyta, Phaeophyceae).

et al. 2010), Sea beans (Braune and Guiry 2011); French: Spaghetti de mer; German: Riementang (Braune 2008); Portuguese: Cintas, Cordas, Corriolas, Esparguete-do-mar (Oliveira 1990, Pereira 2009b, Pereira 2010a); Spanish: Espagueti de mar (Martinez et al. 2014).

Description: Common brown seaweed, with two stage morphology. Small button-like thalli are first produced, from which long strap-like reproductive fronds (receptacles) are formed in autumn. The strap-like reproductive fronds grow quickly between February and May, reaching a length of up to 3 m. The plant releases gametes from June until the winter when it starts to decay. Thalli commonly live for two-three years and reproduce once before dying (Pereira 2010a).

Geographic distribution: NE Atlantic (from Norway to Portugal); Baltic Sea.

Uses: Less known in Asian countries, it is increasingly valued in Europe, both in restaurants and in specialty bakeries. For several years they have manufactured specialty pies, pizzas, pastas, pates, breads, and snacks, since its taste is similar of some cephalopods (squid and cuttlefish). This species is characterized in particular by its high iron content (59 mg per 100 g of algae) and the simultaneous presence of vitamin C, which facilitates the absorption of this trace element. Sea spaghetti is rich in phosphorus, a mineral known to enhance brain function, helping to preserve memory, concentration and mental agility (Michanek 1975, Guiry and Hession 1998, Pereira 2011).

It can be sold and eaten fresh when in season, but it is typically sold dried or pickled, and is eaten most commonly in France and Ireland. It is used as an alternative to both traditional semolina spaghetti and green beans. It is found in popular tartar (sauce), tahini, pâté and cream products, and also in sea vegetable mixes.

See recipes with *Himanthalia elongata* in Annex I.

Laminaria digitata (Hudson) J.V. Lamouroux (Fig. 18)

Common names: English: Tangle (Duddington 1966), Sea girdles, Tangle tail, Wheelbangs, Sea wand, Sea ware (Ohmi 1968), Sea Tangle, Horsetail Kelp, Kelp (Madlener 1977), Strap wrack, Oarweed (Santhanam 2015), Oar weed, Horsetail tangle, Sea Girdle (Arasaki and Arasaki 1983), Kombu Breton; French: Fouet de sorcier (Boisvert 1984); Spanish: Kelp cola de caballo (Radulovich et al. 2013).

Description: Thalli brown to dark-brown, differentiated into a basal holdfast, a strong, slightly flattened stipe merging into the blade. The blade is shaped like a hand with finger-like sections, leathery-firm, with a smooth and slippery surface; stipe smooth, flexible, usually lacking epiphytes, oval section, gradually enlarging towards the lamina (Braune and Guiry 2011).

Figure 18. *Laminaria digitata* (Hudson) J.V. Lamouroux (Ochrophyta, Phaeophyceae).

Geographic distribution: N Atlantic; Atlantic Is (Canary Is, Greenland, Iceland); Baltic Sea.

Uses: Used as food in Ireland and Iceland (Hallsson 1961, Zemke-White and Ohno 1999, Morrissey et al. 2001, Roo et al. 2007), used in W Europe and the UK (Chapman and Chapman 1980), and potential use in Norway (Mæhre et al. 2014), *L. digitata* is imported in Japan and China for making dashi, a soup stock, and for other culinary purposes, such as accelerating the cooking time of vegetables such as beans and lentils.

Laminaria hyperborea (Gunnerus) Foslie (Fig. 19)

Synonym: *Laminaria cloustonii*

Common names: English: Mirkle, Kelpie, Liver weed, Pennant weed (Ohmi 1968), Strapwrack, Cuvie, Tangle, Split whip wrack (Arasaki and Arasaki 1983), Tangleweed (Mourisen 2013), Cuvie (Dickinson 1963), May-weed, Sea rods, Forest Kelp, Northern Kelp (Bunker et al. 2010); Portuguese: Chicote,

Figure 19. *Laminaria hyperborea* (Gunnerus) Foslie (Ochrophyta, Phaeophyceae).

Folha-de-maio, Taborro-de-pé (Pereira 2009b, Pereira 2010a), Rabo-negro (Oliveira 1990, Pereira 2009b).

Description: The blade is broad, large, tough, flat and divided into 5–20 straps or fingers (digitate). The blade is glossy, golden brown to very dark brown in color. The holdfast is large, conical and branched with conspicuous haptera. The stipe is stiff, rough textured, thick at the base and tapers towards the frond. The stipe stands erect when out of water. The stipe is often covered with numerous epifauna and epiflora (Pereira 2010a, Braune and Guiry 2011).

Geographic distribution: European N Atlantic cold-temperate species which does not extend into areas influenced by Arctic waters; its range is the NE Atlantic Ocean, from Scandinavia south to Spain and the Canary Is, the Baltic Sea and the N Sea.

Uses: In Iceland, *L. hyperborea* has traditionally been eaten dried and toasted (Mourisen 2013); is eaten also in Ireland, Norway, Spain, and the UK

(Black 1952, Hallsson 1961, Arasaki and Arasaki 1983, Zemke-White and Ohno 1999, Mæhre et al. 2014, Martinez et al. 2014).

L. hyperborea is one of the two kelp species commercially exploited by the hydrocolloid industry, the other being *L. digitata*. *L. hyperborea* is also utilized by the cosmetic and agrochemical industries and for biotechnological applications, and by the food industry for emulsifiers and gelling agents. Drift kelp has long been collected as an agricultural fertilizer and soil conditioner, and is *L. hyperborea* is still harvested and used in popular kelp meal fertilizer products (Pereira 2015a).

Laminaria ochroleuca Bachelot de la Pylaie (Fig. 20)

Common names: English: Golden Kelp (Bunker et al. 2010), Atlantic Kombu; Portuguese: Folha-de-carriola (Oliveira 1990, Pereira 2010a), Kombu-atlântico, Fitas, Taborrão (Pereira and Correia 2015).

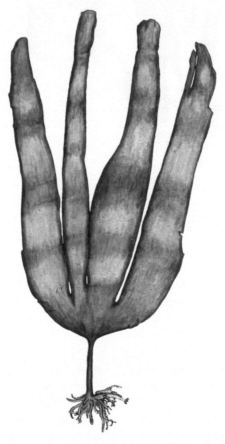

Figure 20. *Laminaria ochroleuca* Bachelot de la Pylaie (Ochrophyta, Phaeophyceae).

Description: Glossy, yellow-brown kelp that is prevalent along the intertidal zones. This kelp is quite conspicuous as it grows quite large under the right conditions. The maximum length recorded is 4 m long, but this length is rarely attained and occurs only in specific areas. Under normal conditions *L. ochroleuca* is more likely to reach a maximum length of about 2 m. *L. ochroleuca* has a large heavy holdfast made up of thick haptera (up to 18 cm in diam.) that supports the plant and anchors it to rocks. This holdfast gives rise to a fairly long, rigid, round, epiphyte-free stipe that tapers somewhat as it approaches the blade. This stipe is so strong and stiff that it stands erect when the plant is out of the water. The blade of this kelp is large, flat, and leathery and divided into five to 20 strap-like digits. This kelp is easily distinguished by the distinct yellow area at the junction of the stipe and the blade. The entire plant actually has a very lovely yellowish hue to its smooth, bright, glossy tissue (Pereira 2015a).

Geographic distribution: Warm-temperate species of kelp, and is most common in the NE Atlantic from the British Isles to the Sahara and the Atlantic Is; Mediterranean.

Uses: Used as food (Kombu) in Europe (Pereira 2010b, Martinez et al. 2014, Pereira 2015a).

Pelvetia canaliculata (Linnaeus) Decaisne & Thuret (Fig. 21)

Common names: English: Cow tang (Dickinson 1963), Channelled wrack and Channel wrack (Bunker et al. 2010); Portuguese: Botelho-bravo (INII 1966).

Description: Common brown seaweed found high on the shore. It is very tolerant of desiccation surviving up to eight days out of the water. *P. canaliculata* lives for about four years and grows up to 15 cm long. The fronds of the algae are curled longitudinally forming a channel (Pereira 2010a).

Geographic distribution: NE Atlantic, from the Arctic Ocean to the Iberian Peninsula, in the English Channel, and in the N Sea. It is common on the Atlantic shores of Europe from Iceland to Spain and Portugal (southern limit of this species).

Uses: Usually cooked, in small quantities, with traditional vegetables; for food, it is sold in different dried forms, and as a seasoning mix (Pereira 2015a). Like a natural multi-vitamin/mineral food supplement and can be dried and sprinkled over food (Green 2014, Hairon 2014, Mæhre et al. 2014).

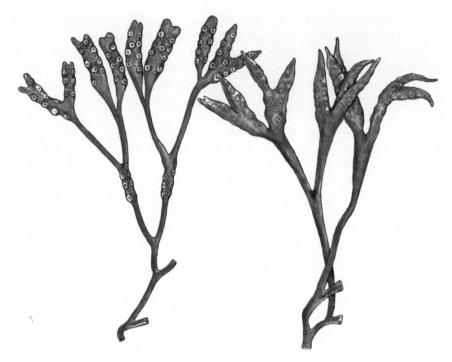

Figure 21. *Pelvetia canaliculata* (Linnaeus) Decaisne & Thuret (Ochrophyta, Phaeophyceae).

Petalonia fascia (O.F. Müller) Kuntze (Fig. 22)

Synonyms: *Ilea fascia*, *Phyllitis fascia*

Common names: English: Sea petals (Chapman and Chapman 1980), Broad leaf weed (Bunker et al. 2010); Japanese: Haba-nori (Yendo 1902, Smith 1904), Seiyohabanori (Madlener 1977), Hondawara (Chapman and Chapman 1980).

Description: *P. fascia* is a small brown alga, typically brown in color but which can also be olive green. Usually it is seen between late autumn and early summer. The alga has a small discoid attachment base, from which emerge fronds some 20 cm long and 4–6 cm wide that are flattened. Being common in winter months it used to be collected extensively by tidal communities during the "hungry gap" that marks the end of winter, just before the beginning of spring (Pereira 2015a).

Geographic distribution: N Atlantic (Greenland to Canary Is); Mediterranean; NW Atlantic (Arctic to New Jersey); SE Atlantic (Senegal, Namibia, S Africa); SW Atlantic (Brazil, Uruguay); Indian Ocean (Pakistan, S Africa); NW Pacific (Japan, China); NE Pacific (Alaska to California); SE Pacific (Chile); Australia, New Zealand; Antarctica, and Sub-Antarctica.

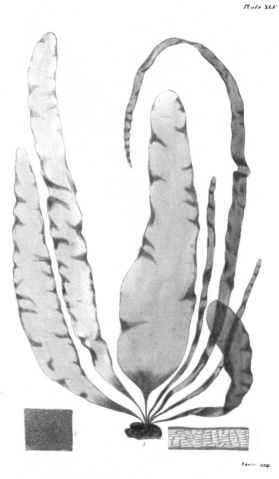

Figure 22. *Petalonia fascia* (O.F. Müller) Kuntze (Ochrophyta, Phaeophyceae). Illustration from William H. Harvey (Plate XLV, Phycologia Britannica, 1846–1951).

Uses: Used for foodstuff in Japan and Korea (Yendo 1902, Smith 1904, Bonotto 1976, Madlener 1977, Chapman and Chapman 1980), this species is high in sugar and starch and has an agreeable flavor which means it can be used cooked or raw. It can be dried and re-constituted (Green 2014, Pereira 2015a).

Saccharina latissima (Linnaeus) C.E. Lane, C. Mayes, Druehl & G.W. Saunders (formerly *Laminaria saccharina*) (Fig. 23)

Common names: English: Sea belt, Poor man's weather glass (Duddington 1966), Sweet wrack, Sugar wrack, Sugar tang, Oarweed, Tangle, Kelp, Sugar

Figure 23. *Saccharina latissima* (Linnaeus) C.E. Lane, C. Mayes, Druehl & G.W. Saunders (Ochrophyta, Phaeophyceae).

sea belt, Sweet tangle (Madlener 1977), Sugarwrack (Rhoads and Zunic 1978), Zuckertang (Braune 2008), Royal or Sweet Kombu (MacArtain et al. 2007); French: Laminaire sucrée (Boisvert 1984); German: See-Palme (Braune 2008); Japanese: Karafuto Kombu, Karafuto Tororo Kombu, Kan-Hoa (Madlener 1977); Portuguese: Rabeiro (Oliveira 1990, Pereira 2009b), Kombu-real (Pereira 2009b, Pereira 2010a).

Description: Sporophyte differentiated into a basal holdfast, a small but firm, cylindrical stipe slightly flattened at the top and a single, undivided blade (phylloid), light brown to deep chocolate brown, that is highly variable in structure: paper-thin to leathery-coarse, shape elongated, ribbon-like to lanceolate (widest part towards the base), margin entire but usually undulating to heavily creased, middle section smooth or more commonly wrinkled, creased or crinkled (Braune and Guiry 2011).

Geographic distribution: N and NE Atlantic (Greenland to Portugal, N Sea, Baltic); NW Atlantic (Canadian Arctic to Massachusetts); NE Pacific (Alaska to California).

Uses: Edible and used in France, Ireland, Spain and Portugal in cooking (Hallsson 1961, Chapman and Chapman 1980, Arasaki and Arasaki 1983, Zemke-White and Ohno 1999, MacArtain et al. 2007, Pereira 2010b, Braune and Guiry 2011, Pereira 2015a); is also consumed in N America (Harrison 2013).

See recipes with *Saccharina latissima* in Annex I.

Saccharina longicruris (Bachelot de la Pylaie) Kuntze (formerly *Laminaria longicruris*)

Common names: English: Kelp, Oarweed (Madlener 1977), Atlantic Kombu (Chapman and Chapman 1980), Atlantic kelp.

Description: The hollow, cylindrical stipe itself can reach up to 10 m long, plus the frond which adds another 1 to 2 m; the large, branched holdfast grips firmly to the rocky substratum allowing a single long, thin, olive-brown, leafy blade to float near the surface. The midsection of these blades is somewhat thicker, but the edges spread and thin and become wide and ruffled (Pereira 2015a).

Geographic distribution: Arctic (Canada); is common also in the N Atlantic, especially around the coasts of New England, US.

Uses: These kelps are a prominent source of algin and food in US and in the Oriental market. Traditionally they have been a source of iodine and potash. Their stipes were used to open wounds, aid in cervical dilation, and induce abortions. Oarweed is harvested in Maine for health food stores where it is sold as "Kombu". Prepared plants may be cooked as a vegetable or added to soups. As with the other kelps, Oarweed is a natural source of glutamic acid (Madlener 1977, White and Keleshian 1994, Zemke-White and Ohno 1999, Roo et al. 2007, Harrison 2013, Pereira 2015a).

Saccorhiza polyschides (Lightfoot) Batters (Fig. 24)

Common names: English: Furbelows (Arasaki and Arasaki 1983), Furbellows (Bunker et al. 2010, Pereira 2010a); Portuguese: Caixeira, Carocha, Cintas, Golfe, Golfo, Limo-correia, Limo-corriola (INII 1966).

Description: Species with a distinctive large warty holdfast and a flattened stipe with a frilly margin. The stipe is twisted at the base and widens to form a large flat lamina, which is divided into ribbon-like sections. Presence of a large bulbous holdfast with warty appearance (Pereira 2010a).

Geographic distribution: NE Atlantic; Atlantic Is (Canary Is).

Uses: Considered an edible species in North of Portugal and Galicia (Spain), and used as food in cold Atlantic regions (Arasaki and Arasaki 1983, Lodeiro et al. 2012, Green 2014, Martinez et al. 2014, Rodrigues et al. 2015b).

Figure 24. *Saccorhiza polyschides* (Lightfoot) Batters (Ochrophyta, Phaeophyceae).

Sargassum vulgare C. Agardh (Fig. 25)

Common names: English: Gulf weed (Braune and Guiry 2011); German: Beerentang (Braune 2008); Portuguese: Sargasso (Neto et al. 2005, Pereira 2009b).

Description: Much branched, bushy plants that grow to 50 cm tall and are attached by a discoid holdfast; thalli pseudo-parenchymatous. Primary and secondary branches are cylindrical and bear lanceolate foliar branches (4 cm long, 3 mm wide) with serrate margins. Bladders are formed on short pedicels. Receptacles are small and often divided and are also formed on pedicels in the axils of the foliar branches; they are often covered with small dots (pits) containing the reproductive organs (Neto et al. 2005).

Geographic distribution: NE Atlantic; Mediterranean; Caribbean Sea; SE Atlantic; Indian Ocean; W Pacific (the Philippines).

Figure 25. *Sargassum vulgare* C. Agardh (Ochrophyta, Phaeophyceae).

Uses: Some *Sargassum* specimens are consumed fresh (Gupta and Abu-Ghannam 2011, Sidik et al. 2012), others cooked in coconut milk or a little vinegar or lemon juice. It is smoked-dried to preserve it. *Sargassum* is also eaten by itself or added to fish and meat dishes. If not strong it can be added to salads after washing, or it can be cooked in water like a vegetable. If the *Sargassum* is strong flavored it can be boiled in two changes of water. Some recipes then call for it to be mixed with brown sugar and used as a filling in steamed buns but it can also be eaten as it is (Nisizawa 2006, Radulovich et al. 2013).

Scytosiphon lomentaria (Lyngbye) Link (Fig. 26)

Common names: English: Leather tube (Arasaki and Arasaki 1983), Chipolata Weed (Bunker et al. 2010), Soda straws, Whip tube; German:

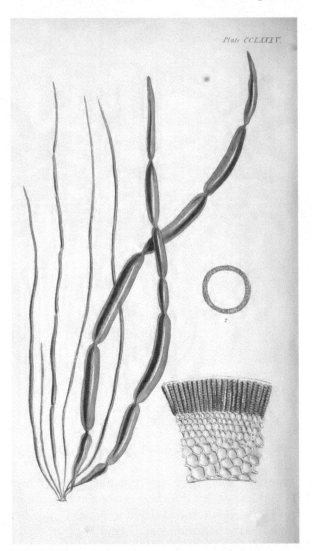

Figure 26. *Scytosiphon lomentaria* (Lyngbye) Link (Ochrophyta, Phaeophyceae). Illustration from William H. Harvey (Plate CCLXXXV, Phycologia Britannica, 1846–1951).

Geschnürter schlauchtang (Braune 2008); Japanese: Kayamo-nori, Sugara, Mugiwara-nori (Madlener 1977); Korean: Korimae (Sohn 1998); Swedish: Korvsnöre (Tolstoy and Österlund 2003).

Description: A light to dark brown, sometimes hollow, cylindrical seaweed, that grows to 33 cm tall. Fronds may occur in groups or singly, are unbranched to 2.3 mm wide, tapering at both ends and arising from a short stipe and small attachment disk (Neto et al. 2005).

Geographic distribution: Cosmopolitan in temperate and cold seas.

Uses: Used for foodstuff in Asia (Tokida 1954, Kang 1968, Bonotto 1976, Arasaki and Arasaki 1983, Tseng 1983, Hotta et al. 1989, Zemke-White and Ohno 1999, Mori et al. 2004). There are numerous local methods of preparing this popular alga. Fujian (China) people boil it until it is tender and then add fish stock to make a soup; or, after it is boiled, soy sauce is added, along with "cellophane noodles" or "long rice," a clear vermicelli-like noodle made of bean flour. Guangdong people prepare it in ways similar to those for *Laminaria* (Chapman and Chapman 1980). In Taiwan, the young plant is mixed with pork and fish and then fried. A thick rice soup is made with the addition of dried, powdered blades of *Endarachne* (Bangmei and Abbott 1987). In Korea is used in various seasoning soup, and fried in oil (Sohn 1998, Roo et al. 2007); is used also as food in France (Zemke-White and Ohno 1999, Roo et al. 2007, Harrison 2013).

Used in the treatment of dry coughs, laryngitis, lymphatic tuberculosis, and has antimitotic action (Chénieux et al. 1980, Demirel et al. 2012).

Taonia atomaria (Woodward) J. Agardh (Fig. 27)

Common name: English: Dotted Peacock weed (Bunker et al. 2010).

Description: Thallus erect, flat ribbon-shaped, paper-like thin, irregularly branching into wedge-shaped segments and/or deeply dividing into narrower, almost linear bands, the upper margin appearing frayed; without midrib; dark, slightly undulating zones of transverse stripes on both sides (hairs and reproductive structures). Stipe-like at the bottom anchored by a dense rhizoidal felt (Braune and Guiry 2011).

Geographic distribution: NE Atlantic (Britain to Mauretania), Atlantic Is (Azores, Canary Is, Madeira, Selvage Is); Adriatic Sea; Mediterranean; Persian Gulf.

Uses: Used as functional foods ingredient (Cornish and Garbary 2010, Pereira et al. 2012, Pereira 2015a).

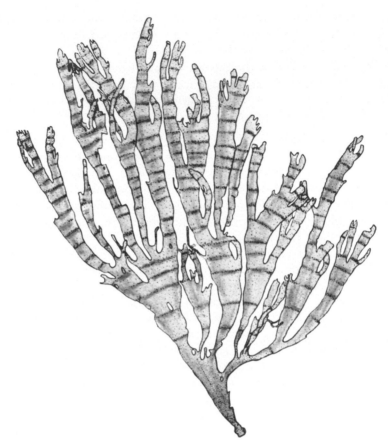

Figure 27. *Taonia atomaria* (Woodward) J. Agardh (Ochrophyta, Phaeophyceae).

2.1.3 Rhodophyta

Ahnfeltia plicata (Hudson) E.M. Fries (Fig. 28)

Common names: English: Bushy Ahnfelt's seaweed (Arasaki and Arasaki 1983), Black scour weed (Bunker et al. 2010), Landlady's wig; French: Fil de fer (Dickinson 1963); Japanese: Netsuki-itanigusa (Tokida 1954), Itanigusa (Arasaki and Arasaki 1983), Kanten (McConnaughey 1985); Swedish: Havsris (Tolstoy and Österlund 2003).

Description: Perennial red seaweed which forms dense, tangled tufts; the fronds are very fine, tough, and wiry with irregular or dichotomous branching and up to 21 cm in length. The holdfast is disk-like or encrusting, 0.5–2 cm in diam. The fronds are dark brown when moist and appear almost black when dry. The uppermost branches are often green (Rayment 2004).

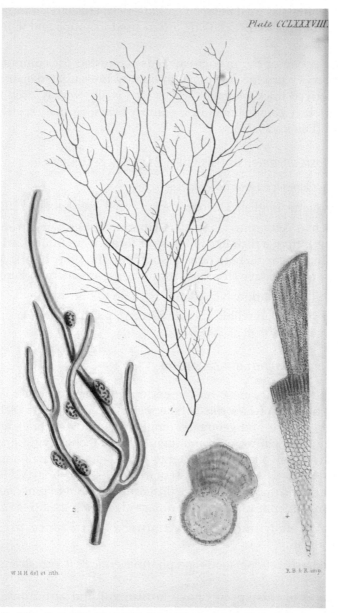

Figure 28. *Ahnfeltia plicata* (Hudson) E.M. Fries (Rhodophyta). Illustration from William H. Harvey (Plate CCLXXXVIII, Phycologia Britannica, 1846–1951).

Geographic distribution: N and NE Atlantic (Greenland to Azores); SW Atlantic (Uruguay); NE Pacific (Alaska to Mexico); NW Pacific (Russia); SE Pacific (Chile); Indian Ocean; Asia (Japan, Russia); SE Asia (Vietnam); Australia (Tasmania); Antarctic and subantarctic Is.

Uses: Consumed as food in Japan (Tokida 1954, Hu 2005); it forms a major component of paleo diet cooking, often used in soups and jellies (Green 2014, Pereira 2015a).

A. plicata is one of the major seaweeds harvested for commercial agar production as it produces a very high-quality, low-sulfated agar. As a non-animal derived thickening agent, *A. plicata* is also widely used for vegan products (such as marshmallows, gummy bears, cosmetics, etc.) in place of gelatin (Pereira 2015a).

Ahnfeltiopsis devoniensis (Greville) P.C. Silva & DeCew

Synonym: *Gymnogongrus devoniensis*

Common name: English: Devonshire fan weed (Bunker et al. 2010).

Description: *A. devoniensis* is a small red marine alga growing to only several cm in length from a disk-like holdfast. It forms a medium-sized flattened frond with regular dichotomous branching. The branches have parallel sides. The reproductive structures (cystocarps) are internal (Pereira 2004).

Geographic distribution: NE Atlantic (Britain to Portugal).

Uses: Considered an edible species (Green); produces iota-kappa hybrid carrageenan (Pereira et al. 2009b).

Amphiroa cryptarthrodia Zanardini

Description: Thallus erect, bushy; endophytic base on other calcareous algae (particularly on *Lithophyllum* species). Size: 2–4 cm high. Dichotomous branching, regular and geometric (angle of about 90°); branch junctions usually not coinciding with intergenicula. Branches lying on different planes, slightly swollen apices. Intergenicula cylindrical with rare annular swellings, tapered on the upper part of fronds 1–4 mm long, 0.25–0.6 mm in the lower diam., upper diam. 0.15–0.18 mm; dark red to pink violet (rare), apart from lighter apices due to annular ridges (Pereira 2014a, 2015a).

Geographic distribution: NE Atlantic (from France to Senegal); Mediterranean.

Uses: Used in functional foods and pharmaceuticals (Cornish and Garbary 2010, Pereira 2015a).

Extracts of this species have antifungal and antimitotic activity (Ballesteros 1992).

Asparagopsis armata Harvey

Common name: English: Harpoon weed, Harpoon-alga.

Description: In North-eastern Europe, gametophyte plants occurring from June or July–August or September (sometimes overwintering), pale purplish-red, quickly degenerating when removed from the water and becoming distinctly orange; fronds bushy, with a cylindrical axis to 1 mm wide and 200 mm long, arising from bare, creeping stolons; irregularly branched, with four rows of branchlets, simple, short, branchlets alternating with longer ones with four rows of simple filamentous ramuli; lower branchlets unbranched, long, tapered, with harpoon-like barbs. Tetrasporophyte (*Falkenbergia*-phase) occurring all year round, but most obvious in October–March, brownish-red, much branched, filamentous, in dense cotton-wool-like tufts to 15 mm in diam. (Braune and Guiry 2011).

Geographic distribution: *A. armata* is native to the Southern Hemisphere (Australia and New Zealand) but has been introduced to the Northern Hemisphere, first recorded in Europe in 1925 and since spreading throughout the Channel Is and all around Great Britain. *A. armata* is now found globally, from the Canary Is to Morocco and throughout the Pacific and Indian Oceans.

Uses: Considered an edible species (Green 2014).

Bangia atropurpurea (Mertens ex Roth) C. Agardh

Common name: Swedish: Purpurtråd (Tolstoy and Österlund 2003).

Description: Gelatinous, unbranched, blackish-purple filaments, at first uniseriate later multiseriate, attached by rhizoidal outgrowths from basal and adjacent cells; cells with central star-shaped chloroplast with pyrenoid (Guiry and Guiry 2014a).

Geographic distribution: *B. atropurpurea* has a widespread amphi-Atlantic range (Kipp et al. 2014, Tittley and Neto 2005).

Uses: Used for foodstuff in Korea (Madlener 1977, Tseng 1983).

Bangia fuscopurpurea (Dillwyn) Lyngbye

Common names: Chinese: Hangmaocai (Montagne 1946), Tou mau tsai (Madlener 1977); English: Velvet thread weed (Bunker et al. 2010); Japanese: Ushike-nori (Tokida 1954, Madlener 1977).

Description: Gelatinous, unbranched, blackish-purple filaments, at first uniseriate later multiseriate, attached by rhizoidal outgrowths from basal and adjacent cells. Cells present central star-shaped rhodoplast with pyrenoid (Guiry 2014a).

Geographic distribution: NE Atlantic; Tropical and subtropical W Atlantic; NE Pacific; Mediterranean Sea; SW Asia (Arabian Gulf, Israel); Asia (China, Japan, Korea, Taiwan); SE Asia (Thailand, Vietnam); Australia; Pacific Is (Hawaiian Is); Antarctic Is.

Uses: Consumed as vegetable in Japan, Taiwan and Thailand (Tokida 1954, Lewmanomont 1978, Arasaki and Arasaki 1983, Hu 2005). In Fujian (China), fishermen often roll this stringy seaweed into cylinders or rolls more than 10 cm long; the fresh or dried rolls are then sold in the market. Preparation consists of cutting the roll into small pieces and frying in a small amount of oil until crisp. Also it used to make soup in Hawaii (Montagne 1946, Bangmei and Abbott 1987).

Extracts of this species have antioxidant, antibacterial, antiviral, cytotoxic (Cornish and Garbary 2010, Stefanov et al. 2009) activity.

Callophyllis laciniata (Hudson) Kützing

Common names: English: Beautiful fan weed (Bunker et al. 2010), Fanweed.

Description: Thick, flattened, sub-cartilaginous, ± palmate, deep crimson fronds, to 200 mm high, with short wedge-shaped stipe from small discoid base. Frond ± deeply cleft into wedge-shaped segments, divide themselves sub-dichotomously, apices rounded, margins smooth or fringed with minute proliferations (Braune and Guiry 2011).

Geographic distribution: NE Atlantic; Mediterranean.

Uses: Considered an edible species (Green 2014).

Catenella caespitosa (Withering) L.M. Irvine (Fig. 29)

Common name: English: Creeping chain weed (Bunker et al. 2010).

Description: A moss-like plant with an irregularly and highly branched frond. The frond is constricted at intervals into different sized segments and the holdfast is a mass of tangled fibers. The plant is small (growing up to 2 cm tall) and is dark purple (Pizzola 2003).

Geographic distribution: NE Atlantic; Atlantic Is (Ascension, Canary Is, Madeira); Tropical and subtropical Atlantic; Caribbean Sea; Indian Ocean; Asia (Japan, Malaysia, the Philippines, Singapore, Thailand).

Uses: Considered an edible species Pakistan (Rahman 2002).

Ceramium diaphanum (Lightfoot) Roth (Fig. 30)

Description: Thallus of numerous erect axes, pinkish-red, 6–11 cm high, thickly or sparsely branched; lower part 0.4–0.5 mm diam., decreasing to 0.1–0.3 mm diam. in outer parts; nodes and internodes clearly marked; branching regularly dichotomous or repeatedly pseudo-dichotomous, occasional short supplementary lateral branchlets; branches divaricated at base, the divergence angle decreases at apex, apices incurved; segments four-five times as long as wide, corticated nodes slightly bulged, broad and thick; internodes cylindrical (Milchakova 2011).

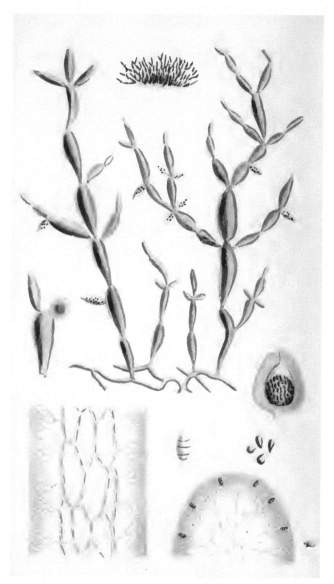

Figure 29. *Catenella caespitosa* (Withering) L.M. Irvine (Rhodophyta). Illustration from William H. Harvey (Phycologia Britannica, 1846–1951).

Geographic distribution: NE Atlantic; Arctic Canada; Adriatic Sea; Baltic Sea; Black Sea; Mediterranean; Atlantic Is (Ascension; Canary Is, Cape Verde Is, Madeira, Selvage Is); Caribbean Sea; Tropical and subtropical Atlantic; India Ocean; Asia (China, Taiwan); SE Asia (Indonesia, the Philippines, Vietnam); Australia; Pacific Is (Micronesia, Mariana Is, Polynesia, Hawaiian Is); Antarctic Is.

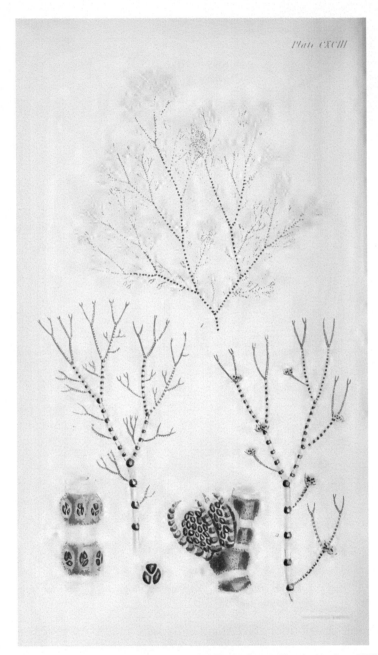

Figure 30. *Ceramium diaphanum* (Lightfoot) Roth (Rhodophyta). Illustration from William H. Harvey (Plate CXIII, Phycologia Britannica, 1846–1951).

Uses: This seaweed could be used in dermatology, cosmetics, medicine and food industry (Milchakova 2011). This species is a source of polysaccharides, BASs with antimicrobial, antiviral and antitumor activity (Milchakova 2011).

Chondracanthus acicularis (Roth) Fredericq (Fig. 31)

Common names: English: Creephorn (Bunker et al. 2010); Portuguese: Cabeça-de-preto, Meruge, Musgo-da-pedra (INII 1966), Barranha (INII 1966, Pereira 2010a), Musgos (Pereira and Correia 2015).

Description: Cartilaginous, cylindrical, purple-red or blackish fronds, to 100 mm long, irregularly bipinnately branched, branches curved, sharply

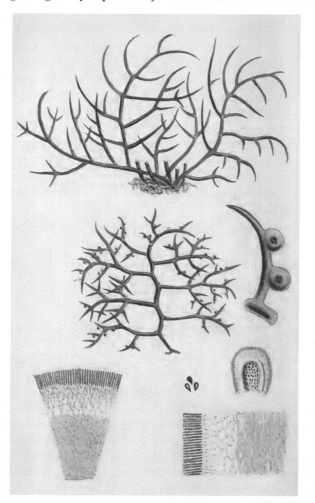

Figure 31. *Chondracanthus acicularis* (Roth) Fredericq (Rhodophyta). Illustration from William H. Harvey (Phycologia Britannica, 1846–1951).

pointed. Base discoid, often stoloniferous and entangled (Braune and Guiry 2011).

Geographic distribution: E Atlantic; Mediterranean; NW and SW Atlantic; Caribbean Sea; Indian Ocean; Australia; Pacific Is (Hawaiian Is).

Uses: Raw material for carrageenan-production (Portugal and W Africa) (Braune and Guiry 2011, Pereira 2013). Used as an ingredient in salads, rice, pasta, scrambled eggs, etc. They are also used in the preparation of soups, creams and sauces. Accompanying both seafood dishes, the fish meat; and even in making desserts and teas (Gómez-Ordóñez and Rupérez 2011).

Chondria coerulescens (J. Agardh) Falkenberg

Common name: English: Iridescent cartilage weed (Bunker et al. 2010).

Description: *C. coerulescens* has bluish or yellowish fronds with blue iridescence. The fronds are flexible and cartilaginous in texture, turning black when dry. Young axes show a striking turquoise iridescence when alive. The thalli consist of cylindrical erect axes or trailing tufts, and are 3–8 cm high when erect. The distinct main axis is 0.4–0.5 mm in diam., branching sparsely at irregular intervals in a spiral pattern to one-three orders of branching. Branches are linear, often long and curve downwards gradually tapering to a slender point, and reattach by secondary holdfast. The morphology shows relatively little variation except that some thalli consist only of inconspicuous isolated erect axes whereas others form dense tufts (Edwards 2005).

Geographic distribution: NE Atlantic (Ireland and Britain to Senegal); Mediterranean.

Uses: Considered and edible species (Green 2014).

Chondrus crispus Stackhouse (Fig. 32)

Common names: English: Irish Moss, Carragheen, Carragheen Moss, Dorset weed, Pearl Moss, Sea Moss, Sea Pearl Moss, Jelly Moss, Rock Moss, Gristle Moss, Curly Moss, Curly Gristle Moss (Madlener 1977), Carrageen, Carraghean, Carrageenin (Rhoads and Zunic 1978); French: Goemon blanc, Mousse d'Irlande (Madlener 1977), Mousse d'Irlande; German: Irischmoos, Irisches moos (European Communities 1998); Italian: Muschio Irlandese (European Communities 1998); Japanese: Tsuno-mata, Hosokeno-mimi (Smith 1904); Portuguese: Musgo-gordo, Botelho (Sanches 1989), Botelha, Cuspelho, Musgo, Limo-folha (Oliveira 1990), Musgo gordo, Folha-de-alface, Musgo da Irlanda (Pereira 2010a), Crespo, Musgo-irlandês, Folhinha (INII 1966, Pereira and Correia 2015); Spanish: Condrus (European Communities 1998); Swedish: Karragener (European Communities 1998).

Figure 32. *Chondrus crispus* Stackhouse (Rhodophyta). Illustration from William H. Harvey (Plate LXL, Phycologia Britannica, 1846–1951).

Description: Highly variable (polymorphous) thalli may reach 15 cm long, cartilaginous consistency and reddish-pink or brown color and iridescent in water. These algae are fixed by a disk that start as a unbranched stipe gradually expanding into fan-like blade, repeatedly dichotomously divided, with ends rounded or truncated. On the surface of the blades may appear small dilations (2–3 mm in diam.) which are the reproductive structures (Pereira 2010a).

Geographic distribution: *C. crispus* has a wide distribution. It includes the NW Atlantic from Labrador and the Maritime Provinces in Canada southward to New Jersey and Delaware in the United States. It can be found to a limited extent in the western Baltic Sea, from northern Russia and Norway to southern Spain, spanning the NE Atlantic, the N Sea, and the English Channel. *C. crispus* can be found in the Mediterranean, Portugal, the Azores, the Faroes, and West Africa, as well as in the Bering Sea, from Russia to Alaska.

Uses: Irish moss or carrageen moss, by far the tastiest seaweed, continues to be lightly harvested for food in Ireland, Scotland, on the coasts of England, Iceland, and more recently Canada and the US (Hallsson 1961, Arasaki and Arasaki 1983, Zemke-White and Ohno 1999). In days past it was recommended for delicate people, as it was claimed to be strengthening and easily digested. It was made into jelly or blancmange, being boiled for some time in water, strained and boiled up again with milk, spice, sugar and lemon peel. It thickens milk as gelatin or corn-flour does, and at one time was ordered by doctors in cases of throat and chest complaints. Its reddish-purple or green fronds, forked and frilled, were spread in the sun to dry and bleach, and afterwards packed in bags and sent off for sale. Masses of Irish moss bleached and dried white by sun and strong sea air, were often found spread on the grassy headlands along the coasts of Scotland and Ireland (Arasaki and Arasaki 1983, Roo et al. 2007, Worthington 2014, Pereira 2015a).

Quite early in New England's history articles were published mentioning Irish moss as a common household remedy. By the mid 1800s it was commonly used as a food additive. It also was eaten as a relish; consumed fresh or dried; boiled with milk, sugar and spices as a pudding or healthy drink; cooked in butter and served as a vegetable with meat or fish; or added to stews. It was used as an emulsifier for cod fish oil, and made into jellies for sick patients, as well as fillings, icings, meringues, glazes and marshmallows, and as a clarifying agent in beer. Today it is a source of carrageenan, commonly used as a thickener and stabilizer in milk products like ice cream and luncheon meats (Worthington 2014).

"Hana nori", a yellow strain of *C. crispus*, that resembles another traditional Japanese seaweed, first introduced to the Japanese market in 1996; the dried product, to be reconstituted by the user, was reported to be selling well at the end of 1999, with forecasts of a market valued at tens of millions of US dollars. It is used in seaweed salads, sashimi garnishes and as a soup ingredient (McHugh 2003). Acadian Seaplants Limited (Nova Scotia, Canada) has a culture of a unique strain of edible *C. crispus*, Hana Tsunamata™ (Fig. 33), for the Asian (mostly Japanese) human food market (kaiso salads, sashimi garnishes and soups) by manipulating the color and the texture of selected isolates (Chopin and Ugarte 2006, Acadian Seaplants 2014). This species is used also as food in Korea (Bonotto 1976, Madlener 1977).

See recipes with *Chondrus crispus* in Annex I.

Figure 33. Edible *Chondrus crispus*, Hana Tsunamata™, from Acadian Seaplants Limited (Nova Scotia, Canada).

Chordaria flagelliformis (O.F. Müller) C. Agardh

Common names: English: Slimy Whip Weed (Bunker et al. 2010); German: Geisseltang (Braune 2008); Japanese: Naga-matsumo (Tokida 1954).

Description: Dark brown seaweed, brushy with branches whip-like and with few secondary branches, often longer than the main stem; thallus string-like, with divaricated branches, slippery, solid, 10–30 cm high; branching irregular, alternate from all sides; branches cylindrical, or slightly compressed, to 3 mm wide, slightly tapering towards the base. Branch apices blunt (Gosner 1978).

Geographic distribution: Arctic (Canada, Svalbard, White Sea); N Atlantic; Baltic Sea; Atlantic Is (Greenland, Iceland); Tropical and subtropical W Atlantic; Asia (Japan, Korea, Russia); Antarctic Is (Campbell Is).

Uses: Is eaten in Japan, fresh or dried (Tokida 1954, Levring et al. 1969, Michanek 1975, Hoppe 1979, Rao et al. 2010); also eaten in Korea (Bonotto 1976).

Extracts of this species have antimicrobial (Homsey and Hide 1974), anticoagulant (Hoppe 1979), and agglutinin (Shiomi 1983) activity.

Corallina officinalis Linnaeus (Fig. 34)

Common name: English: Common coral weed (Bunker et al. 2010); French: Coralline rose.

Description: Whitish-pink to lilac, calcified, articulated fronds, 60–120 mm high, axis cylindrical to compressed, repeatedly pinnate from and expanded discoid base, branching often irregular. Growth form very variable often stunted. In unfavorable habitats erect system vestigial, but extensive base may be present (Brodie et al. 2013).

Geographic distribution: Worldwide.

Uses: Edible seaweed (Dominguez 2013, Gray 2015).

Extracts of this species have antibacterial (Taskin et al. 2007), and antihelminthic (Pereira 2015a) activity. *C. officinalis* is a very popular ingredient among cosmetics and health and personal care companies. There are known sellers of *C. officinalis*-based products in the US, China, Italy,

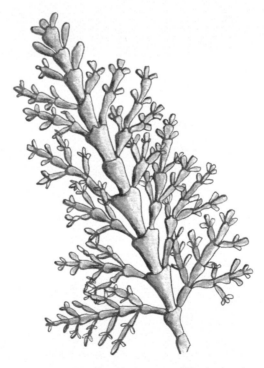

Figure 34. *Corallina officinalis* Linnaeus (Rhodophyta).

France, Switzerland, and Germany. These are products for men and women, and include toners, moisturizers, cleansers, emulsions, essences, astringents, eye creams, wash gels, shower gels, shaving balm, hydration sprays and creams, and masks (Milchakova 2011, Pereira 2014d, Pereira 2015a).

Cordylecladia erecta (Greville) J. Agardh

Common name: Erect clublet (Bunker et al. 2010).

Description: Narrow cylindrical, cartilaginous, brownish-red fronds, to 100 mm high, in tufts from thin, spreading, discoid base often hidden below sand; Sparingly sub-dichotomously branched, branches usually tapering to fine point; dioecious. Tetrasporangia and spermatangia in terminal, deciduous, spindle-shaped pods, cystocarps clustered in swollen areas near apices. Reproduction takes place in the depths of winter (Braune and Guiry 2011).

Geographic distribution: NE Atlantic; Atlantic Is (Azores).

Uses: Considered an edible species (Green 2014).

Cystoclonium purpureum (Hudson) Batters

Common name: English: Purple claw weed (Bunker et al. 2010).

Description: Rather soft, cylindrical, dull purplish pink fronds, 3 mm wide, to 600 mm long. Branches numerous, alternate, branchlets tapered at both ends; branches sometimes drawn out into long twisting tendrils; multiaxial, medulla a cordlike strand of loosely interwoven, narrow filaments, surrounded by large, rounded cells, with outer layer of small, angular, assimilatory cells; known to have an onion-like smell shortly after collection (Braune and Guiry 2011).

Geographic distribution: N Atlantic (Arctic Canada to Spain); North Sea (Helgoland); Baltic, Mediterranean (Greece), Indian Ocean (Pakistan), Australia (Queensland).

Uses: Considered and edible species (Green 2014); is also a source of carrageenan (Braune and Guiry 2011).

Devaleraea ramentacea (Linnaeus) Guiry

Description: Thallus hollow, saccate or tubular, flattened or terete, membranaceous or coriaceous when old, one or several fronds arising from a small basal disk (Guiry 1982).

Geographic distribution: Atlantic Is (Greenland); N Atlantic (Arctic Canada, Nova Scotia); NE Pacific (Alaska); Asia (Japan).

Uses: Used as food in Nova Scotia, Canada (Michanek 1975).

Dilsea carnosa (Schmidel) Kuntze (Fig. 35)

Synonyms: *Dilsea edulis, Iridaea edulis, Schizymenia edulis*

Common names: English: Dulse (Dickinson 1963), Red rags (Bunker et al. 2010).

Description: Dark red, frequently becoming yellow above, thickest of the foliose red algae in the N Atlantic, flattened cartilaginous fronds, arising in groups of small, medium and large from an thick, discoid holdfast, obtuse, ovate with tapered base, to 500 mm long, 250 mm broad (Pereira 2010a).

Geographic distribution: NE Atlantic; Atlantic Is (Azores); Baltic Sea.

Uses: Possibly eaten in Britain, Scotland, Ireland and Iceland (Chapman and Chapman 1980, Zemke-White and Ohno 1999, Roo et al. 2007), but reports may have confused species with *Palmaria palmata* (Guiry and Garbary 1991). This perennial seaweed is not recommended for eating these days as it contains suspect compounds; the name change itself again indicative of the changing face of marine botany as modern science comes to better understand seaweeds. However this common, perennial, deep red seaweed

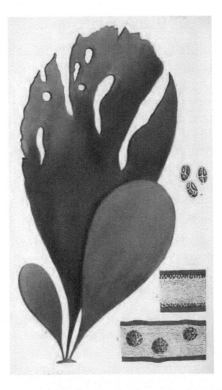

Figure 35. *Dilsea carnosa* (Schmidel) Kuntze (Rhodophyta). Illustration from William H. Harvey (Phycologia Britannica, 1846–1951).

was generally eaten raw but sometimes fried. Perhaps keep this one as emergency food only! (Harrison 2008).

Dumontia contorta (S.G. Gmelin) Ruprecht (Fig. 36)

Synonym: *Dumontia incrassata*

Common names: English: Dumont's tubular weed (Bunker et al. 2010); Japanese: Ryûmonsô (Tokida 1954).

Description: Erect, gelatinous, pale yellowish to purple fronds, solitary or tufted from discoid base, tubular, cylindrical or compressed, irregularly branched, to 500 mm long, filled with watery mucilage. Axis and branches tapered at both ends, irregularly swollen, often twisted, to 5 mm or more broad. Basal disk to 20 mm diam., persistent, erect fronds initiated internally (Braune and Guiry 2011).

Figure 36. *Dumontia contorta* (S.G. Gmelin) Ruprecht (Rhodophyta). Illustration from William H. Harvey (Plate LIX, Phycologia Britannica, 1846–1951).

Geographic distribution: Artic Canada; N Atlantic; Atlantic Is (Azores Is); Baltic Sea; White Sea; Asia (Japan, Russia); NE Pacific (Alaska).

Uses: Used as food in Japan (Tokida 1954); unique properties of *D. contorta* have been identified as potent anti-viral agents, effective against herpes. Red algae also have many wonderful health benefits when eaten or taken as dietary supplements as they posses many micro-nutrients and minerals as well as components that are known to be good for soft, healthy skin and hair (Gómez-Ordóñez and Rupérez 2011).

Ellisolandia elongata (J. Ellis & Solander) K.R. Hind & G.W. Saunders

Synonym: *Corallina elongata*

Description: Whitish pink to reddish lilac, calcified, articulated fronds, fish-bone-like arrangement, to 50 mm high, axis compressed, repeatedly pinnate from discoid base, more abundantly and regularly branched than *C. officinalis*; articulations small (Pereira 2010a, Pereira 2015a).

Geographic distribution: NE Atlantic (Ireland and Britain to Senegal); Atlantic Is (Azores, Canary Is, Madeira, Selvage Is); Adriatic Sea; Mediterranean; Asia (Korea); Pacific Is (Hawaiian Is, Polynesia).

Uses: Used for R-phycoerythrin extraction and for functional foods (Rossano et al. 2003).

C. elongata is used for producing bio-energy on an industrial scale, contain antitumor, antifungal, antiviral, and antimicrobial BASs, terpens and phenols, PUFA, α-tocopherol, xylogalactans of agar group and alginic acids; used in medicine and dermatology (Ballesteros et al. 1992, Osman et al. 2010, Milchakova 2011).

Furcellaria lumbricalis (Hudson) J.V. Lamouroux

Synonym: *Furcellaria fastigiata*

Common names: English: Black carrageen (Brennan 1950), Clawed fork weed (Bunker et al. 2010); Swedish: Kräkel (Tolstoy and Österlund 2003).

Description: A reddish brown to brownish black seaweed with glossy, cartilaginous, cylindrical fronds, branching dichotomously six to 11 times. The fronds rise from a much branched holdfast up to 25 mm in diam. The reproductive bodies occur as pod-like structures at the ends of the branches. The seaweed grows up to about 30 cm in length (Rayment 2008).

Geographic distribution: Northern European waters and reaching to Canada.

Uses: *F. lumbricalis* is an economically important species around the Baltic Sea region where it is harvested as a raw material in the production of carrageenans, a compound that is used extensively in the food industry as a stabilizer and thickener in products like ice cream, pudding, and gelatinized items to name a few. Extracts of *F. lumbricalis* are also utilized in cosmetic products for thickening and stabilization as well as added minerals and skin softening agents (Mouritsen 2013, Green 2014).

Gelidium corneum (Hudson) J.V. Lamouroux (Fig. 37)

Synonyms: *G. attenuatum, G. sesquipedale*

Common names: English: Atlantic agar (Sáa 2002); Japanese: Tengusa, Tokorotengusa; Portuguese: Ágar, Cabelo, Cabelo-de-cão, Febra, Francelha,

Figure 37. *Gelidium corneum* (Hudson) J.V. Lamouroux (Rhodophyta). Illustration from William H. Harvey (Plate LIII, Phycologia Britannica, 1846–1951).

Garagar, Guia, Limo-encarnado, Limo-fino, Limo-preto, Pelinho, Pêlo, Ratanho, Sedas (INII 1966); Spanish: Ocle, Caloca (Martinez et al. 2014).

Description: *G. corneum* is a beautiful, red seaweed that grows in thick bushes along European coasts. Fronds of this algae are deep red to purple in color (bleaching to yellow when dried), thick and hard in texture, cylindrical at the base, but flattening and tapering towards the pointed tip with irregular, fine branches. Erect fronds (up to 40 cm in length) are supported by basal rhizoids and prostrate branches. *G. corneum* is usually found attached to rocks in shallow subtidal regions, sometimes in the lower intertidal zone, but usually at a depth where the basal portion of the plant remains submerged (Pereira 2010a).

Geographic distribution: Found from Europe through N Africa and the Mediterranean, as well as Australia, New Zealand, Hawaii, and the Caribbean.

Uses: This is the highest quality of sea gelatin. It is the soluble fiber extracted from this species and has a neutral flavor. It is used as a thickening agent and gelatin in desserts (cakes, compotes, jams, custards, puddings, sorbets, juices) (Sáa 2002, Pereira 2010b, Lodeiro et al. 2012). It is also used to thicken sauces, soups, stews and purées. Due to its great purity it is used in microbiological research. A soupspoon of agar flakes can thicken one liter of liquid by boiling for 8 min (Sáa 2002). This species is also eaten in the Hawaiian Is (Reed 1906).

Gigartina pistillata (S.G. Gmelin) Stackhouse (Fig. 38)

Common names: English: Pestle Weed (Bunker et al. 2010); Portuguese: Borracha, Botelho-borriço, Botelho-riço, Corno-de-veado, Pinheirinho (INII 1966), Musgos (Pereira and Correia 2015).

Description: *G. pistillata* is the type species of the genus *Gigartina* and their thalli are erect, up to 20 cm tall, dark-red or red-brown, cartilaginous, elastic, dichotomously branched, attached to the substratum through a small disk (Pereira 2010a).

Geographic distribution: NE Atlantic (Ireland and Britain to Senegal); SE Atlantic (S Africa); SE Asia (Malaysia).

Uses: Edible seaweed (Gómez-Ordóñez et al. 2010, Gómez-Ordóñez and Rupérez 2011, Sidik et al. 2012), carrageenan producer (Amimi et al. 2001, Pereira et al. 2009), and with antioxidant activity (De Souza et al. 2007, Jimenez-Escrig et al. 2012).

Figure 38. *Gigartina pistillata* (S.G. Gmelin) Stackhouse (Rhodophyta).

Gracilariopsis longissima (S.G. Gmelin) M. Steentoft, L.M. Irvine & W.F. Farnham

Synonyms: *Gracilaria verrucosa, G. confervoides*

Common names: Chinese: Hai mien san, Fen tsai, Hunsai, Hai tsai, Hoi tsoi (Madlener 1977), Hai-meín-san (Tseng 1933); English: Thin dragon beard plant, Ceylon moss (Madlener 1977); Fijian: Lumiwawa, Lumiyara (Ostraff 2003); Filipino: Susueldot-baybay (Galutira and Velasquez 1963), Gulaman (Tseng 1933); Hawaiian: Ogo (McConnaughey 1985); Japanese: Ogo-nori (Smith 1904), Ogo, Ogonori (Tokida 1954, Madlener 1977), Ogo-nori (Chapman and Chapman 1980); Korean: Kkoshiraegi (Sohn 1998); Vietnamese: Nuoc-mam, Rau-cau, Xoa xoa (Madlener 1977).

Description: Color purplish red to gray to translucent green; plant body (thallus) very bushy, becoming less bushy with age; branches fleshy, repeatedly dividing, occasionally dichotomously (into two branches) up to 2 mm in width, round in cross-section, tapering distally; swollen bump-like fruiting bodies (cystocarps) may be numerous, scattered all over the branches (Bowling 2012a).

Geographic distribution: NE Atlantic; Baltic Sea; Mediterranean; SW and SE Atlantic; Indian Ocean; SW Asia (Iran, Israel, Sri Lanka); SE Asia (Vietnam); NE and E Pacific; Australia; Pacific Is (Hawaiian Is, Polynesia, Samoan Is).

Uses: Consumed in NE Atlantic (Zemke-White and Ohno 1999, Harrison 2012); in Japan and Korea is used as food in soups and for agar production (Tokida 1954, Kang 1968, Okazaki 1971, Bonotto 1976, Madlener 1977, Mshigeni 1983, Tseng 1983, Hotta et al. 1989, South 1993, Sohn 1998, Hu 2005, Lee 2008, Sidik et al. 2012). This seaweed is used as a food in two ways: (1) washed free from salt and eaten as a salad raw or boiled, and (2) sun-bleached, dried, and marketed as a cheap substitute for gelatin. In Manila the sea weeds are brought into the markets during the rainy season; they are frequently used as food in Fiji, the Philippines, France, Malaysia, Indonesia, Thailand, China, Japan and Korea (Zaneveld 1959, Galutira and Velasquez 1963, Lewmanomont 1978, Simoons 1990, South 1993a, Ostraff 2003, Ohno and Largo 2006, Roo et al. 2007, Ahmad et al. 2012, Irianto and Syamdidi 2015); used as food in Hawaii, but not common (MacCaughey 1918).

Extracts of this species have antimicrobial activity (Homsey and Hide 1974). This species has been used in traditional medicine to treat pulmonary tuberculosis, stomach disorders (Dawson 1966, Dawes 1981), urinary diseases, dropsy, and goiter (Tseng and Zhang 1984, Tokuda et al. 1986).

Gracilaria multipartita (Clemente) Harvey (Fig. 39)

Common name: English: Cleaved wart weed (Bunker et al. 2010).

Description: Thalli translucent, dull purple or reddish-brown algae which has fronds that measure up to 25 cm long. It is cartilaginous, very brittle, and has a compressed stipe (Mayhew 2002).

Geographic distribution: NE Atlantic; Mediterranean.

Uses: Used as food and for agar extraction (Senate 1869, Givernaud et al. 1999); extracts of this species have antibacterial (Etahiri et al. 2003, Farid et al. 2009, Shelar et al. 2012, Omar et al. 2012), and antifungal (Omar et al. 2012) activity.

Gymnogongrus griffithsiae (Turner) Martius (Fig. 40)

Common name: Japanese: Ito-okitsunori (Tokida 1954).

Description: Thallus stiff-erect, wiry-cartilaginous, cylindrical to compressed, brown-red to blackish purple fronds, to 75 mm high, from an expanded discoid base; repeatedly dichotomous, fastigiated, with rounded, somewhat flattened apices (Braune and Guiry 2011, Pereira 2015a).

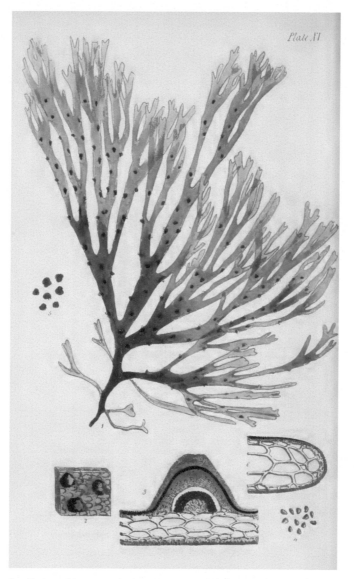

Figure 39. *Gracilaria multipartita* (Clemente) Harvey (Rhodophyta). Illustration from William H. Harvey (Plate XI, Phycologia Britannica, 1846–1951).

Geographic distribution: NE and E Atlantic; Mediterranean; NW and SW Atlantic; Caribbean; Asia (Japan); Australia.

Uses: Used as food in Japan (Tokida 1954). Produces sulfated galactans—carrageenans with antioxidant, and antiviral activity (Pereira 2015a).

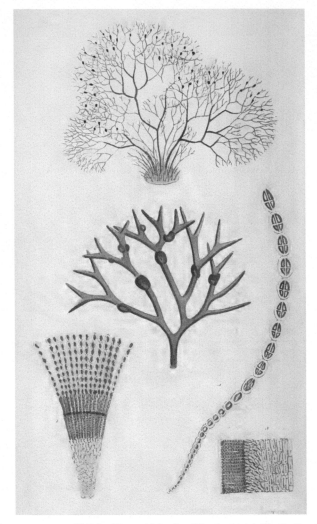

Figure 40. *Gymnogongrus griffithsiae* (Turner) Martius (Rhodophyta). Illustration from William H. Harvey (Phycologia Britannica, 1846–1951).

Halurus equisetifolius (Lightfoot) Kützing

Common names: English: Sea Tail (Dickinson 1963), Sea Horsetail (Bunker et al. 2010).

Description: Erect plants consisting of up to seven irregularly branched main axes growing 6 to 22.5 cm high and 2 to 3 mm wide, resembling a horsetail fern. The main axes branch to four orders and are clothed (either sparsely or densely) in whorls of between five and eight short incurved branchlets which divide di- to trichotomously. The branchlets are 1.4 to 2.4 mm long and consist of four to seven cells of which the terminal one is mucronate. Without the branchlets, the main axis is 200 to 350 µm wide. Plants are soft and spongy when young becoming more cartilaginous when older and are a dark red color.

Geographic distribution: NE Atlantic; Atlantic Is (Azores, Canary Is); Mediterranean; Asia (Japan).

Uses: Considered an edible species (Green 2014).

Halurus flosculosus (J. Ellis) Maggs & Hommersand

Common name: English: Mrs. Griffith's little flower (Bunker et al. 2010).

Description: Rather rigid, tufted, uniseriate, ecorticate, bright red filaments, to 200 mm long. Repeatedly ± dichotomously branched, with frequent short branchlets, axils very acute. Articulations cylindrical, five- six times as long as broad in the lower parts, and shorter distally (Braune and Guiry 2011).

Geographic distribution: NE Atlantic; Adriatic Sea; Atlantic Is (Azores, Canary Is, Cape Verde Is); Mediterranean; Auckland Is.

Uses: Considered an edible species (Green 2014).

Laurencia viridis Gil-Rodríguez & Haroun

Common name: Portuguese: Erva-malagueta (Neto et al. 2005).

Description: Greenish-red, rigid, erect, terete, thalli forming matted tufts that grow to 2–30 cm tall and arise from a basal disk. Thalli are pseudo-parenchymatous in structure. Main axes of the thalli are 1–2 mm diam., profusely, variably branched with an acute angle of branching (Neto et al. 2005).

Geographic distribution: Atlantic Is (Azores, Canary Is, Cape Verde Is, Madeira, Selvage Is).

Uses: On some islands of Azores (Portugal) it is collected, cleaned, and pickled in vinegar and eaten with fried food (Neto et al. 2005).

Mastocarpus stellatus (Stackhouse) Guiry (Fig. 41)

Synonyms: *Gigartina mamillosa, G. stellata*

Common names: Danish: Vortetang; English: Grape pip weed (Bunker et al. 2010); Portuguese: Corninho, Folhinha, Crespo (INII 1966, Oliveira 1990, Pereira 2009b), Alface-miúda, Botelha (INII 1966, Pateira 1993), Limo-musgo, Musgo (INII 1966, Pereira and Correia 2015).

Figure 41. *Mastocarpus stellatus* (Stackhouse) Guiry (Rhodophyta).

Description: A small red alga (up to 17 cm in length), the fronds are channeled with a thickened edge and widen from a narrow stipe with disk-like holdfast. The channeling is often slight and is most noticeable at the base of the frond. Mature plants have conspicuous growths of short, small papillae (reproductive bodies) on the fronds. The plant is dark reddish-brown to purple in color and may be bleached. The common name false Irish moss is used as it may be confused with *Chondrus crispus* (Irish moss) (Pereira 2010a).

Geographic distribution: NE Atlantic (Scandinavia to Mauretania, N Sea); NW Atlantic (Newfoundland and Nova Scotia to Rhode Is).

Uses: Considered and edible species in several European countries, namely in Spain (Galicia) (Gómez-Ordóñez and Rupérez 2011, Lodeiro et al. 2012, Abreu et al. 2014); in Iceland this species is eaten mainly during adverse seasons (Hallsson 1961, Zemke-White and Ohno 1999).

M. stellatus is a source of carrageenan, like *Chondrus crispus* (Braune and Guiry 2011). This extract is used widely in a variety of industries from food to pharmaceuticals to cosmetics. In Ireland, traditionally, both *C. crispus* and *M. stellatus* have been collected, dried, and used for cooking soups and jellies as well as for making a drink to ward off respiratory illness (Roo et al. 2007).

Extracts of this species have antioxidant and anticoagulant (Jimenez-Escrig et al. 2012, Gómez-Ordóñez et al. 2014) activity.

Mesophyllum lichenoides (J. Ellis) Me. Lemoine

Common name: English: Pink plates (Bunker et al. 2010).

Description: Pale to dark purple thin, brittle, leafy calcified fronds, attached at base, margins free, lobed. Fronds semicircular, concentrically banded; reproduction takes place in winter and spring in small, wart-like conceptacles; gametophytes and sporophytes similar and provided of very characteristic hemispheric, bulky conceptacles (Braune and Guiry 2011, Pereira and Correia 2015).

Geographic distribution: NE Atlantic (Ireland to Mauritania); Mediterranean.

Uses: Considered an edible species (Green 2014).

Osmundea osmunda (S.G. Gmelin) K.W. Nam & Maggs
(formerly *Laurencia osmunda*)

Common names: English: Pepper dulse (Kirby 1953), Royal Fern-weed (Bunker et al. 2010).

Description: Cartilaginous, usually markedly compressed, dark purplish-brown to pale yellow fronds, to 100 mm or more long, from a discoid holdfast; very variable in size and form; main axis usually simple, branching alternate distichous, repeatedly pinnate. Ultimate ramuli short, blunt (Braune and Guiry 2011).

Geographic distribution: NE Atlantic; Atlantic Is (Madeira); Tropical and subtropical WE Atlantic; Asia (Korea).

Uses: Considered an edible species (Green 2014).

Osmundea pinnatifida (Hudson) Stackhouse
(formerly *Laurencia pinnatifida*) (Fig. 42)

Common names: English: Pepper Dulse (Bunker et al. 2010); Hawaiian: Limu maneoneo, Limu olipeepee, Limu lipee (Reed 1906); Portuguese: Argacinho-das-lapas, Botelho-preto, Pele-de-lapa (INII 1966, Oliveira 1990, Pereira 2009b); Erva-malagueta (Neto et al. 2005, Pereira 2010a).

Description: Dark-red (sometimes bleached to yellow-red), cartilaginous, fleshy, erect, markedly compressed and densely branched fronds that grow to 2–6 cm tall and arise from a stoloniferous base. The thalli are pseudo-parenchymatous, variable in form and size, and may form extensive turfs or occur as single large plants. The main axis grows to 2.5 mm wide and has alternate, distichous, repeatedly pinnate branching; tips have a longitudinal groove (seen with an x10 hand lens) (Neto et al. 2005).

Geographic distribution: Can be found in the N Atlantic, Atlantic Is, N Sea, English Channel, and Mediterranean; Pacific Is (Hawaiian Is).

Figure 42. *Osmundea pinnatifida* (Hudson) Stackhouse (Rhodophyta).

Uses: This aromatic seaweed is dried and used as a pepper- or curry-flavored spice in Scotland and Ireland. In some islands of Azores (Portugal) it is collected, cleaned, and pickled in vinegar and later eaten with fried food (Chapman and Chapman 1980, Zemke-White and Ohno 1999, Neto et al. 2005, Roo et al. 2007, Harrison 2013, Rodrigues et al. 2015a, 2015b). Recently, students from the Professional School of Praia da Vitória (Azores) developed some innovative dishes with algae, including the *O. pinnatifida*, called "Tuna Tataki Rolled in Pepper-Dulse and Laver Dumpling", "Flan Pudding with Pepper-Dulse Caramel", "Lemon Sorbet with Strawberry and Pepper-Dulse Crisp" (Matos 2013); is also eaten in the Hawaiian Is (Reed 1906).

Osmundea truncata (Kützing) K.W. Nam & Maggs

Description: *O. truncata* is less common than *O. pinnatifida*, is more pinnately branched, and generally grows epiphytically.

Geographic distribution: NE Atlantic; Atlantic Is (Canary Is, Selvage Is); Adriatic Sea; Baltic Sea; Black Sea; Mediterranean.

Uses: Considered an edible species in Denmark (Wegeberg 2010).

Phymatolithon calcareum (Pallas) W.H. Adey & D.L. McKibbin

Synonym: *Lithothamnion calcareum*

Common name: English: Calcified seaweed (FDA 1999), Mäerl (CEVA 2014).

Description: Usually unattached, sparsely or densely branching, heavily calcified, chalky reddish-violet, brittle, branches sometimes fusing, slender or robust, cylindrical or flattened; endings rounded, surface smooth, matte; very variable in form (Braune and Guiry 2011).

Geographic distribution: NE Atlantic; Atlantic Is (Faroe Is, Iceland).

Uses: Authorized as vegetable and condiment in France (CEVA 2014); Mäerl is used in dietetics as a food complement, sold either directly in a powdered form or as capsules or tablets, and plays a part in the treatment of mineral deficiencies in food (FDA 1999); is also used in paleo diet cooking (Green 2014).

Plocamium cartilagineum (Linnaeus) P.S. Dixon (Fig. 43)

Synonym: *Plocamium coccineum*

Common names: English: Cockscomb (Dickinson 1963), Cock's comb, Branched cock's comb (Bunker et al. 2010), Red comb weed; French: Boucle de cheveux; Portuguese: Botelho-melado, Roseta (INII 1966).

Description: Narrow, compressed, cartilaginous, bright scarlet fronds, to 300 mm long, tufted, much divided; branching irregularly alternate, pinnules alternately second in twos to fives, with acute apices, lowest of each set a simple spur, others increasingly strongly pectinate (Braune and Guiry 2011).

Geographic distribution: NE Atlantic (Scandinavia to Senegal, North Sea), SE Atlantic (Namibia); Mediterranean; Indian Ocean (Pakistan, Mauritius); NW Pacific (Japan); Pacific Is; NE Pacific (Alaska to California); SE-Pacific (Chile); Australia, New Zealand; Antarctica.

Uses: Considered an edible species (Green 2014).

Polysiphonia fucoides (Hudson) Greville (Fig. 44)

Common names: English: Black siphon weed (Bunker et al. 2010); Swedish: Fjäderslick (Tolstoy and Österlund 2003).

Description: Cartilaginous, cylindrical, tufted, brownish purple fronds, to 300 mm long (more usually about 70 mm long), from branched rhizoidal holdfast; branching ± alternate, tripinnate, ramuli with terminal tufts of colorless dichotomous fibrils. Plants perennial, ramuli and branchlets shed in winter, leaving jagged stumps. Large central siphon surrounded by 12–20 pericentral siphons, corticated only at base. Articulations as long as broad in older parts, to 1.5 times as long as broad distally (Braune and Guiry 2011).

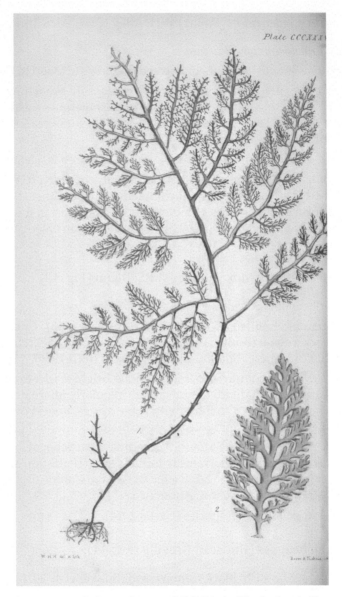

Figure 43. *Plocamium cartilagineum* (Linnaeus) P.S. Dixon (Rhodophyta). Illustration from William H. Harvey (Plate CCXXXV, Phycologia Britannica, 1846–1951).

Geographic distribution: Arctic; White Sea; N Atlantic; Adriatic Sea; Black Sea; Atlantic Is (Azores, Canary Is, Greenland, Iceland, Madeira, Selvage Is); Tropical and Subtropical Atlantic.

Figure 44. *Polysiphonia fucoides* (Hudson) Greville (Rhodophyta). Illustration from William H. Harvey (Plate CCIX, Phycologia Britannica, 1846–1951).

Uses: *P. fucoides* is the source of antimicrobial BASs, phenol derivatives, essential amino acids; used in food and as herbal medicine in some Asian countries (Milchakova 2011).

Palmaria palmata (Linnaeus) Weber & Moh (Fig. 45)

Common names: English: Dillisk, Dillesk, Crannogh, Water leaf, Sheep dulse (Madlener 1977), Dried dulse (McConnaughey 1985), Dulse (Lembi and Waaland 1988), Shelldulse (Anon 2000); French: Goéman à Vache (Madlener 1977); Icelandic: Saccha, Sol (Hallsson 1961, Madlener 1977); Japanese: Darusu (Madlener 1977); Norwegian: Sou sol (Madlener 1977); Portuguese: Botelho-comprido (INII 1966, Oliveira 1990, Pereira 2009), Dulse, Folha (Pereira and Correia 2015).

Figure 45. *Palmaria palmata* (Linnaeus) Weber & Moh (Rhodophyta). Illustration from William H. Harvey (Plate CCXVII, Phycologia Britannica, 1846–1951).

Description: A foliose red algae with a tough flat frond usually between 20 and 50 cm in length, but sometimes up to 1 m. The algae grow directly from a small discoid holdfast gradually widening and subdividing. The stipe is inconspicuous, rarely to 5 mm long. Older parts may have small "leaflets" along the margin especially where damaged; dark red, with purple tints under water (Pereira 2010a, Pereira 2015a).

Geographic distribution: North coasts of the Atlantic and Pacific Oceans, as far north as Arctic Canada and Russia, and as far south as Portugal in Europe, and New Jersey and California in the United States (US). In the W Pacific, the southern range of *P. palmata* includes Japan and Korea.

Uses: Of all the marine algae, "dulse" probably has the taste that is most agreeable to the western palate. When fresh, dulse can be eaten as a salad, possibly after being soaked in fresh water; that causes its cells to burst. It is often too tough to eat raw, but drying makes it easier to chew and brings out its pleasant salty and nut-like taste. "Dulse" should not be cooked as this causes it to break up (Mouritsen 2013). This species is eaten in Nova Scotia (Canada), France, Iceland, Ireland, Portugal, the UK, US, and Indonesia (Hallsson 1961, Arasaki and Arasaki 1983, Kuhnlein and Turner 1996, Zemke-White and Ohno 1999, Roo et al. 2007, Pereira 2010b, Harrison 2013, Mouritsen et al. 2013, Irianto and Syamdidi 2015, Pereira 2015a).

Cultivation from vegetative thalli of *P. palmata* in open sea ropes has been successfully developed, at a commercial scale, in Asturias (Spain). Biomass obtained from a 2.400 m^2 farm is canned and sold for human consumption in a still growing market (Sosa et al. 2006).

This species is a good source of dietary requirements; it is rich in potassium, iron, iodine and trace elements, and relatively low in sodium. A small amount can provide more than 100% of the daily amount of vitamin B_6, 66% of Vitamin B_{12}, iron and fluoride (Pereira 2011).

See recipes with *Palmaria palmata* in Annex I.

Porphyra dioica J. Brodie & L.M. Irvine

Common name: English: Black laver (Bunker et al. 2010).

Description: Membranous, monostromatic, olive-green to brown-purple or blackish fronds, to 500 mm long and 200 mm wide, from short stipe and basal holdfast (Braune and Guiry 2011).

Geographic distribution: NE Atlantic; Atlantic Is (Faroe Is, Iceland).

Uses: Authorized as vegetables and condiments in France (CEVA 2014).

Porphyra linearis Greville

Common names: English: Winter Laver (Bunker et al. 2010); Portuguese: Erva-do-calhau, Erva-patinha (Neto et al. 2005, Matos 2013).

Description: With narrow stem that attaches the base and appears mainly in winter (Pereira 2010a).

Geographic distribution: NE and NW Atlantic; Atlantic Is, Mediterranean; Adriatic Sea; Baltic Sea; NE Pacific (Alaska); Asia (Japan).

Uses: The red seaweed *Porphyra* is collected, then fried or incorporated into a soup or an omelet, called "torta de erva-patinha" or "tortas-do-calhau", in the Azores Is (Neto et al. 2005, Pereira 2010b, Matos 2013, Pereira 2015a).

Porphyra purpurea (Roth) C. Agardh (Fig. 46)

Synonyms: *Porphyra nereocystis, P. vulgaris*

Common names: Chinese: Chi Choy (Madlener 1977); English: Purple laver (Bunker et al. 2010), Purple vegetable (Braune and Guiry 2011), Red Laver,

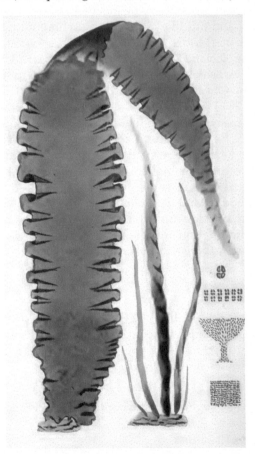

Figure 46. *Porphyra purpurea* (Roth) C. Agardh (Rhodophyta). Illustration from William H. Harvey (Phycologia Britannica, 1846–1951).

Purple laver (Madlener 1977), Rose nori (Chapman and Chapman 1980), Red nori (McConnaughey 1985); Japanese: Amanori (TFJ 1893), Asakusa nori, Hoshinori (Kirby 1953), Nori (Madlener 1977).

Description: *P. purpurea* is usually purple-red, crimson, olive-brown, or greenish-purple to olive-black color. Fronds typically grow up to 20 cm long and are irregularly lobed with a central holdfast. Fronds have a double layer of tissue, but range from being thin to broad, and are usually divided into lobes; *P. purpurea* occurs throughout the year (Braune and Guiry 2011).

Geographic distribution: Currently growing across the majority of the Northern Hemisphere, likely due to its preference for low temperature water.

Uses: In China, practically all species are collected and used as food ("Zicai", "purple vegetable"), and some cultivation takes place; the weed is seldom eaten fresh in Japan and nearly always is made into thin sheets like paper, and dried for preservation; this is heated over the fire and eaten with broth, or by itself with a little soy sauce; it is also used for various other culinary purposes (TFJ 1893); in Japan and Korea they are also extensively cultivated and commercially exploited in large quantities (Braune and Guiry 2011). Used also as food in the US, and NW Atlantic (Yanovsky 1936, Chapman and Chapman 1980, Redmond et al. 2014).

Porphyra umbilicalis Kützing (Fig. 47)

Common names: English: Laverbread (Ohmi 1968), Purple laver, Sloak, Slook, Laver (Madlener 1977), Tough Laver (Bunker et al. 2010); French: Laitue rouge (Boisvert 1984); Japanese: Chishima-kuronori (Tokida 1954, Arasaki and Arasaki 1983), Nori (McConnaughey 1985); Portuguese: Erva-patinha (Neto et al. 2005), Folhuda (Morton et al. 1998), Nori-Atlântico (Pereira and Correia 2015).

Description: A red alga (up to 40 cm across) with a circular, broad frond that is membranous but tough. The plant attaches to rocks via a minute discoid hold-fast, is greenish when young becoming purplish-red and has a polythene-like texture (Pereira 2010a).

Geographic distribution: Occurs in the N Atlantic. In the east, it is found in Iceland and has been recorded from Norway to Portugal and in the W Mediterranean. In the west, *P. umbilicalis* is found from Labrador in Canada to the mid-Atlantic coast of the US; Asia (Japan).

Uses: Collected in France, the US (NW Atlantic), Japan and on various Atlantic Is, *P. umbilicalis* is used in the preparation of various dishes such as soups and "tortas"; consumed also in Japan (Tokida 1954, Hallsson 1961, Arasaki and Arasaki 1983, Zemke-White and Ohno 1999, Neto et al. 2005, Roo et al. 2007, Pereira 2010b, Harrison 2013, Redmond et al. 2014).

Figure 47. *Porphyra umbilicalis* Kützing (Rhodophyta).

P. umbilicalis is rich in protein, vitamins A, C, E, and B, and trace minerals, and also rich in omega-3 polyunsaturated fatty acids (EPA and DHA) (Pereira 2011).

See recipes with *Porphyra* in Annex I.

Pyropia leucosticta (Thuret) Neefus & J. Brodie
(formerly *Porphyra leucosticta*)

Common names: English: Pale patch laver (Bunker et al. 2010); Hawaiian: Limu lua'u Limu lipahee (Reed 1906), Limu luan, Limu lua'u (MacCaughey 1918); Portuguese: Erva-do-calhau (Matos 2013), Erva-patinha (Neto et al. 2005, Pereira 2010a, Matos 2013); Spanish: Cochayuyo (Polo 1977).

Description: Delicate membranous monostromatic reddish-brown fronds, becoming pink on drying, to 150 mm long, with very short stipe from basal holdfast (Braune and Guiry 2011).

Geographic distribution: NE Atlantic; Atlantic Is; N Sea; Baltic Sea; Mediterranean; Black Sea; NW Atlantic (Maine to Florida); Antarctic and the subantarctic Is.

Uses: Used as food in NW Atlantic and Azores, Portugal (Zemke-White and Ohno 1999, Neto et al. 2005, Roo et al. 2007, Pereira 2010b, Pereira 2012, Matos 2013, Redmond et al. 2014); used also as food in Japan, China, and

Israel; can be used as a fodder additive (Milchakova 2011); a very highly prized delicacy in Hawaii (Reed 1906, MacCaughey 1918); is also consumed in Peru (Polo 1977).

Extracts of this species have antioxidant activity (Zubia et al. 2009).

Rhodymenia pseudopalmata (J.V. Lamouroux) P.C. Silva (Fig. 48)

Common names: English: Dulse (Gaia 2009), Rosy fan weed (Bunker et al. 2010).

Description: Flattened, fan-shaped, rather stiff, rose-red fronds, to 100 mm high, with long or short stipes arising from a discoidal base. Fronds repeatedly dichotomously lobed, axils wide, apices rounded, margin smooth (Guiry 2015a).

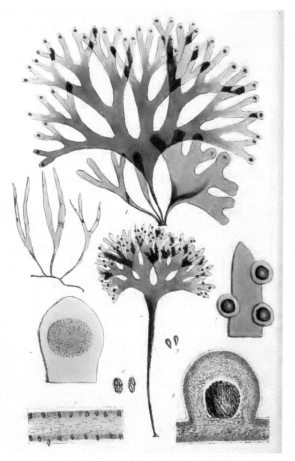

Figure 48. *Rhodymenia pseudopalmata* (J.V. Lamouroux) P.C. Silva (Rhodophyta). Illustration from William H. Harvey (Phycologia Britannica, 1846–1951).

Geographic distribution: NE Atlantic; Adriatic Sea; Atlantic Is (Azores, Canary Is, Cape Verde Is, Madeira, Selvage Is); Gulf of Mexico; Caribbean Sea; Tropical and subtropical Atlantic; E Pacific (Galapagos Is).

Uses: Considered an edible species in Ireland (Gaia 2009).

Rhodothamniella floridula (Dillwyn) Feldmann

Common name: English: Sand binder (Bunker et al. 2010).

Description: *R. floridula* is perennial brownish-red seaweed found on the lower shores. It usually covers large areas of rock in sandy habitats. At the base of the seaweed, filaments bind with sand to form a spongy, carpet like mass. The filaments are well-spaced and branch out up to 3 cm in length. Upright filaments of the seaweed uncovered by the ebbing tide appear as tufts of hair; when plants dry out they have a purplish tinge (Riley 2005).

Geographic distribution: NE Atlantic (Ireland and Britain to Portugal); SE Atlantic (Namibia, S Africa); SW Atlantic (Argentina).

Uses: Used for the production of pharmaceuticals and functional foods. Extracts have antioxidant activity (Cornish and Garbary 2010, Pereira 2015a).

Solieria filiformis (Kützing) P.W. Gabrielson

Description: Thalli smooth, mucousy and cartilaginous alga with cylindrical branches that are very flexible and branch dichotomously, usually murky beige-brown (Einav 2014).

Geographic distribution: N and S Atlantic; Mediterranean.

Uses: Considered an edible species (Gray 2015), and used also for carrageenan extraction (de Araujo et al. 2012, Einav 2014); can be used as food in soups and other dishes (Radulovich et al. 2013).

Extracts of this species have anti-inflammatory and anti nociceptive (de Araujo et al. 2011, de Araujo et al. 2012), and anticoagulant (Rodrigues et al. 2010, de Araujo et al. 2012) activity.

Sphaerococcus coronopifolius Stackhouse (Fig. 49)

Common name: English: Berry wart cress (Bunker et al. 2010).

Description: Narrow, compressed, two-edged, cartilaginous, scarlet fronds, main axes dark brownish-red, to 300 mm long; branching abundant, distichous, sub-dichotomous or alternate, terminal branchlets acute, fringed

Figure 49. *Sphaerococcus coronopifolius* Stackhouse (Rhodophyta). Illustration from William H. Harvey (Plate LXI, Phycologia Britannica, 1846–1951).

with short marginal proliferations. Tetrasporophyte is *Haematocelis fissurata*, a thick crust with oily fissures, with which it is sometimes found (Guiry and Guiry 2014j).

Geographic distribution: NE Atlantic; Atlantic Is (Azores); Mediterranean; SW Asia (Turkey).

Uses: Considered an edible species (Patarra et al. 2011, Gray 2015).

Extracts of this species have antiviral (Bouhlal et al. 2011), antifouling (Piazza et al. 2011), antibacterial (Etahiri et al. 2001, Smyrniotopoulos et al. 2010), antitumor (Smyrniotopoulos et al. 2010), antioxidant (Fino et al. 2014), antibiotic and antimalaria (Etahiri et al. 2001) activity.

Spyridia griffithsiana (J.E. Smith) G.C. Zuccarello, Prud'homme van Reine & H. Stegenga

Similar species: *Spyridia filamentosa*

Description: This species is distinguished from *S. filamentosa* by having two (or occasionally one) significantly smaller lower axial cells on determinate branches than more distal axial cells. These small cells also lack a (or have a very reduced) cortical ring of cells. In contrast, *S. filamentosa* has basal determinate branches with axial cells of approximately the same size as the more distal cells, and all possess a corticating ring of cells (Zuccarello et al. 2004).

Geographic distribution: NE Atlantic; Mediterranean Sea.

Uses: Considered an edible species (Green 1966).

Vertebrata lanosa (Linnaeus) T.A. Christensen

Common name: English: Wrack siphon weed (Bunker et al. 2010, Mæhre et al. 2014).

Description: Cartilaginous, cylindrical, densely tufted, dark reddish-purple fronds, to 75 mm long, attached by creeping rhizoids with branches penetrating the host fronds of *Ascophyllum nodosum*. Repeatedly pseudo-dichotomously branched, apices pointed, widely forked; large central siphon with 12–24 pericentral siphons, ecorticate, articulations shorter than broad (Guiry 2015b).

Geographic distribution: N Atlantic; Atlantic Is (Iceland).

Uses: Potential use as food in Norway (Mæhre et al. 2014).

2.2 South Atlantic and Caribbean Sea

2.2.1 Chlorophyta

Anadyomene stellata (Wulfen) C. Agardh

Common name: Spanish: Repollo de mar (Radulovich et al. 2013).

Description: Thallus bright green; leaf-like sheets rather stiff, erect, not limp and mucilaginous as in *Ulva* or *Monostroma*, often several emerging from a common branching rhizoid; stipe almost invisible and sharply wedge-shaped, merging into the fan-shaped, irregularly lobed, slightly crinkled thallus with undulated margins (Braune and Guiry 2011).

Geographic distribution: Warmer parts of the NE Atlantic (Azores, Canary Is); Mediterranean; Caribbean; Indian Ocean; W Pacific (the Philippines, Indonesia); S Australia.

Uses: Used as food in the Caribbean Is and Costa Rica, cooked fresh or dried; used also as flour or meal (Radulovich et al. 2015).

Chaetomorpha intestinalis Kützing

Description: Uniseriate, unbranched filaments, singular or united in tufts, attached by lobed or discoid basal cells (Braune and Guiry 2011).

Geographic distribution: Tropical and subtropical W Atlantic; Caribbean Sea; Indian Ocean; Asia (Taiwan).

Uses: Used in the Caribbean Is as food (Radulovich et al. 2013).

Chaetomorpha crassa (C. Agardh) Kützing

Common names: Chinese: Tai tyau (Madlener 1977); English: Hair-shaped green algae, Big green algae (Madlener 1977); Filipino: Kauat-kauat, Riprippiis (Zaneveld 1950, Madlener 1977, Agngarayngay et al. 2005), Pansit-pansit (Domingo and Corrales 2001); Lukotlukot (Leyte 2015).

Description: The thalli consist of unbranched filaments forming loose clumps entangled with other algal species; The clumps are bright green with some greenish white portions; fragments consist of cylindrical cells, 600–700 µ in diam., sometimes long as broad but usually twice to two and half as long as broad (Madlener 1977).

Geographic distribution: NE Atlantic; Mediterranean; Tropical and subtropical W and E Atlantic; Indian Ocean; SW Asia (India, Kuwait, Pakistan, Sri Lanka, Turkey, Yemen); Asia (China, Japan, Korea, Taiwan); SE Asia (Indonesia, the Philippines, Singapore, Thailand, Vietnam); Australia; Pacific Is (Micronesia, Mariana Is, Solomon Is).

Uses: Direct use as food in China, Korea, Taiwan, Thailand and the Philippines to make salads (Zaneveld 1959, Madlener 1977, Lewmanomont 1978, Arasaki and Arasaki 1983, Lee 2008, Agngarayngay et al. 2005, Sidik et al. 2012, Leyte 2015). The plants are commonly prepared to make a gelatin-like sweet meat from them (Zaneveld 1950, 1959). Cooked fresh similar to spinach (Radulovich et al. 2015).

Chaetomorpha javanica Kützing

Common name: Indonesian: Lumut-laut (Zaneveld 1950).

Description: Thalli with rigid filaments, pale green, wide.

Geographic distribution: Tropical W Atlantic; Indian Ocean; Asia (China); SE Asia (Indonesia, Vietnam); Australia; Pacific Is (Solomon Is).

Uses: Used as food in Indonesia and Malaysia, dried and boiled (Zaneveld 1950, 1959, Davidson 2004, Sidik et al. 2012). After being soaked at night in fresh water they are dried in the sun. The dried weeds are boiled and eaten with bacon (Levring et al. 1969, Zaneveld 1969).

Cladophora laetevirens (Dillwyn) Kützing (Fig. 50)

Common names: Spanish: Pelillo, Espinaca de mar (Radulovich et al. 2013).

Description: Thallus light to medium green, forming dense tufts 2–10 cm high with pseudo-dichotomous axes from a small group of rhizoids from cells in the basal region; epilithic; filaments tapering only slightly, branched from almost every cell above often with somewhat falcate and unilateral branchlets (van den Hoek and Womersley 1984).

Geographic distribution: NE Atlantic; Adriatic Sea; Black Sea; Mediterranean; Atlantic Is (Azores, Bermuda, Canary Is, Cape Verde Is, Madeira, Selvage Is); Tropical and subtropical W and E Atlantic; Caribbean Sea; Mexican Gulf; Arabian Gulf; Indian Ocean; Asia (China, Japan, Taiwan); SE Asia (Indonesia, Malaysia, the Philippines, Thailand); Australia and New Zealand; Pacific Is (Ryukyu Is, Solomon Is).

Uses: Used in Thailand and the Caribbean Is as food. Especially in Thailand the *Cladophora* genus is consumed just like noodles and is also common to use it in soups. It is very appetizing cooked (boiled) and tastes just like spinach (Radulovich et al. 2013).

C. laetevirens constitutes a potential source of BASs for pharmaceuticals, and contain PUFA (Milchakova 2011).

Cladophora prolifera (Roth) Kützing

Description: Unattached or basally attached coarse filaments those are usually less than 0.5 mm wide and 3–5 cm long. The filaments are formed of a single row of often swollen cells; if attached then by a discoid base or by rhizoidal outgrowths (Pereira 2015a).

Geographic distribution: NE Atlantic; Adriatic Sea; Black Sea; Mediterranean; Atlantic Is (Azores, Bermuda, Canary Is, Cape Verde Is, Madeira, Selvage Is); Tropical and subtropical W and E Atlantic; Caribbean; Mexican Gulf; Arabian Gulf; Indian Ocean; Asia (China, Japan, Taiwan); SE Asia (Indonesia,

Figure 50. *Cladophora laetevirens* (Dillwyn) Kützing (Chlorophyta). Illustration from William H. Harvey (Phycologia Britannica, 1846–1951).

Malaysia, the Philippines, Thailand); Australia and New Zealand; Pacific Is (Ryukyu Is, Solomon Is).

Uses: Used in the Caribbean Is as food (Radulovich et al. 2013).

Codium decorticatum (Woodward) M.A. Howe

Description: It is a little branched subtidal species, which can reach 1 m long. The thallus is flattened at the points where the branches are formed; utricles not mucronate and larger than those of other *Codium* species.

Geographic distribution: Adriatic Sea, Mediterranean Sea; Atlantic Is (Canary Is); Caribbean Sea; Tropical and subtropical W Atlantic; Indian Ocean; Asia (China, Indonesia, the Philippines, Australia; Pacific Is (Hawaiian Is); Antarctic Is.

Uses: Used as food in India (Rao 1970), and in the Caribbean Is (Michanek 1975).

Codium isthmocladum Vickers

Description: Thalli with green color and 8–10 cm in height, erect, spongy consistency, branched dichotomically four to six times, fixed to the substratum through more or less extended basal portion; made up abundantly branched filaments coenocytic and densely interwoven; upright cylindrical portions, 3–4 mm in diam.

Geographic distribution: Tropical and subtropical W Atlantic; SE Asia (Indonesia).

Uses: Used as food in Yucatan (Mexico) (Robledo and Pelegrín 1997, 2009).

Codium taylorii P.C. Silva

Description: Erect trunk of dark-green coloration and of firm consistency; sub-dichotomous ramification with wide angles. Cylindrical, flattened branches at the base and with rounded apex; oval and cylindrical utricle; oval gametangia (1–2 per utricle) up to 50 µ in diam. and 730 µ in length; the thickness of the utricle apex (30 µ) clearly differentiate this species from the species *C. decorticatum* (4–8 µ) (Salas 2008).

Geographic distribution: Atlantic Is (Ascension, Azores, Bermuda, Canary Is, Cape Verde Is, Madeira, Selvage Is); Tropical and subtropical W Atlantic; E Mediterranean (Israel); India Ocean; SW Asia (India, Oman); SE Asia (Indonesia, Thailand).

Uses: In Israel this species is used as food (Zemke-White and Ohno 1999, Lipkin and Friedlander 2006, Roo et al. 2007, Harrison 2013); in the Caribbean Is is cooked fresh, or fried with egg batter (Radulovich et al. 2015).

Gayralia brasiliensis Pellizzari, M.C. Oliveira & N.S. Yokoya

Description: Single foliaceous monostromatic thallus, bright green and becoming dark olive green after drying; liberation of zooids from specimens starts with marginal zone discoloration, disintegration of the cell wall, and the release of four biflagellate zooids. Zooid fusions and a *Codiolum* phase were not observed; therefore, reproduction is asexual. After zooid release and a short swimming period, one of the flagella attaches to the substratum, while the zooid spins a little longer until settling; after zooid attachment, cell divisions starts, giving rise to a uniseriate filament, which becomes multiseriate without formation of tubular or saccate stages. A small foliaceous thallus develops and expands into a monostromatic plantlet after 30 to 40 days in culture, around 300 μ broad. The mature thallus size is usually 7 ± 2.6 cm, with cell size and thallus thickness of about 8 ± 3 μ and about 25 ± 1.8 μ, respectively. Cells are uninucleate, with a large central vacuole, parietal chloroplast, and one or two pyrenoids (Guiry and Guiry 2014e).

Geographic distribution: SW Atlantic (Brazil).

Uses: During the last two decades, the monostromatic green seaweed *Gayralia* sp. has been harvested sporadically by local fishermen on the Paraná coast of southern Brazil and sold to Japanese restaurants (Pellizzari et al. 2006).

Gayralia oxysperma (Kützing) K.L. Vinogradova ex Scagel et al.

Common name: Swedish: Ljus havssallt (Tolstoy and Österlund 2003).

Description: Thallus light to medium green, delicate but fairly firm, somewhat shiny when partly dried, 2–8 cm high and usually as much across, irregularly foliose, often with ruffled marginal parts but usually without lobes, usually epilithic, attached by rhizoids from the basal cells; thallus 10–20 μ thick. Cells of the expanded blade irregularly polygonal to rounded, 7–15 μ across with gelatinous walls, mostly irregularly arranged but sometimes in groups or in rows; chloroplast filling most of the cell, with one or two prominent pyrenoids (Womersley 1984).

Geographic distribution: N Atlantic; Arctic Canada, Caribbean; Atlantic Is (Azores, Bermuda, Canary Is, Iceland); SW Atlantic; Adriatic Sea; Baltic Sea; Mediterranean; Indian Ocean; Asia (Japan, Taiwan, Myanmar, the Philippines); New Zealand; Pacific Is (Micronesia, Hawaiian Is); NE Pacific (Alaska to California).

Uses: Used as food in the Caribbean and Central America (Harrison 2013).

Halimeda incrassata (J. Ellis) J.V. Lamouroux (Fig. 51)

Description: Heavily calcified algae that is abundant in shallow habitats to 13 m deep. It features stiff, segmented fronds that are irregular and flat shaped. It grows in different forms, some forming large clumps and others with just five-six branches.

Geographic distribution: Tropical and subtropical W Atlantic; Caribbean; Indian Ocean; Asia (China, Japan, Taiwan); SE Asia (Indonesia, the Philippines, Singapore, Thailand, Vietnam); Papua New Guinea; Australia;

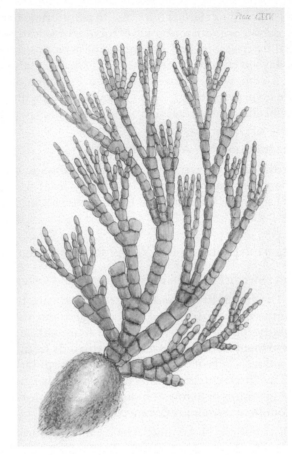

Figure 51. *Halimeda incrassata* (J. Ellis) J.V. Lamouroux (Chlorophyta). Illustration from William H. Harvey (Plate CXXV, Phycologia Britannica, 1846–1951).

Pacific Is (Samoa, Polynesia, Micronesia, Fiji, Hawaiian Is, Mariana Is, Republic of Palau, Solomon Is, Tahiti).

Uses: Seaweed with a high concentration of calcium that can serve as food supplements in the Caribbean Is (Radulovich et al. 2013).

Halimeda opuntia (Linnaeus) J.V. Lamouroux

Description: This species forms loose clumps and the rhizoids attach to the surface at various points wherever segments of the plant body come in contact with the substrates. The branch segments are flat, kidney-shaped to fan-shaped with distinct mid ribs, and are significantly calcified. There is no stem segment. The plants are found on rocks in subtidal zones along moderately wave exposed to calm shorelines (JIRCAS 2012d).

Geographic distribution: Tropical and subtropical W Atlantic; Caribbean; Indian Ocean; Asia (China, Japan, Taiwan); SE Asia (Indonesia, Malaysia, Myanmar, the Philippines, Singapore, Thailand, Vietnam); Papua New Guinea; Australia; Pacific Is (Samoa, Polynesia, Micronesia, Fiji, Hawaiian Is, Mariana Is, Marshall Is, Republic of Palau, Solomon Is).

Uses: Seaweed with a high concentration of calcium that can serve as food supplements in the Caribbean Is (Radulovich et al. 2013).

Ulva clathrata (Roth) C. Agardh (Fig. 52)

Common names: Arabic: Tahalib (FAO 1997); Chinese: Tai tyau (Madlener 1977), Taitiao (Tseng 1983); English: Spiky tendrils (Bunker et al. 2010);

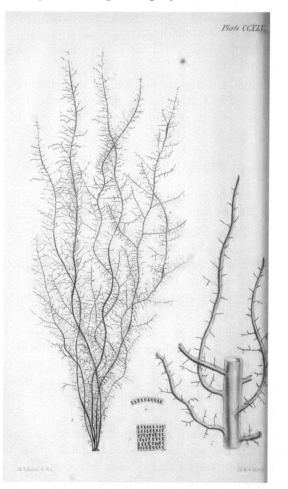

Figure 52. *Ulva clathrata* (Roth) C. Agardh (Chlorophyta). Illustration from William H. Harvey (Plate CCXLV, Phycologia Britannica, 1846–1951).

Stone hair (Madlener 1977), Bright green nori (FAO 1997); Japanese: Aonori (Pereira 2015a); Korean: Parae (Sohn 1998).

Description: This species forms tufts, bright green, composed of branched axes, which can reach several centimeters long (20–30 cm). The main axis and branches are covered with very characteristic conical branchlets (Pereira 2010a).

Geographic distribution: *U. clathrata* is widely distributed in warm seas from the W Pacific through the Atlantic Ocean, Gulf of Mexico, Mediterranean, Indian Ocean and SE Pacific.

Uses: Used in the US, Caribbean, China, Japan and Taiwan for foodstuff; *U. clathrata* is a highly nutritious seaweed, with high protein content. It is eaten by humans (algae tortillas!), used as an animal feed in shrimp farms, etc. (Kang 1968, Arasaki and Arasaki 1983, Tseng 1983, Pereira 2011, Farasat et al. 2013, Pereira 2015a). In Korean cuisine is used in various seasoning with soy sauce (Kang 1968, Sohn 1998, Zemke-White and Ohno 1999, Roo et al. 2007, Harrison 2013).

U. clathrata has anti-fertility, hypoglycemic and diuretic properties and favors cardiovascular and nervous activity; antibacterial, antiviral, antifungal, and antiprotozoal activities are also known (Milchakova 2011, Pereira 2015a).

Ulva linza Linnaeus (Fig. 53)

Synonym: *Enteromorpha linza*

Common names: English: Slender sea lettuce (Chapman and Chapman 1980), Doubled Ribbon Weed (Bunker et al. 2010), Green string lettuce; Chinese: Hai-cai (Arasaki and Arasaki 1983); Hawaiian: Limu ele-ele (MacCaughey 1918); Japanese: Awo-nori (Smith 1904), Usuba-awonori (Tokida 1954), Usaba-aonori (Arasaki and Arasaki 1983); Korean: Parae (Sohn 1998), Ipparae (Song et al. 2013); Swedish: Fingrenig tarmalg (Tolstoy and Österlund 2003).

Description: Large, ribbon-like species of green seaweed that may reach up to 30 cm in length. The thalli are unbranched and often have a frilled margin. The thalli taper into a distinct stipe below and are highly compressed. The width of the thallus is greater in the middle than at the base and may reach 5 cm in width. *U. linza* is bright light to dark green in coloration (Pereira 2014h, 2015a).

Geographic distribution: NE Atlantic; Atlantic Is (Azores Is, Bermuda, Canary Is, Iceland, Madeira, Selvage Is); Tropical and subtropical W and E Atlantic; NE Pacific (Alaska); SW Asia (Arabian Gulf, Cyprus, Israel, Turkey); Asia (China, Japan, Korea, Russia, Taiwan); SE Asia (Indonesia); Australia and New Zealand; Pacific Is (Hawaiian Is).

Figure 53. *Ulva linza* Linnaeus (Chlorophyta). Illustration from William H. Harvey (Phycologia Britannica, 1846–1951).

Uses: *U. linza* is used as an edible seaweed in many cultures for its high nutrient content and silky texture. Used also for making a cooling tea as well as for a vegetable in China (Bangmei and Abbott 1987); in Japanese and Korean cuisine is used in various seasoning with soy sauce (Kang 1968, Bonotto 1976, Arasaki and Arasaki 1983, Sohn 1998, Zemke-White and Ohno 1999, Roo et al. 2007, Song et al. 2013); and has been known for its consumption in Israel (Lipkin and Friedlander 2006). This species is also consumed in N America (Harrison 2013), and in the Hawaiian Is (Reed 1906, MacCaughey 1918).

Green algae extracts are also very nutrient rich and make a beneficial addition to natural cosmetic products (Pereira 2014h). Extracts have antiviral activity to *influenza A* (Fassina and Berti 1962).

Ulva rigida C. Agardh

Common names: Green laver, Lattuga marina, Lechuga de mar, Meersalat, Sea lettuce, Stijve zeesla, Ulve rigide; Portuguese: Alface-do-mar (Neto et al. 2005).

Description: Thallus a flat cellular membrane or frond variable in shape and size that may form tufts 2 cm tall although individuals can grow to sheets to 15 cm tall. The thallus is firm in texture, two cell layers thick, and arises from a small stipe and discoid holdfast. The margin of the frond may have small denticulations near the base (Neto et al. 2005).

Geographic distribution: *U. rigida* has a worldwide distribution in temperate and warm seas.

Uses: *U. rigida* is often utilized as a fresh sea vegetable by many island cultures for its high nutrient content and fresh taste. In Azores Is (Portugal), this species is used in cooking to prepare soups and "tortas" (Neto et al. 2005, Matos 2013); is also consumed in the Mediterranean Basin (Harrison 2013); fresh and boiled (at 100°C for 2 min) *U. rigida* are marinated with two different formulations by using 2% lemon salt and 2% vinegar; the marination of *U. rigida* is made at room temperature for 20 days. Marinated fresh and boiled *U. rigida* by using lemon salt and vinegar can be an alternative for human foods (Kılın et al. 2013).

U. rigida is source of nitrate, sulfate, antimicrobial and antivirus BASs, vitamins, gibberellins; contain cytostatic substances; plants are used in the treatment for boils, dropsy, urinary diseases and nose-bleeds; forage additive in mollusk aquaculture; a promising source of biogas; post-fermentation waste products are used as fertilizers, also raw material for cosmetics (refreshing liquid, shampoo, lotion and skin powder); in some countries used as food and in the pulp-paper industries (Milchakova 2011, Pereira 2015a).

Valonia aegagropila C. Agardh

Common names: Spanish: Bolitas de mar, Bio-bolas (Radulovich et al. 2013).

Description: Thallus green, composed of compressed or loosely interwoven siphonous cells forming hemispherical domes or irregular cushions of indeterminate size; 2–4 cm tall and sometimes reach 15 cm or more in diam. Sub-dichotomous or irregular branching to several orders; septate only at branch points; cells elongate, 2–4 mm broad, 3–20 mm long; basal cells in contact with substratum serving as rhizoidal cells; cells multinucleate; chloroplasts numerous per cell and discoid, each with single pyrenoid (Huang 2015c).

Geographic distribution: Adriatic Sea; Mediterranean; Atlantic Is (Canary Is, Selvage Is); Tropical and subtropical W Atlantic; Caribbean; Mexican Gulf; Arabian Gulf; Indian Ocean; Asia (China, Japan, Taiwan); SE Asia (Indonesia, Malaysia, the Philippines, Singapore, Vietnam); Australia; Pacific Is (Samoa, Polynesia, Micronesia, Fiji, Hawaiian Is, Mariana Is, Marshall Is, Solomon Is).

Uses: Used in the Caribbean Is as food (Radulovich et al. 2013).

Valonia macrophysa Kützing (Fig. 54)

Common name: Japanese: Tamago Valonia (Oshie-Stark et al. 2003); Spanish: Bolitas de mar, Bio-bolas (Radulovich et al. 2013).

Description: Thallus olive-green forms bladder- to club-shaped multinucleate "cells" called coenocytes, which colonize the substratum in abstruse irregular blistered carpets (Braune and Guiry 2011).

Geographic distribution: NE Atlantic; Adriatic Sea; Mediterranean; Atlantic Is (Ascension, Azores, Bermuda, Canary Is, Madeira, Selvage Is); Tropical and subtropical W Atlantic; Caribbean; Indian Ocean; Asia (Japan); SE Asia (Indonesia, the Philippines, Vietnam); Australia; Pacific Is (Samoa, Polynesia, Fiji).

Uses: Considered an edible species (Radulovich et al. 2013, Subsea 2015).

Figure 54. *Valonia macrophysa* Kützing (Chlorophyta).

2.2.2 Ochrophyta – Phaeophyceae

Brassicophycus brassicaeformis (Kützing) Draisma, Ballesteros, F. Rousseau & T. Thibaut

Common name: Hanging wrack (Heyden 2013).

Description: Plants with many terete erect axes and bilaterally, alternately arising branches. Branches become terminally modified into siliqua-shaped receptacles; unisexual conceptacles in two rows on the margins of the flattened receptacles. Perennial holdfast is a spreading rhizome that forms large mats.

Geographic distribution: S Africa; Indian Ocean.

Uses: Used as food in S Africa, delicious and crunchy, used mixed with couscous and coleslaw (Heyden 2013).

Canistrocarpus cervicornis (Kützing) De Paula & De Clerck

Description: Thallus tufted, yellow-brown, erect, somewhat intertwined; branches screw-like twisted, narrow ribbons, mostly forking, but asymmetrically branching (one fork longer than the other), widening at the fork-base (to 4 mm), outer thallus parts narrow (1–2 mm wide), slightly tapering upwards, tips acute, margins smooth; anchored to right substrata by a discoidal holdfast (Pereira 2014c).

Geographic distribution: Warm E Atlantic; Caribbean; SW Atlantic (Brazil); NW Pacific (Japan, China); Indian Ocean.

Uses: Used for foodstuff (Gray 2014); extracts have antiviral (Vallim et al. 2010), anti-snake venom (Domingos et al. 2011, Moura et al. 2011), and antifouling (Bianco et al. 2009) activity.

Colpomenia sinuosa (Mertens ex Roth) Derbès & Solier (Fig. 55)

Common names: English: Papery sea bubble (Novaczek 2001), Silver-ballon, Sea ballon, Sea potato (Santhanam 2015); French: Ballons, Voleuse d'huitres (Dickinson 1963); Japanese: Fukuronori (Tokida 1954).

Description: Thallus bladder-like, smooth, slick, hollow, crisp, spherical to sac-like, irregularly expanded or somewhat lobed, to 30 cm diam., 10 cm high, golden-brown (Pereira 2015a, Santhanam 2015).

Geographic distribution: NE Atlantic; W Mediterranean; Persian Gulf; Indian Ocean; Asia (Japan, the Philippines); NE Pacific (Alaska to California); Australia and New Zealand.

Uses: Edible, used as food in India (Rao 1970), and the Philippines (Trono 1998, Zemke-White and Ohno 1999, Roo et al. 2007, Braune and Guiry 2011,

Figure 55. *Colpomenia sinuosa* (Mertens ex Roth) Derbès & Solier (Ochrophyta, Phaeophyceae).

Harrison 2013). Chop and add to any salad, stir fry, make soup or stew. Used for making dried flavor flakes (Novaczek 2001). In the Persian Gulf, this species is considered a potential edible algae (Mohammadi et al. 2013).

Dictyota ciliolata Sonder ex Kützing

Description: Thalli are erect, to 8 cm tall, attached by means of a single stupose holdfast. Stolonoidal fibers are absent. Straps, 2–3 mm wide, are slender and dichotomously branched. The margins are dentate, rarely smooth, while the surface is always smooth. The apices are rounded. The medulla and cortex are uniformly one-layered (De Clerck et al. 2002).

Geographic distribution: Tropical and subtropical Atlantic; Atlantic Is (Bermuda, Canary Is, Madeira, Selvage Is); Caribbean Is; Indian Ocean; SE Asia (Indonesia, Malaysia, the Philippines, Singapore, Thailand); Asia (Japan, Korea); Australia; Pacific Is (Polynesia, Hawaiian Is, Samoan Archipelago).

Uses: In the Caribbean Is this species is cooked fresh or dried, used as flour or meal, fried, or baked (Radulovich et al. 2013, Radulovich et al. 2015).

Dictyopteris delicatula J.V. Lamouroux

Description: This species has erect, light-brown, strap-shaped blades attached to the substratum at basal holdfast or to adjacent branches, creating a tangled mass 2–8 cm in height. Dichotomous to irregularly branch bi-layered blades, 0.5–5.0 mm wide, have a distinctly raised midrib that may be several cells thick. Cells of blades are arranged in parallel rows at acute angles to midrib. Scattered clusters of hyaline hairs in dense tufts arise along midrib on only one surface of midrib (Schmitt and Mamoozadeh 2014).

Geographic distribution: Atlantic Islands (Bermuda, Canary Is, Cape Verde Is); N America (Florida, Mexico, N Carolina, S Carolina, Texas); Central America (Belize, Costa Rica, Panama); Caribbean Is (Bahamas, Barbados, Cuba, Martinique, the Netherlands Antilles, Puerto Rico, Virgin Is).

Uses: Used for foodstuff (Gray 2015). Heterofucans of this species have anticoagulant, antioxidant and antitumor activities (Magalhaes et al. 2011).

Dictyota caribaea Hörnig & Schnetter

Description: Is an erect olive-brown alga with strap-shaped fronds that measured up to 32 mm long in the studied area. The fronds have unequal dichotomous divisions that range 1–2 mm wide, 90–140 mm thickness and non-constricted branches with sharp apexes (De Oliveira et al. 2009).

Geographic distribution: Tropical and subtropical W Atlantic; Caribbean.

Uses: Used in the Caribbean Is as food (Radulovich et al. 2013).

Dictyota friabilis Setchell

Synonym: *Dictyota pfaffii*

Description: Lower portions of this algae spread over the substrates, but upper portions are slightly erect. The branches are strap-shaped, dichotomously divided, up to 5 mm wide, with entire margins and round tips. The color is greenish brown with concentric lines when it is under water. The plants grow on rocks and gravel in subtidal zones along the shoreline with water movement (JIRCAS 2012a).

Geographic distribution: Atlantic Is (Canary Is); Tropical and subtropical in W Atlantic; Caribbean; Indian Ocean; Asia (China, Japan, Korea, Taiwan); SE Asia (Indonesia, Malaysia, the Philippines, Thailand, Vietnam); Australia; Pacific Is (Samoa, Caroline Is, Polynesia, Micronesia, Fiji, Hawaiian Is, Mariana Is, Palau, Solomon Is, Tahiti).

Uses: Considered an edible species (Gupta and Abu-Ghannam 2011).

Dictyota menstrualis (Hoyt) Schnetter, Hörning & Weber-Peukert

Description: Thallus erect, 15–25 cm high, yellow brown to dark brown, dichotomously branched blades strapshaped, 2–15 mm wide, with smooth margins and blunt apices; surface cells rectangular to polygonal. Medullary layer of large hyaline cells (Parish et al. 2014).

Geographic distribution: N America (Florida, N Carolina, S Carolina), Caribbean, S America (Brazil, Colombia, Venezuela), Atlantic Islands (Azores, Canary Is, Cape Verde Is, Salvage Is).

Uses: Used for foodstuff (Gray 2015); extracts of this species have anticoagulant (Albuquerque et al. 2004), anti-nociceptive and anti-inflammatory (Albuquerque et al. 2013), and antiviral (Pereira et al. 2004) activity.

Dictyota mertensii (Martius) Kützing

Description: Thallus bushy, brown, often iridescent blue-green under water, erect, robust and stately; the flattened fronds show a branching pattern atypical for the genus: distinct main axes branch repeated-alternating, terminal forkings of the latitudinal axes turn into spur-like, 1–2 mm long, pointed or rounded tips (Braune and Guiry 2011).

Geographic distribution: Atlantic Is (Bermuda, Canary Is, Cape Verde Is); Tropical and subtropical W and E Atlantic; Caribbean; SE Asia (the Philippines, Thailand, Vietnam); Pacific Is (Fiji).

Uses: Used in the Caribbean Is as food (Radulovich et al. 2013).

Dictyota stolonifera E.Y. Dawson

Description: Thallus upright near base, parallel to substratum toward apices, 0.5–8 cm high, light brown, with occasional dark striping and light patches at -v- of dichotomies; branching somewhat dichotomous. Blades stubby, strap-shaped, 2–6 mm wide, to 80 µ thick; apices broad, rounded (Littler and Littler 2015a).

Geographic distribution: Caribbean Sea; SE Asia (the Philippines); Pacific Is (Hawaiian Is); NE Pacific (Baja California, Colombia, Nicaragua, Panama).

Uses: Used in the Caribbean Is as food (Radulovich et al. 2013).

Ecklonia maxima (Osbeck) Papenfuss

Common names: English: Sea bamboo (Heyden 2013).

Description: Large seaweed, dark brown, with a very long, hollow stipe tapering slightly towards the base and often becoming inflated at the top like a gas-filled bulb. This acts as a float bladder holding the stipe vertically and the ribbon-like assimilators floating at the sea surface (Braune and Guiry 2011).

Geographic distribution: SE Atlantic (Namibia, S Africa).

Uses: Used as food in Namibia and S Africa (Molloy 1990), candied, fried, pickled. This kelp is harvested commercially for use as a gel in food products, toothpaste, paint and ink, to stabilize embankments and waterproof cement and for use as plant food and fertilizers, being rich in mineral salts. Fisher folk along the Cape coast (S Africa) fill the bulb with seafood, plug the ends with wet newspaper and cook the "food in a bulb" over a fire (Heyden 2013).

Laminaria abyssalis A.B. Joly & E.C. Oliveira

Description: Alternation of large sporophyte bearing unilocular meiosporangia with paraphyses (sori) and microscopic dioecious and oogamous, heteromorphous gametophytes. Thallus divided into blade, stipe and rhizoids.

Geographic distribution: Tropical and subtropical W and E Atlantic; Caribbean Sea.

Uses: Used in the South western Atlantic as food (Radulovich et al. 2013).

Laminaria pallida Greville

Common names: English: Split-fan kelp, Split fan (Heyden 2013).

Description: Sporophyte similar to that of *L. digitata* and *L. hyperborea*, stipe rigid, solid or hollow, round, gradually widening into a large, fan-shaped blade towards the end, light brown color. The blade is deeply longitudinally split into a number of individual, ribbon-like fronds; number of fronds in sheltered conditions lower than when growing in wave-exposed sites. Holdfast made up of branching, rounded haptera (Braune and Guiry 2011).

Geographic distribution: Atlantic Is (Canary Is, Tristan da Cunha Is); SE Atlantic (Namibia, S Africa); Indian Ocean.

Uses: Used as food in S Africa, candied, fried, and pickled (Heyden 2013).

Research has shown *L. pallida* to be an excellent source of alginate, a biological compound that has numerous medical applications. Alginate derived from *L. pallida* is biocompatible, meaning the human body does not recognize it as a threat and will not attack it with an immune response. Because of this, *L. pallida* alginate is used to coat transplants of tissue like bone marrow to avoid rejection of the foreign tissue by the host body. *L. pallida* is also an important source of gelling and thickening agents used in food (Heyden 2013).

Lessonia flavicans Bory de Saint-Vincent

Synonym: *Lessonia vadosa*

Description: This species can grow up to 4 ms tall from a richly branched haptera with a single dichotomously branched stipe up to 5 cm in diam.;

branches of the stipe bear blades anywhere from 6 to 39 cm broad and 17 to 86 cm long. The margins of these blades are entirely to coarsely toothed, with the cortex comprised of a solid, parenchymatous tissue. The range of blade morphology seen in this species alone displays the amount of phenotypic plasticity that exists in plants of this genus in response to different environmental influences. *L. flavicans* is found in deeper water than other *Lessonia* species in S America (up to 20 m deep), often separated from stands of *Lessonia* species by a belt of *Macrocystis pyrifera* (SIA 2014).

Geographic distribution: S America (Argentina, Chile, Falkland/Malvinas Is); Antarctic and the subantarctic Is.

Uses: Used for foodstuff (Gray 2015), and extraction of alginate in colder S Atlantic (Arasaki and Arasaki 1983).

Padina boergesenii Allender & Kraft (Fig. 56)

Common name: English: Leafy rolled-blade alga (Florent 2014).

Description: Thalli light brown to tan, moderately ventrally calcified, 4–6 cm long and wide; blades broadly or narrowly lobed, short-stipitate from a bulbous fibrous holdfast (Kraft 2009).

Geographic distribution: This species has a scattered distribution across tropical, subtropical and warm-temperate regions.

Figure 56. *Padina boergesenii* Allender & Kraft (Ochrophyta, Phaeophyceae).

Uses: Used for foodstuff (Gray 2005). Extracts of this spices have hepatoprotective (Karthikeyan et al. 2010), anti-diabetic (Senthilkumar et al. 2014), and antioxidant (Rajamani et al. 2014) activity.

Padina gymnospora (Kützing) Sonder

Common names: English: Limey petticoat (Santhanam 2015); Spanish: Alcachofa marina (Radulovich et al. 2013).

Description: Thalli brown, darker at the base, robust, up to 200 mm across, of fan-shaped lobes that split into wedge-shaped pieces blade edges are in-rolled upwards and inwards concentric bands of hairs and sporangia prominent (Baldock 2003).

Geographic distribution: Worldwide, in the tropics.

Uses: Used for foodstuff in India (Rao 1970), Malaysia (Gray 2005, Ahmad et al. 2012), and the Caribbean Is (Robledo and Pelegrín 2009, Radulovich et al. 2013).

Extracts of this species have anti-inflammatory (Marques et al. 2012), hemagglutinating and anticancer (Neeta et al. 2012), and antibacterial (Chapman and Chapman 1980) properties.

Padina sanctae-crucis Børgesen

Description: This species forms curled, fanlike branches from a single stalk. The plant is about 15 cm tall. Blades are often irregularly split and branched. The upper surfaces of the fans are calcified and whitened, but the rest of the plant is brownish. All the branches are crossed by closely set growth lines.

Geographic distribution: Atlantic Is (Bermuda); Gulf of Mexico; Caribbean Sea; Tropical W Atlantic; Caribbean Sea; Indian Ocean; Asia (Japan, Taiwan); SE Asia (Indonesia, the Philippines, Thailand); Australia; Pacific Is (Micronesia, Fiji, Hawaiian Is, Mariana Is); Tropical E Pacific.

Uses: With food potential (Radulovich et al. 2013).

Sargassum cymosum C. Agardh

Common names: Hawaiian: Limu kala (Reed 1906); Spanish: Sargazo (Radulovich et al. 2013).

Description: Thallus to 10–200 cm or more in length, with one to a few simple, terete to compressed, stipes 1–20 cm long arising from a discoid-conical holdfast. Stipes bearing radially or distichously borne, long primary branches, produced seasonally from the stipe apices and subsequently deciduous, leaving scars or other residues on the stipe. Primary branches 10 cm to 200 cm or more long.

Geographic distribution: Atlantic Is (Azores, Bermuda, Canary Is, Cape Verde Is, Selvage Is); Tropical and subtropical W and E Atlantic; Caribbean; Indian Ocean; Asia (China, Japan, Taiwan); SE Asia (Vietnam); Pacific Is (Hawaiian Is).

Uses: Used in the Caribbean and Hawaiian Is as food (Reed 1906, Radulovich et al. 2013).

Sargassum filipendula C. Agardh

Common names: English: Gulf weed, Gulfweed, Sargassum grass, Tuna weed; Spanish: Sargazo llorón, Sargazo (Radulovich et al. 2013).

Description: *S. filipendula* is used in traditional cuisines of most of South America and Asia. The nutrient-rich extracts of *S. filipendula* are also used in cosmetic products from lotions to face masks (SIA 2014).

Geographic distribution: Atlantic Is (Bermuda, Canary Is, Madeira, Selvage Is); E Mediterranean; Arabian Gulf; Caribbean Sea; Gulf of Mexico; Tropical and subtropical W Atlantic; SE Asia (Indonesia, Malaysia).

Uses: Used for foodstuff; canned and frozen *S. filipendula* has been consumed in Egypt for its high vitamin, mineral and fiber content; is used also in traditional cuisines of most of S America and Asia (Zemke-White and Ohno 1999, Lipkin and Friedlander 2006, Robledo and Pelegrín 2009, Harrison 2013, Radulovich et al. 2013, Gray 2015).

Extracts of this species have antioxidant and anti-proliferative (Silva et al. 2011) activity.

Sargassum fluitans (Børgesen) Børgesen

Common names: English: Broadleaf gulfweed (Bowling 2012); Spanish: Sargazo (Radulovich et al. 2013).

Description: Free floating; color orange to brown; plant body (thallus) with many branches, rounded in cross-section; branches bear flat, elongate, leaf-like organs (blades) with a prominent midrib and serrated margins on a small rounded stalk; spherical, air-filled sacs (bladders) on short stalks in axils of leafy part; blades 2–6 cm long, 3–8 mm wide; one individual may reach up to 1 m (Bowling 2012b).

Geographic distribution: Atlantic Is (Bermuda); Tropical and subtropical W Atlantic; Caribbean Sea; Gulf of Mexico; SE Asia (Indonesia, the Philippines).

Uses: All parts of this *Sargassum* are edible, including the numerous crustaceans that make this seaweed their home. It has a somewhat bitter taste and is not considered to be as desirable as many of the more N Pacific and Atlantic seaweeds. However, it is quite plentiful and a decent source of calories. Traditionally it is chopped and cooked in many ways including boiled, steamed, and sauteed in hot oil. Experiment until you find a method and flavor you like (Radulovich et al. 2013, Merriwether 2014).

Sargassum liebmannii J. Agardh

Common name: Spanish: Sargazo (Radulovich et al. 2013).

Description: Thallus to 10–200 cm or more in length, with one to a few simple, terete to compressed, stipes 1–20 cm long arising from a discoid-conical holdfast. Stipes bearing radially or distichously borne, long primary branches, produced seasonally from the stipe apices and subsequently deciduous, leaving scars or other residues on the stipe. Primary branches 10 cm to 200 cm or more long (Cortés et al. 2014).

Geographic distribution: Tropical and subtropical Atlantic; Caribbean Is; NE and E Pacific (California, Mexico).

Uses: In the Caribbean Is this species is cooked fresh or dried, used as flour or meal, fried, or baked (Radulovich et al. 2015).

Sargassum natans (Linnaeus) Gaillon

Common names: English: Common gulfweed, Narrowleaf gulfweed, Spiny gulfweed; Spanish: Sargazo (Radulovich et al. 2013).

Description: As the name suggests, Common Gulfweed is the most common *Sargassum* species found in the Sargasso Sea and washed up on Bermuda's beaches. *S. natans* is a bushy seaweed with narrow leaf blades which are golden brown with toothed edges. The rubbery-textured leaves range from 2–6 mm wide and 2–10 cm long. The gas-filled floats are less than 6 mm in diam. and are held on short stalks along the stems among the leaves. The floats of *S. natans* have a single protruding spine 2–5 mm long. *S. natans* does not have a single main stem; instead it grows in many directions forming clumps that can reach 60 cm long. It is these clumps that form together into much larger mats (Bowling 2012c, Karen 2014, Merriwether 2014).

Geographic distribution: NE Atlantic (Portugal, Spain); Atlantic Is (Azores, Bermuda, Canary Is, Cape Verde Is, Madeira, Selvage Is); Tropical and subtropical W Atlantic; Caribbean Sea; SE Asia (Indonesia, the Philippines); Australia.

Uses: Some *Sargassum* are consumed fresh, others cooked in coconut milk or a little vinegar or lemon juice. It is smoked-dried to preserve it. *Sargassum* is also eaten by itself or added to fish and meat dishes. If not strong it can be added to salads after washing, or it can be cooked in water like a vegetable. If the *Sargassum* is strong flavored it can be boiled in two changes of water. Some recipes then call for it to be mixed with brown sugar and used as a filling in steamed buns but it could be eaten as it is (Radulovich et al. 2013, Deane 2014c, Merriwether 2014).

Sargassum platycarpum Montagne

Common name: Spanish: Sargazo (Radulovich et al. 2013).

Description: Thallus up to 25 cm tall, erect, attached to the substratum by a small discoid holdfast up to 3 mm in diam. The main axis is cylindrical, 1 mm diam. Lateral branches slender, smooth and alternate on the main axis. Phylloids scattered or alternate, stalked, lanceolate, up to 2 cm long and 3 mm wide, apex acute, some tapering into a petiole, margins are coarsely serrate. Cryptostomata are large, arranged in a single irregular row on each side of a conspicuous midrib. Vesicles are spherical to ovate, up to 6 mm long and 4 mm wide, stalked, the compressed stalk up to 6 mm long. Receptacles are in auxiliary clusters, crowded, closely cymose, the cluster up to 7 mm long (Abdel-Kareem 2009).

Geographic distribution: Tropical and subtropical Atlantic; Caribbean Is.

Uses: In the Caribbean Is this species is cooked fresh or dried, used as flour or meal, fried, or baked (Radulovich et al. 2013, Radulovich et al. 2015).

Sargassum polyceratium Montagne

Common name: Spanish: Sargazo (Radulovich et al. 2013).

Description: Plants brown, crowded, leathery, densely branched to 60 cm high. Distinct holdfast, firmly attached to rocky substratum; main axis developing from holdfast, and up to 3 cm in length; one to several lateral branches developing from main axis, with secondary branching and sometimes tertiary branching as well. Specimens found do not all have spines on lateral branches, a characteristic of the species; blades closely crowded, curved, with serrated margin, and dispersed surface cryptostomata; younger branches with blades to 2.5 cm long and 1 cm wide.

Geographic distribution: Tropical and subtropical W Atlantic; Caribbean; SE Asia (Indonesia, the Philippines).

Uses: Used in the Caribbean Is as food (Radulovich et al. 2013).

Spatoglossum schroederi (C. Agardh) Kützing

Common name: Spanish: Espatogloso (Radulovich et al. 2013).

Description: Thalli erect, arising from a matted rhizoidal holdfast, up to 80 cm long, complanate, divided into sub-dichotomous to sub-palmate segments, 0.5–5 cm broad, with undulate to dentate margins, lacking midrib or veins. Growth initiated from a short row of apical cells.

Geographic distribution: Atlantic Is (Azores, Bermuda, Canary Is); Tropical and subtropical W and E Atlantic; Caribbean Sea; Mediterranean; Indian Ocean; Australia; Pacific Is (Micronesia); Tropical E Pacific.

Uses: Is a very interesting species, with great potential to produce antioxidants, but with limited use as food (Radulovich et al. 2013).

2.2.3 Rhodophyta

Bryothamnion triquetrum (S.G. Gmelin) M.A. Howe

Common name: Spanish: Pelo de cabra (Radulovich et al. 2013).

Description: Thallus coarse and bushy, to 25 cm high; dark brown to red; branches numerous, irregularly alternate; branchlets stiff to 3 mm long, in three vertical rows, creating triangular branches; apices of branches incurved and pointed; axes polysiphonous with seven–nine pericentral cells, heavily corticated, and generally triangular in transverse section.

Geographic distribution: Atlantic Is (Cape Verde Is); N America (Florida, Mexico); Central America (Costa Rica, Panama); Caribbean Is (Bahamas, Barbados, Cuba, Hispaniola, Jamaica, Lesser Antilles, Netherlands Antilles, Puerto Rico, Trinidad and Tobago, Virgin Is); S America (Brazil, Venezuela); Africa (Angola, S. Tomé and Príncipe).

Uses: With food potential (Radulovich et al. 2013).

Champia feldmannii Díaz-Piferrer

Description: Plants growing in tufts, abundantly and repeatedly irregularly branched, measuring up to 12 cm high. The segments are clearly barrel-shaped measuring from 1 to 8 mm in diam. and each segment is about 5 to 8 mm long, the upper ones being smaller. The tips of the branches are frequently curved. Each segment is a hollow vesicles separated from neighboring segments by a cellular diaphragm (Pereira and Ugadim 1974).

Geographic distribution: Tropical and subtropical W and SW Atlantic (Brazil, Colombia, Venezuela).

Uses: Considered an edible species (Gray 2015).

Extracts of this species have antitumor (Lins et al. 2009), antithrombotic and anticoagulant (Assreuy et al. 2008) activity.

Cryptonemia crenulata (J. Agardh) J. Agardh

Description: Foliose, linear-lanceolate to broad and lobed, entire, proliferous or dissected into numerous cuneate, lanceolate or obovate lobes, occasionally laciniate, usually with distinct stalk or stipe, continuing as a midrib in some species, branched or unbranched, and with small discoid holdfast.

Geographic distribution: Atlantic Is (Bermuda, Canary Is); Caribbean Sea; Gulf of Mexico; Tropical W and E Atlantic; Indian Ocean; SE Asia (Indonesia, the Philippines); Pacific Is (Fiji).

Uses: With food potential (Radulovich et al. 2013).

Eucheuma isiforme (C. Agardh) J. Agardh

Common names: English: Sea moss, Seamoss; Spanish: Eukeuma (Radulovich 2013).

Description: *E. isiforme* is perennial Rhodophyta, pale straw-yellow to yellow-brown or reddish, which looks like tumbleweed. It has multidirectional rubbery, cartilaginous branches, which sometimes grow secondary branches. It reaches a height of 30 to 50 cm with coarse succulent-like branches with bumpy surfaces. *E. isiforme* grows freely in shallow, calm waters around islands and patch reefs (Braune and Guiry 2011).

Geographic distribution: Caribbean, Gulf of Mexico (Braune and Guiry 2011).

Uses: *E. isiforme* is a known local ingredient for families and restaurants in drinks, cakes, and breads, and as a vegetable. It is most popularly blended into a cold shake (used to make Sea moss or Irish moss health drink in Jamaica and Trinidad) (Zemke-White and Ohno 1999).

Some claim it is a natural medical remedy for glaucoma, menopause, arthritis, and tuberculosis, as well as fatigue, headaches, and colds. Reported aphrodisiac and sexually restorative properties contribute to the popularity of *E. isiforme*. This species is a carrageenophyte, a source of iota carrageenan, and was exported at one time to Denmark due to a supply shortage from other areas (Chapman and Chapman 1980, Robledo and Pelegrín 2009, Braune and Guiry 2011, Harrison 2013).

Eucheuma nudum J. Agardh

Description: It was separated from the other species primarily on the basis of its terete, spineless, conical-tipped branches (Cheney 1986).

Geographic distribution: Florida and the Caribbean.

Uses: Used as food in Florida and Bermuda (Arasaki and Arasaki 1983).

Galaxaura rugosa (J. Ellis & Solander) J.V. Lamouroux

Description: Thallus upright, pale-pink, tufted, repeated-regular forking; weakly to moderately calcified (Braune and Guiry 2011, Pereira 2015a).

Geographic distribution: E Atlantic (Madeira, Canary Is, Cape Verde to Gabon); W Atlantic (Brazil); Caribbean Sea; Indian Ocean; NW Pacific (Japan, the Philippines, Indonesia); Pacific Is; Australia, New Zealand.

Uses: With food potential (Radulovich et al. 2013).

Gelidium crinale (Hare ex Turner) Gaillon

Description: Thallus dark red-brown, forming dense turfs or masses 1–3 cm high and 1–3 cm in spread, with basal stolons, usually short and very branched, bearing erect, slender branches; stolons terete, 100–200 µ in diam. Erect branches numerous, terete to slightly compressed, of similar dimensions throughout (Womersley 1994). Forms dense turfs where covered by frequent wave-wash or in pools, on coasts of moderate to strong wave action and often subject to sandy conditions (Womersley and Guiry 1994).

Geographic distribution: Black Sea; Asia (China, Japan, Taiwan); SE Asia (Indonesia, the Philippines, Vietnam); Australia and New Zealand; Pacific Is (Polynesia, Micronesia, Hawaii, Mariana Is, Samoan Archipelago); N Atlantic; W Atlantic (Tropical and subtropical W Atlantic); S Atlantic.

Uses: Used as food in China an SE Asia (Bangmei and Abbott 1987, Milchakova 2011, Gray 2015).

G. crinale is a valuable commercial species, the source of high-quality agar, agarose, agaropectin, polysaccharides and sulfates, vitamins A, B, C, E, and BASs with anticoagulant, antimicrobial, antiviral and antitumor effect (Milchakova 2011, Pereira 2015a).

Gelidium serrulatum J. Agardh

Common name: English: Seamoss (Fournillier 2015).

Description: The algae are attached to the substratum with a tough holdfast that is usually removed when harvested, leaving no base to regenerate new shoots. The agar of this species has high gel strength and is the best quality of any "Seamoss" found in the Caribbean (agar is the carbohydrate produced by "Seamoss" which, when dissolved in hot water, thickens to form a gel on cooling) (Fournillier 2015).

Geographic distribution: Tropical W and E Atlantic; Caribbean Sea; Tropical E Pacific.

Uses: Traditionally "Seamoss" harvested from natural stocks has been used for food as well as medicinal products. These include the popular milk drink that is locally believed to be an aphrodisiac (Radulovich et al. 2013, Fournillier 2015).

Gelidium vittatum (Linnaeus) Kützing

Synonym: *Suhria vittata*

Common name: English: Red ribbons (Chapman 1970, Heyden 2013).

Description: Thallus narrow, deeply crimson, strap-like with a conspicuous midrib, spirally twisted in parts, margins ruffle-like densely covered with delicate small leaflets; usually only loosely and irregularly branching; almost exclusively as epiphyte on the stipes of *Ecklonia maxima* (Phaeophyceae) (Braune and Guiry 2011).

Geographic distribution: Endemic to S Africa; SE Atlantic (Namibia) and Indian Ocean (S Africa, Mauritus).

Uses: On the continent of Africa, seaweed appears to be largely neglected as a direct source of food. Chapman (1970) refers to the traditional use of *G. vittata*, known along the S African coast as "red ribbon", for jelly-making by the early Cape colonists; boiled in water to extract the agar used to make sweet and savory jellies (Chapman 1970, Chapman and Chapman 1980, Heyden 2013).

Gracilaria birdiae E.M. Plastino & E.C. Oliveira

Common name: Portuguese: Macarrão do Mar (Pires et al. 2012).

Description: Thalli cylindrical and branched, when dried resembles spaghetti (macaroni type).

Geographic distribution: W and SW Atlantic.

Uses: Used as food in Brazil (França-Pires et al. 2012, Fidelis et al. 2014).

Gracilaria bursa-pastoris (S.G. Gmelin) P.C. Silva

Synonym: *Gracilaria compressa* (C. Agardh) Greville

Common names: English: Shepherd's Purse Wart Weed (Bunker et al. 2010); Hawaiian: Ogo, Limu manauea (Fortner 1978); Japanese: Ogo (Kent 1986).

Description: Thallus greenish-red, yellow and dark brownish-red, with cylindrical axes, forking and with lateral branching, tufted, erect, branches gradually tapering off, not narrowing at their origins, the cystocarps-bearing plants are densely covered with hemispherical reproductive structures (cystocarps); texture cartilaginous-meaty, stiff, and bristly; attached to the rocks by a discoid holdfast (Braune and Guiry 2011).

Geographic distribution: NE and E Atlantic (Ireland to Morocco, W Africa, Cape Verde Is), Mediterranean, Caribbean, Indian Ocean (India, Kenya, Sri Lanka); W and NW Pacific (Japan, China, the Philippines, Thailand); Pacific Is (Hawaii).

Uses: Economically important in Asia as food, and a source of agar (Johnston 1966, Lewmanomont 1978, Zemke-White and Ohno 1999, Hu 2005, Ohno and Largo 2006, Braune and Guiry 2011, Harrison 2013). The old Hawaiians chopped and salted the "limu", then mixed it with other "limu" or fish or meat. Later, they used it to thicken chicken and pork stews. The range of present-day uses is far greater. "Ogo" can be prepared as a candy, pickle, salad, tempura, soup vegetable, or dip. Because of its mild taste, this seaweed is recommended for the beginning "limu" eater (Fortner 1978). This species is also used as food in England and Wales (Chapman and Chapman 1980, Green 2014). In the Caribbean Is this species is cooked fresh or dried, used as flour or meal, fried, or baked (Radulovich et al. 2015).

Gracilaria damaecornis J. Agardh

Description: Erect thallus arise from a small discoid holdfast, the thalli are cylindrical, depressed or blade-shaped, with lateral, alternate or sub-dichotomous branches, the style of the apex and the base of branches are different with species, has a vegetative thalli, thallus has a cortex and medulla.

Geographic distribution: Atlantic Is (Bermuda); Gulf of Mexico; Tropical W Atlantic; Caribbean Sea; SE Asia (Indonesia, the Philippines).

Uses: With food potential (Radulovich et al. 2013).

Gracilaria domingensis (Kützing) Sonder ex Dickie

Description: Plants bushy, 10 (–35) cm tall, rose-red to green brown. This flattened species with a dentate or pinnulate thallus margins, has cortical cells of 5–10 µ in diam., medullary cells of 150–170 µ in diam., and subsurface cells of intermediate size (Oliveira et al. 1983, Hadad and Fricke 2008).

Geographic distribution: Tropical and subtropical W Atlantic; SE Asia (Indonesia).

Uses: *G. domingensis* has been collected in Brazil and exported to Japan for food (Oliveira 2006, Calado et al. 2012); used as food in the Caribbean and Central America (Zemke-White and Ohno 1999, Roo et al. 2007, Harrison 2013, Radulovich et al. 2013).

Gracilaria dura (C. Agardh) J. Agardh

Description: Thallus bushy, cartilaginous, dark-purple, 3–20 cm high, 0.3–0.9 mm diam., attached and unattached forms known; branching numerous, dichotomous and trichotomous, alternate and unilateral; branches often equal in length, sometimes connivent, with small bunches at apices; upper branches slightly narrower, obtuse; unattached thalli broader and coarser, often giving rise to cauliflower-shaped knobby galls (Milchakova 2011).

Geographic distribution: Atlantic Is (Canary Is); Adriatic Sea; Mediterranean; Black Sea; Indian Ocean; Asia (China).

Uses: *G. dura* is a valuable commercial species, the source of agar, microelements, PUFA, α-tocopherol, R-phycoerythrin, arachidonic acid, gibberellins and prostaglandins; also used as an agricultural fertilizer, in the fragrance industry and food in some countries (Milchakova 2011).

Gracilaria foliifera (Forsskål) Børgesen

Common names: Portuguese: Botelho, Golfinho, Limo-folha (INII 1966); Tamil: Cigarette pasi (Kaliaperumal et al. 1995).

Description: Plants sparingly bushy, 10–30 cm tall, dull purple or faded, the lower part relatively slender, above more coarse, thick, sub-terete to compressed or expanded and 2–15 mm or somewhat more in width, sub-linear or laciniate, the margin often prolific, branching of one to several degrees, usually in the plane of the blade, di- or trichotomous or alternate, tetrasporangia 20–35 µ diam. long, 30–45 µ long, formed in the upper mature branches just below the surface, pericarps projecting strongly on faces or margins of the blades (Paula 2005).

Geographic distribution: Tropical and subtropical W Atlantic; Caribbean; Mediterranean; Indian Ocean; Asia (Indonesia, the Philippines, Vietnam); Australia.

Uses: Used as food in Gulf of Mexico and Cuba (Arasaki and Arasaki 1983), and for foodstuff in India, as potential raw material for agar production (Chapman and Chapman 1980, Kaliaperumal et al. 1995).

Gracilaria gracilis (Stackhouse) M. Steentoft, L.M. Irvine & W.F. Farnham (Fig. 57)

Common names: English: Slender wart weed (Bunker et al. 2010); Portuguese: Cabelo-da-velha, Carriola (Oliveira 1990, Pereira 2009b, Pereira and Correia 2015).

Description: Cartilaginous, cylindrical, dull purple fronds, to 500 mm long, one or several rising from small, fleshy, perennial discoid holdfast. Branching very irregular, sparse or profuse, branches to 2 mm diam., apices pointed; intertidal tissue of large thin-walled cells with narrow outer cortical zone of small colorless cells. The color of the thallus may turn from dark red in winter into pale orange in summer. This Rhodophyta is usually found in shallow waters, growing on rocks, gravel and sand. Species of the

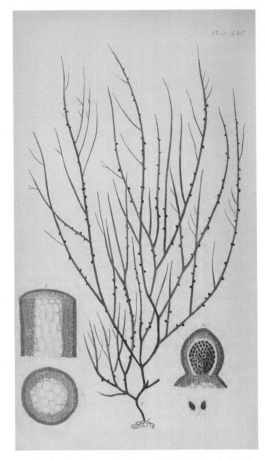

Figure 57. *Gracilaria gracilis* (Stackhouse) M. Steentoft, L.M. Irvine & W.F. Farnham (Ochrophyta, Phaeophyceae). Illustration from William H. Harvey (Plate LXV, Phycologia Britannica, 1846–1951).

genus *Gracilaria* are characterized by their capacity for assimilating nitrogen rapidly and keeping it in stock.

Geographic distribution: NE Atlantic; Atlantic Is (Canary Is, Madeira, Selvage Is); Black Sea; Tropical and subtropical Atlantic, SW Asia (India, Oman, Turkey); Asia (Russia, Taiwan); SE Asia (Indonesia, the Philippines, Singapore); Pacific Is (Polynesia, Fiji).

Uses: Used as food in Vietnam, and feed for the culture of abalone (*Haliotis midae*), in S Africa (Critchley et al. 1998, Roo et al. 2007, Milchakova 2011); considered an edible species in Portugal and other European countries (Green 2014, Rodrigues et al. 2015b).

Gracilariopsis tenuifrons (C.J. Bird & E.C. Oliveira) Fredericq & Hommersand

Description: Largest specimens reach 40 cm in length. Thalli are terete and erect from a discoid holdfast.

Geographic distribution: SW Atlantic (Brazil and Venezuela); Caribbean Is (Cuba and Guadalupe).

Uses: Used for foodstuff and agar extraction (Serra 2013).

Hydropuntia caudata (J. Agardh) Gurgel & Fredericq

Synonym: *Gracilaria caudata*

Description: The cartilaginous plants may consist of a few or many equally-sized terete axes that are sparingly or densely branched; branches typically terminate in acute tips. Cystocarps are characterized by multinucleate tubular nutritive cells that originate from the lowermost spore-bearing cells and fuse with gametophytic cells at the base of that fruiting body. Male structures are organized within deep pits with the spermatangial parent cells connecting to neighboring vegetative cells (Schmitt 2008).

Geographic distribution: Tropical and subtropical W Atlantic; Caribbean.

Uses: Considered an edible species (Costa et al. 2010, Costa et al. 2014).

Hydropuntia cornea (J. Agardh) M.J. Wynne

Synonym: *Gracilaria cornea*

Description: Plants of the largest species reach 30 cm in length (Xia and Abbott 1987). Thalli are terete to slightly compressed, and cylindrical to irregularly articulated, arising from a crustose holdfast or a creeping system of rhizomes.

Geographic distribution: Caribbean and S Atlantic (Brazil).

Uses: Used for food and agar production in the Caribbean Is (Zemke-White and Ohno 1999, Roo et al. 2007, Pereira-Pacheco et al. 2007, Robledo and Pelegrín 2009, Calado et al. 2012, Harrison 2013, Gray 2015).

Extracts of this species have antioxidant (Souza et al. 2011), anti-nociceptive and anti-inflammatory (Coura et al. 2012) activity.

Hydropuntia crassissima (P.L. Crouan & H.M. Crouan) M.J. Wynne (formerly *Gracilaria crassissima*)

Common name: English: Wild seamoss.

Description: Plants of the largest species reach 30 cm in length. Thalli are terete to slightly compressed, and cylindrical to irregularly articulated, arising from a crustose holdfast or a creeping system of rhizomes (Schneider and Wynne 2007).

Geographic distribution: Atlantic Is (Bermuda); Tropical and subtropical W Atlantic; Asia (Japan).

Uses: This species has been used for at least a century in fishing communities since migration or English colonization of the Caribbean Is. A hot drink called "Atole" is prepared with this algae, made from dried seaweed boiled water to release their phycocolloids; later added to evaporated milk and vanilla. Other local communities prepare this drink with coconut milk. *H. crassissima* is also used in the preparation of soups, puddings and other dishes (Zemke-White and Ohno 1999, Ortega et al. 2001, Roo et al. 2007, Harrison 2013).

Hydropuntia secunda Gurgel & Fredericq

Synonym: *Gracilaria secunda*

Description: Seaweed with cylindrical structure and firm texture, with alternating and irregular branches (Radulovich et al. 2013).

Geographic distribution: Gulf of Mexico; Tropical W Atlantic; Caribbean Sea.

Uses: It is used in the preparation of a drink made based on condensed milk and which some populations of the Caribbean attribute aphrodisiac effects (Radulovich 2013).

Laurencia obtusa (Hudson) J.V. Lamouroux (Fig. 58)

Common names: English: Rounded brittle fern weed (Bunker et al. 2010), Corsican moss; Hawaiian: Limu Ho`onunu (Abbott 1984); Indonesian: Sangan (Chapman and Chapman 1980); Sangau (Zaneveld 1959).

Description: Globular tufts of brittle, cartilaginous, narrow, cylindrical, reddish brown to yellowish red fronds, 150 mm long, from small discoid base, bushy, with green or yellow axes and rose branchlets, main stems long, which are sparingly alternately branched, 0.75–1.50 diam. the smallest branches and the short, truncate, ultimate branchlets opposite or sub-verticillate, 0.5–0.75 mm diam.; tetrasporangia in a band below the apex of the hardly modified branchlets (Pereira 2014g).

Geographic distribution: Worldwide in warm temperate to tropical seas.

Uses: In Indonesia, Thailand and Vietnam is used as food and for agar extraction (Zaneveld 1959, Lewmanomont 1978, Arasaki and Arasaki 1983, Istini et al. 1998, Zemke-White and Ohno 1999, Hong et al. 2007, Roo et al. 2007, Milchakova 2011, Sidik et al. 2012, Pereira 2014g, Irianto and Syamdidi

Figure 58. *Laurencia obtusa* (Hudson) J.V. Lamouroux (Rhodophyta). Illustration from William H. Harvey (Phycologia Britannica, 1846–1951).

2015). Also used as food in the Mediterranean Basin (Harrison 2013) and in the Hawaiian Is (Reed 1906).

L. obtusa presents a complex agar-like sulfated galactan. These polysaccharides belong to the agar group, being agarose derivatives with a rather high content sulfate group and with a reduced amount of 3,6-anhydro-L-galactose residues (700–950 cm^{-1}) (Usov and Elashvili 1991, Pereira et al. 2013).

Nemalion elminthoides (Velley) Batters (Fig. 59)

Synonyms: *Nemalion helminthoides, N. lubricum*

Common names: English: Threadweed (Madlener 1977), Sea noodles (Chapman and Chapman 1980), Sea noodle (Bunker et al. 2010); Japanese: Umi-zomen (Yendo 1902), Umisomen, Tsukomo nori (Madlener 1977); Portuguese: Esparguete-da-costa (Neto et al. 2005, Pereira 2009b).

Description: Cylindrical, reddish to purple brown, softly gelatinous thallus, simple or sparingly dichotomously branched, 2–3 mm wide, 250 mm long, with discoid base (Braune and Guiry 2011).

Geographic distribution: NE Atlantic (Scandinavia to Canary Is); SE Atlantic; SW Atlantic (Brazil, Uruguay); Mediterranean; NW Pacific (Japan); NE Pacific (Alaska to Mexico); Australia and New Zealand.

Uses: Used in Japan and Italy, either raw, in soups and in salads (Chapman and Chapman 1980, Gómez-Ordóñez and Rupérez 2011, Milchakova 2011, Gray 2015). In some parts of Japan this plant is dried and bleached; it is then eaten with soy or vinegar, but its use is not common (Yendo 1902).

Figure 59. *Nemalion elminthoides* (Velley) Batters (Ochrophyta, Phaeophyceae).

Plocamium cartilagineum (Linnaeus) P.S. Dixon (Fig. 60)

Synonym: *Gelidium cartilagineum*

Common names: English: Cockscomb (Dickinson 1963), Cockscombe (Braune and Guiry 2011), Red comb weed (SIA 2014); French: Boucle de cheveux (SIA 2014).

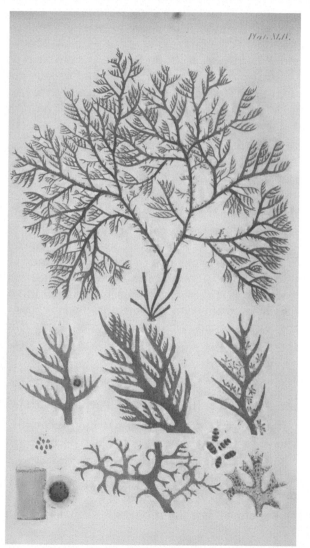

Figure 60. *Plocamium cartilagineum* (Linnaeus) P.S. Dixon (Rhodophyta). Illustration from William H. Harvey (Plate XLIV, Phycologia Britannica, 1846–1951).

Description: *P. cartilagineum* is scraggly-looking red alga that grows up to 15 to 30 cm long, with a transparent dark red to purple color. Arising from a discoid holdfast, the fronds are flattened, crisp, and cartilaginous in texture, branching irregularly with smaller branches curving at the tips and forming comb-like groupings (pectinate). These comb-like terminal branchlets are the distinguishing characteristic for *P. cartilagineum*, making it easily recognizable in the field. This branching pattern gives the overall plant a tufted, feathered appearance. When reproductive the female gametophytes can be recognized by a covering of cystocarps (bumps) 1 mm in diam. over the entire frond. *P. cartilagineum* is found on temperate coasts with moderate to strong exposure from 2 to 26 m deep. This plant usually grows on rocky substrates, but is also often found growing attached to the stipes of larger kelps such as *Laminaria hyperborea* (Phaeophyceae). *P. cartilagineum* can be an abundant alga in temperate areas (Braune and Guiry 2011).

Geographic distribution: NE Atlantic (Scandinavia to Senegal, N Sea), SE Atlantic (Namibia); Mediterranean; Indian Ocean (Pakistan, Mauritius); NW Pacific (Japan); Pacific Is; NE Pacific (Alaska to California); SE Pacific (Chile); Australia, New Zealand; Antarctica.

Uses: *P. cartilagineum* is harvested commercially as the main raw material for agar production on the Pacific coast of North America. Agar is widely used not only in laboratories as a growth medium for bacteria and other cultures, but in food and cosmetics as a gelling agent and stabilizer. It is a very pure, natural, firm gel. *P. cartilagineum* extracts are also high in lypolitic (lipid-digesting) sterols and as such are useful as an additive in slimming applications such as creams and massage products where they are able to provoke the release of fatty acids and eliminate surface fat, acting as a skin-firmer (Pereira 2015a).

Porphyra capensis Kützing

Common name: Cape laver (Bianchi 1999), Purple laver (Heyden 2013).

Description: Thallus thin, with varying shape, but generally ovoid, dark purple-black to yellow, with slippery texture, that becomes crisp and brittle when dry (Bianchi 1999).

Geographic distribution: SE Atlantic (Angola, Namibia, S Africa); India Ocean, Antarctic Is; SE Pacific (Chile).

Uses: Fried with oatmeal is a traditional dish in Wales and the Hebrides (Heyden 2013).

Porphyra laciniata C. Agardh (Fig. 61)

Common names: English: Laver sloke, Laver, Laver slack (Kirby 1953), Red Laver, Purple laver; Japanese: Hoshinori, Asakusa-nori, Asakusa nori (Kirby 1953).

Description: Fronds 5–20 cm long, purple, clustered together, expanded, delicately membranaceous, pellucid, very irregularly divided into several lobes; the point of attachment frequently within the frond, which is then peltate. Margin wavy, entire or irregularly cut; apices often truncate (Harvey 1864).

Geographic distribution: NE Atlantic; Atlantic Is (Cape Verde Is); SW Atlantic (Brazil); Indian Ocean; Asia (Japan).

Uses: Used as food in Alaska and several European countries; miners of south Wales were the biggest consumers of laver in Great Britain (Chapman

Figure 61. *Porphyra laciniata* C. Agardh (Rhodophyta). Illustration from William H. Harvey (Phycologia Britannica, 1846–1951).

and Chapman 1980, CEVA 2014); this species is used in bread "laver bread", may be fried with butter and in Ireland it is made into jelly by stewing or boiling, then used later (Lembi and Waaland 1988).

See recipes with *Porphyra* in Annex I.

Pyropia acanthophora (E.C. Oliveira & Coll) M.C. Oliveira, D. Milstein & E.C. Oliveira (formerly *Porphyra acanthophora*)

Description: Thalli with monostromatic blades, arising from a single discoid attachment, and sexual thalli monoecious with spermatangia and zygote-sporangia in patches.

Geographic distribution: Atlantic Is (Canary Is); Tropical and subtropical W Atlantic.

Uses: Used as food in Brazil (Zemke-White and Ohno 1999, Oliveira 2006, Roo et al. 2007, Harrison 2013).

Pyropia spiralis (E.C. Oliveira & Coll) M.C. Oliveira, D. Milstein & E.C. Oliveira (formerly *Porphyra spiralis*)

Description: Thin, nearly transparent sheets grow clinging on rocks; tiny holdfast; can appear ruffly.

Geographic distribution: Tropical and subtropical W Atlantic.

Uses: Used as food in the Caribbean, Central America, and Brazil (Zemke-White and Ohno 1999, Oliveira 2006, Roo et al. 2007, Harrison 2013).

Schizymenia apoda (J. Agardh) J. Agardh

Common name: Orange sheets (Heyden 2013).

Description: Thalli are medium to dark red, membranous, circular or ellipsoid, rarely irregularly shaped, up to 8 cm in length and 4 cm in width, expanding from a stipe. Blades are thickened along the margin and beset with numerous, small protuberances, looking toothed (Kim et al. 2012).

Geographic distribution: Atlantic Is (Azores, Madeira, Tristan da Cunha Is); SE Atlantic (Namibia, Somalia, S Africa); Asia (China, Korea, Taiwan).

Uses: In S Africa is used to wrap roasted chicken (Heyden 2013).

Yonagunia formosana (Okamura) Kawaguchi & Masuda

Synonyms: *Carpopeltis formosana, Prionitis formosana*

Description: Thalli grow gregariously on bedrocks in the upper subtidal zone at 1–3 m in depths. Attachment to the substratum is by a discoid holdfast from which three to five erect blades are formed, the blades being terete and 1–1.5 mm in diam. for 2–3 mm basally and then gradually becoming compressed to uniformly flattened and 2–3 mm wide and 350–500 mm thick. They reach 3–8 cm in length at maturity and are rigidly cartilaginous and dark red or purplish red (occasionally bluish gray) in color. Branching is five–11 times dichotomous or sub-dichotomous with wide axils (Kawaguchi et al. 2004).

Geographic distribution: Caribbean Is; Indian Ocean; Asia (China, Japan, Taiwan, the Philippines, Vietnam); Pacific Is (Samoa, Polynesia, Fiji).

Uses: Used as food in the Caribbean—Tobago Is (Hansen et al. 1981). Setchell (1924) recorded this alga as edible from Samoa, although to date no record of it being eaten is known. It is rarely collected, probably because of its habitat (exposed places) and cryptic habit (Skelton 2003).

2.3 East Asia (China, Japan, Korea)

2.3.1 Chlorophyta

Capsosiphon fulvescens (C. Agardh) Setchell & N.L. Gardner

Common names: Korean: Maesaengui (Sohn 1998); Swedish: Gyllentarmalg (Tolstoy and Österlund 2003).

Description: This species is rare. They are small plants, green-yellow color. Upper thalli are in the form of tubes, while the base consists of only one or two rows of cells. The cells are arranged in very pronounced longitudinal cross series loosely bound, sometimes producing false branches. Cell walls are gelatinous.

Geographic distribution: N Atlantic; Canadian Arctic; Adriatic Sea; Baltic Sea; Atlantic Is (Greenland, Iceland); Indian Ocean; Asia (Japan, Korea); NE Pacific (Alaska, California).

Uses: Used as food in Korea, in boiled soup with oysters (Kang 1968, Sohn 1998, Zemke-White and Ohno 1999, Roo et al. 2007, Harrison 2013, Islam et al. 2014).

Caulerpa cupressoides (Vahl) C. Agardh

Common names: English: Cactus tree alga (De Kluijver et al. 2014); Pacific name (Tonga Is): Kaka (Ostraff 2003).

Description: Erect fronds arise from stolons of up to several diameters in length, anchored to the substratum by rhizoidal structures. Fronds are usually rich forked toward the apex, and at the base are naked (usually three) rows of rounded or flattened, ovoid to conically pointed, short branchlets with a distinctly pointed tip (Braune and Guiry 2011, De Kluijver et al. 2014, Pereira 2015a).

Geographic distribution: NE Atlantic (Canary Is); Caribbean; NW Pacific (Japan, China); W Pacific (the Philippines, Indonesia); Pacific Is; Australia.

Uses: Commonly eaten species in Pacific Is (Deane 2014a).

Caulerpa okamurae Weber-van Bosse

Description: Thallus composed of horizontal stolon anchored by colorless rhizoids, bearing erect photosynthetic fronds, vesicular structures with soft and succulent texture. Radial branching regarded as primitive, bilateral as more recent, interpretations which are supported by ultrastructure of chloroplasts. Reduction of light also results in change of erect portions from radial to bilateral symmetry.

Geographic distribution: Asia (China, Japan, Korea); Australia (Queensland).

Uses: Used for foodstuff in Japan and Korea (Johnston 1966, Kang 1968, Bonotto 1976, Arasaki and Arasaki 1983, Hotta et al. 1989).

Codium cylindricum Holmes

Description: Plant erect, slender, 15–40 cm in length, with a discoid holdfast; branches cylindrical, dichotomously, 2.5–5.5 mm in diam.; utricles clavate, cylindrical, 850–1200 µ in length, 175–500 µ in diam., with truncate or sub-truncate apices; hairs and hair scars abundant; gametangia ovoid or elliptic, 220–310 µ long, 110–130 µ broad (Chang 2015b).

Geographic distribution: Asia (China, Japan, Korea, Taiwan, the Philippines); Australia and New Zealand; Pacific Is (Micronesia).

Uses: Considered an edible species in Korea (Arasaki and Arasaki 1983).

Codium fragile (Suringar) Hariot

Common names: Chinese: Shui song (Lembi and Waaland 1988); English: Fleece, Sponge Tang (Madlener 1977), Sponge seaweed, Sponge weed, Green sea-velvet, Green sponge fingers, Fragile green sponge fingers (Bunker et al. 2010); Japanese: Miru (Madlener 1977); Korean: Chonggak (Madlener 1977, Sohn 1998).

Description: Appears as a fuzzy patch of repeatedly branching tubular fingers; these formations hang down from rocks during low tide, which serve as the inspiration for a few of its common names. The color of *C. fragile* ranges from medium green to dark green to blackish green. The entire thallus is velvety and spongy in texture, relatively soft, and is sometimes tomentose with profuse hairs. The erect, dichotomously branched fronds, or fingers, are up to 1 cm wide, and can extend to lengths of over 30 cm. The branches of *C. fragile* are round in cross section but may be flattened beneath dichotomies of branches. The fronds usually arise from a spongy disk-shaped holdfast, which resembles a small, broad cushion. A pungent odor is often associated with this species, which is less common for seaweeds. The utricles hairs ends with a characteristic peak, the apical mucronate is visible with a magnifying glass.

Geographic distribution: N Atlantic; Atlantic Is (Canary Is); Adriatic Sea; Baltic Sea; Asia (China, Japan, Taiwan, Indonesia, the Philippines); NE Pacific (Alaska); Australia (Tasmania) and New Zealand; Antarctic Is.

Uses: Used as food in China (Bangmei and Abbott 1987, Harrison 2013); in Korean cuisine is used in various seasoning with soy sauce and vinegar, mixed in Kimchi (Sohn 1998). Used as food in the form of vinaigrette and winter vegetables (Lembi and Waaland 1988), dried or salt cured (Levring et al. 1969). Used in sweets, salads and soups in Japan, Korea and the Philippines (Chapman and Chapman 1980, Arasaki and Arasaki 1983, Zemke-White and Ohno 1999, Roo et al. 2007, Harrison 2013); considered an edible species in Tasmania (Sanderson and Benedetto 1988) and Denmark (Wegeberg 2010).

See recipes with *Codium* in Annex I.

Codium platylobium Areschoug

Synonym: *Codium lindenbergii*

Common name: Miru (Smith 1904).

Description: Thallus spongy, anchored to rocks or shells by a weft of rhizoids.

Geographic distribution: SE Atlantic (Senegal, S Africa); Asia (Japan, Korea).

Uses: In Japan, after drying they are preserved in salt; they are prepared for food by boiling or baking in water, and are put in soups; or, after washing, by mixing with soy-bean sauce and vinegar (Smith 1904).

Codium subtubulosum Okamura (formerly *Codium divaricatum*)

Description: Thallus spongy, anchored to rocks or shells by a weft of rhizoids, varying in size from 1 cm to 10 m long; thalli dichotomously branched; erect or repent; branches wholly terete or variously flattened, at times anastomosing.

Geographic distribution: Asia (China, Japan, Korea, Taiwan); Pacific Is (Hawaiian Is).

Uses: Used as food in Japan and Korea (Bonotto 1976, Arasaki and Arasaki 1983).

Codium vermilara (Olivi) Delle Chiaje

Description: Thallus 20–40 cm long, 0.3–0.8 cm thick, deep-green, large and porous, attached with basal disk-like holdfast giving rise to several erect thalli; regularly or irregularly branched, sometimes proliferous, terete with short terminal segments; utricles obovate to oblong, square ended, with appreciable apical thickening; morphologically diverse: prostrate or erect, cylindrical or flat, cushion-like or globular; branches cylindrical, gradually tapering to apices; basal part incrassate; apical branches relatively thin, the apices blunt (Milchakova 2011).

Geographic distribution: NE Atlantic; Adriatic Sea; Black Sea; Mediterranean; Atlantic Is (Canary Is, Selvage Is); Asia (Japan, Korea).

Uses: Used as traditional food in Japan and other Asian countries (Milchakova 2011).

C. vermilara is the source of BASs, valuable polysaccharides, PUFAs and cytostatic drugs for medicine (Milchakova 2011).

Monostroma antarcticum V.J. Chapman

Description: Thalli forming leafy monostromatic blades 2–30 cm (or more) in length. Typical blade ontogeny involves spore germination to prude a filamentous prostrate cushion which up-heaves to form a saccate stage which then opens to produce a blade.

Geographic distribution: Asia (Japan); New Zealand.

Uses: Used as food in Japan (Tokida 1954).

Monostroma crassidermum Tokida

Common name: Japanese: Atsukawa-hitoe (Tokida 1954).

Description: Thallus green, consisting of flat, spread-out, leaf-like, soft-skinned, irregularly shaped sheets made up of one only cell layer (Braune and Guiry 2011).

Geographic distribution: Asia (Japan, Korea).

Uses: Used as food in Japan (Tokida 1954).

Monostroma kuroshiense F. Bast

Common name: Japanese: Shimanto nori (Bast 2014).

Description: The frond is blade-like with eponymous one-cell thickness and therefore it is also known as "Slender sea-lettuce" (Bast 2014).

Geographic distribution: Asia (Japan).

Uses: Among green seaplants, M. *kuroshiensis* is the most intensely cultivated genus, constituting about 90% of the total green algal cultivation (Nisizawa et al. 1987), almost exclusively for the Japanese produce "hitoegusa" (Bast 2014).

Prasiola japonica Yatabe

Common names: Japanese: Kawa-nori (Namikawa 1906), Datyagawa-nori (Skvortzov 1919), Nikko-nori, Kawanori (Arasaki and Arasaki 1983).

Description: Thalli forming monostromatic blades generally expanded above and narrowing to short stipitate region at base. Blade shape varies from orbicular, to irregular or spatulate to long linear; marine species generally 0.5 to 20 mm in high.

Geographic distribution: Asia (China, Japan).

Uses: Used as food in Japan (Namikawa 1906, Levring et al. 1969, Watanabe 1970, Arasaki and Arasaki 1983, Lembi and Waaland 1988).

Ulvaria splendens (Ruprecht) Vinogradova

Synonym: *Monostroma fuscum*

Common name: Japanese: Kuro-hitoegusa, O-hitoegusa (Tokida 1954).

Description: Young thallus starts out as a sac and then splits into a blade expanding to least 30 cm in diam. The blade is bright green, one cell layer thick, and almost opaque.

Geographic distribution: NE Atlantic; Atlantic Is (Greenland); Baltic sea; Asia (Japan, Korea, Russia); NE Pacific (Alaska, British Columbia, California); SE Pacific (Chile).

Uses: Used as food in Japan (Tokida 1954).

Umbraulva japonica (Holmes) Bae & I.K. Lee (formerly *Ulva japonica*)

Description: Thallus expands flabellately with cuneate and pinnate segments and has an olive-green color (Bae and Lee 2001).

Geographic distribution: Asia (Japan, Korea, Taiwan).

Uses: Used as food in Korea (Arasaki and Arasaki 1983).

Ulva compressa Linnaeus (formerly *Enteromorpha compressa*) (Fig. 62)

Common names: English: Thread weed, Tape weed (Bunker et al. 2010); Filipino: Bagisbagis (Agngarayngay et al. 2005), Lumot (Velasquez 1972);

Figure 62. *Ulva compressa* Linnaeus (Chlorophyta). Illustration from William H. Harvey (Plate CCCXXXV, Phycologia Britannica, 1846–1951).

Hawaiian: Limu ele-ele (MacCaughey 1918); Japanese: Awo-nori (Smith 1904); Korean: Parae (Sohn 1998); Swedish: Tarmalg (Tolstoy and Österlund 2003).

Description: Plants attached, light or bright-green in color; adult plants usually tubular; more or less compressed, dilated towards the apex, tapering below, giving several branches from the gradually contracted stalk like base; branches similar to the main frond; fronds up to 1.5 cm high (Kaliaperumal et al. 1995).

Geographic distribution: Cosmopolitan.

Uses: Used for foodstuff in Indonesia and Korea (Kang 1968, Bonotto 1976, Zemke-White and Ohno 1999, Irianto and Syamdidi 2015), and for animal fodder in Korea (Bonotto 1976, Harrison 2013); used fresh or dried for both human and animal consumption for its high nutrient levels and good taste. Is often collected as food by people in the Yellow Sea and Gulf of Chihli areas of Northeastern China and by the aboriginal Yemei tribe in Taiwan (Bangmei and Abbott 1987); also eaten in Hawaii, Pakistan, Japan, the Philippines and in the Malay Peninsula (MacCaughey 1918, Zaneveld 1955, 1959, Velasquez 1972, Chapman and Chapman 1980, Rahman 2002, Agngarayngay et al. 2005, Harrison 2013). In Korean cuisine is used in various seasoning with sesame oil, and sometimes vinegar (Sohn 1998); in Indonesia used to make salad, and vegetable soup (Istini et al. 1998, Roo et al. 2007). In India, it is used as a vegetable and also in the form of salad, jam and powder (Kaliaperumal et al. 1995).

Many benefits have been associated with *U. compressa* consumption such as antimicrobial, antiviral, and antioxidant properties. Extracts of this species have also hypocholesterolemic activity (Nisizawa 1979).

2.3.2 Ochrophyta – Phaeophyceae

Acrothrix pacifica Okamura & Yamada

Common name: Japanese: Nise-mozuku (Tokida 1954).

Description: Frond epiphytic on *Chorda filum*, over 30 cm, in height, cylindrical fistulose and scarcely attaining 1 mm in diam. below, gradually tapering above, repeatedly branched, more or less regularly alternate, branches gradually shortening from the base upwards; is morphologically an alternation of macroscopic sporophytes with microscopic gametophytes or, karyologically, a diploid sporophyte alternates with a haploid gametophyte. Furthermore, unfused gametes developed into haploid sporophytes (Tokida 1954, Umezaki 1990).

Geographic distribution: Asia (China, Japan, Korea, Russia).

Uses: Used as food in Japan (Tokida 1954).

Agarum clathratum Dumortier

Synonym: *Agarum cribrosum*

Common name: Japanese: Aname (Tokida 1954).

Description: Thalli sieve-like, profusely perforated, with numerous small holes (Stearn 1973).

Geographic distribution: Asia (China, Japan, Korea, Russia); N Atlantic (Arctic, Greenland); NW Pacific (Alaska).

Uses: In Japan and Korea is used for foodstuff, especially for the manufacture of potash salts (Tokida 1954, Kang 1968), and for production of alginates (Kang 1968, Bonotto 1976).

Alaria crassifolia Kjellman

Common names: English: Lesser kelp (Hu 2005); Japanese: Chigaiso (Subba Rao 1965, Airanthi et al. 2011).

Description: Purplish brown seaweed growing on subtidal rocks, the lower portion of the thalli pinnately deeply-lobed, the apical portion oblong, coriaceous, very thick along the midrib (Hu 2005).

Geographic distribution: Asia (China, Japan).

Uses: Used as food in China, Japan and Korea (Subba Rao 1965, Johnston 1966, Kang 1968, Chapman and Chapman 1980, Zemke-White and Ohno 1999, Hu 2005, Roo et al. 2007, Airanthi et al. 2011, Harrison 2013).

Analipus japonicus (Harvey) M.J. Wynne

Synonyms: *Chordaria abietina*, *Heterochordaria abietina*

Common names: English: Bottlebrush seaweed, Far needle (Arasaki and Arasaki 1983), Fir needle, Sea fir needle (Madlener 1977), Sea fir (Chapman and Chapman 1980); Japanese: Matsuma (Smith 1904), Matsumo (Subba Rao 1965, Madlener 1977, Arasaki and Arasaki 1983).

Description: Thallus is light to dark brown with numerous radially-arranged flattened lateral branches, which are somewhat curled at the ends and hollow. This species has a distinctive, perennial, crustose base up to 5 cm in diam. and erect axes up to 35 cm tall that die back in the fall. Habitat: This species occurs in the mid to low intertidal on rocks and prefers semi-protected to exposed habitats (Lindeberg and Lindstrom 2014a).

Geographic distribution: Asia (Japan, Korea, Russia); NE Pacific (Alaska, California).

Uses: Used as food in China and Japan (Tokida 1954, Subba Rao 1965, Johnston 1966, Madlener 1977, Chapman and Chapman 1980, Arasaki and Arasaki 1983, Simoons 1990, Kim 2011). This is particularly abundant in northern Japan, where it is collected and packed in salt; it is cooked with soy and is a common food of the poorer classes in the north (Yendo 1902, Smith 1904).

Arthrothamnus bifidus (S.G. Gmelin) J. Agardh

Common names: Japanese: Mekoashi-kombu chigaiso (Subba Rao 1965), Nekoashi-kombu (Arasaki and Arasaki 1983).

Description: Life history diplohaplontic with alternation of large sporophyte bearing unilocular meiosporangia with paraphyses (sori) and microscopic dioecious and oogamous, heteromorphous gametophytes (Yamada 1938, Parker and Guiry 2014a).

Geographic distribution: NE Pacific (Alaska); Asia (Japan, Russia).

Uses: Used as food in Japan and cold N Pacific, utilized for Kombu production (Subba Rao 1965, Johnston 1966, Levring et al. 1969, Chapman and Chapman 1980, Arasaki and Arasaki 1983).

Arthrothamnus kurilensis Ruprecht

Common names: Japanese: Chishima-nekoashikombu, Kidachi-mimikombu (Tokida 1954), Chishimanekoashi (Subba Rao 1965).

Description: Life history diplohaplontic with alternation of large sporophyte bearing unilocular meiosporangia with paraphyses (sori) and microscopic dioecious and oogamous, heteromorphous gametophytes; sporophytes with stem eight times dichotomously branched (Tokida 1954, Parker and Guiry 2014a).

Geographic distribution: Asia (Japan, Russia).

Uses: Used as food in Japan, as raw material for Kombu (Tokida 1954, Subba Rao 1965, Johnston 1966, Levring et al. 1969, Chapman and Chapman 1980).

Cladosiphon okamuranus Tokida

Common names: English: Slender slippery weed (Novaczek 2001); Japanese: Mozuku (McHugh 2003, Ohno et al. 1998), Okinawa-mozuku (Toma 1997).

Description: This sea vegetable is slender, pale brown, sparsely branched, slippery, floppy (not stiff) and jelly-like. Branches are 1–1.5 mm wide and the plant can reach a length of at least 30 cm. The branching pattern may be alternate (first to one side, then to the other) or irregular (Novaczek 2001).

Geographic distribution: Asia (Japan).

Uses: After being washed to remove the salt, it is used as a fresh vegetable, eaten with soy sauce and in seaweed salads (McHugh 2003). This sea vegetable is highly prized in Japan, and consumed with soy sauce and vinegar (Toma 1997, Ohno et al. 1998, Zemke-White and Ohno 1999, Novaczek 2001, Roo et al. 2007, Saito et al. 2010, Harrison 2013).

Desmarestia dudresnayi subsp. *tabacoides* (Okamura) A.F. Peters,
E.C. Yang, F.C. Küpper & Prud'Homme van Reine (formerly
Desmarestia tabacoides)

Description: Thalli are erect and can be terete, compressed or ligulate,
simple or branched, reaching several meters in length. Compared with
the Laminariales (i.e., *Laminaria* spp.), they differ in having a thallus of
pseudo-parenchymatous construction and trichothallic growth (adapted
from Altamirano et al. 2014).

Geographic distribution: Asia (Japan and Korea); W Pacific (California and
Mexico).

Uses: In Korea is used for foodstuff (Madlener 1977), and animal fodder
(Hotta et al. 1989).

Ecklonia bicyclis Kjellman

Synonym: *Eisenia bicyclis*

Common names: Japanese: Arame, Kajimi, Sagarame (Smith 1904, Suba
Rao 1965, Madlener 1977, Mouritsen 2013).

Description: *E. bicyclis* is a rather small kelp with a stiff, woody stipe up to
1 m tall and two flattened, oval fronds or lobes, with many lateral blades.
Each year fronds are shed and new ones develop creating a branched and
feathery plant. This kelp is native to warm-temperate waters of the Pacific,
specifically around Japan where it displays a distinct seasonality of growth
and reproduction (Smith 1904).

Geographic distribution: *E. bicyclis* species is native to Japan, but may be
cultivated elsewhere. It is limited to warm-temperate waters.

Uses: Used for preparation of miso soup, nori-jam, salads, and seaweed
powder for various foods in the Japanese cuisine (Smith 1904, Suba Rao
1965, Takagi 1975, Arasaki and Arasaki 1983, Ohno and Largo 1998). It
contains a fair amount of calcium, which is present in chelated form, that
is, bound to an organic or amino acid, which permits the body to absorb
more of the mineral (Pereira 2011, Mouritsen 2013); used also as food in
Korea (Takagi 1975).

The extracts of *E. bicyclis* have antihypertensive (Takagi 1975) activity.
This species is also used for alginates production (Kang 1968, Bonotto 1976,
Madlener 1977, Hotta et al. 1989).

Ecklonia cava Kjellman

Common names: English: Paddle weed; Japanese: Kajime, Noro-kajime (Arasaki and Arasaki 1983).

Description: *E. cava* a large brown algae is an important kelp, forming vast underwater forests with plants growing up to 3 m in length. The large strong holdfast is composed of many branched haptera which give rise to a single plant with a long cylindrical stipe of 1 to 2 m. Many long, smooth, leathery blades emerge from the stipe, forming a clump at the top of the plant, very reminiscent of a palm tree. *E. cava* grows exclusively in subtidal of deep pools where they are anchored directly to the rocky substratum.

Geographic distribution: Asia (China, Japan, Korea).

Uses: Most of the harvest of *E. cava* is used for food in China, Japan and Korea, where it is boiled with soy sauce (Chapman and Chapman 1980, Arasaki and Arasaki 1983, Ohno et al. 1998, Zemke-White and Ohno 1999, Ohno and Largo 2006, Roo et al. 2007, Harrison 2013). However, the western world has caught on to the many health benefits of the seaweed and now *E. cava* extract (ECE) is being sold in capsule or powdered form for dietary supplements.

This species has antihypertensive action (Takagi 1975), and is used for alginates extraction (Kang 1968, Bonotto 1976, Hotta et al. 1989).

Ecklonia kurome Okamura (Fig. 63)

Common names: Chinese: Kunbu, Miangichai (Tseng 1983).

Description: Central lamina width (5.1–67.8 cm), the median fascia thickness (0.9–3.3 mm), the primary pinnae width (2.2–24.0 cm), the ratio of width to length of the primary pinnae (0.07–0.56) and the primary pinnae number index (PPNI; 0.24–2.21). The results suggest that the morphological variations observed within this species are related to their habitat. A thick and distinct median fascia and an undulate central lamina margin are found on plants from moderate wave-exposed locations. Plants having narrow primary pinnae with indistinct ruga are observed on shores facing the open sea. Wide and thin plants grow in wave-sheltered habitats. Flat plants are found on a shore exposed to continuous strong currents throughout the year (Tsutsui et al. 1996).

Geographic distribution: Asia (China, Japan, Korea).

Uses: Used as food in Japan (Johnston 1966), and China (Levring et al. 1969, Tseng 1983, Bangmei and Abbott 1987, Gupta and Abu-Ghannam 2011, Zhao et al. 2015).

Figure 63. *Ecklonia kurome* Okamura (Ochrophyta, Phaeophyceae).

Ecklonia stolonifera Okamura

Common names: Korean: Kompi (Sohn 1998); Japanese: Kizame arame (Chapman and Chapman 1980).

Description: This species usually possesses a massive holdfast that extends into a long, hollow, gas-filled stipe up to 15 m in length. The stipe ends in a gas-filled bladder that extends into a flat, solid primary frond from which secondary fronds emerge. These secondary fronds can quite reach 3 m in length. Because of its hollow stipe and bladder this genus generally floats at the sea surface.

Geographic distribution: Asia (Japan and Korea).

Uses: In Korea is used as food in salads (Zemke-White and Ohno 1999); also eaten in Japan (Johnston 1966, Kang 1968, Chapman and Chapman 1980, Arasaki and Arasaki 1983, Hotta et al. 1989, Sohn 1998, Roo et al. 2007, Harrison 2013).

Extracts of *E. stolonifera* have phlorotannins with antioxidant and hepatoprotective properties (Lee et al. 2012).

Eckloniopsis radicosa (Kjellman) Okamura

Description: Large sporophyte bearing unilocular meiosporangia with paraphyses (sori) and microscopic dioecious and oogamous, heteromorphous gametophytes. Sporophytes composed of holdfast with branched, filamentous haptera, a short stipe and a ruffled blade with slight or no marginal pinnate lobes.

Geographic distribution: Asia (Japan).

Uses: Used as food in Japan (Johnston 1966).

Eudesme virescens (Carmichael ex Berkeley) J. Agardh

Common names: Japanese: Yezo-mozuku, Nise-futomozuku (Tokida 1954); Swedish: Olivslemming (Tolstoy and Österlund 2003).

Description: Thallus cylindrical, yellowish brown, to 500 mm long; frond solid but soft, gelatinous, 1–4 mm wide with numerous irregular branches. Pseudo-parenchymatous, easily squashed to reveal a multiaxial medulla of numerous colorless filaments and a cortex of stalked fan-like tufts of club-shaped pigmented assimilatory filaments 15–28 cells long. Assimilatory filaments grade into axial cells. Hair is present (Guiry 2014b).

Geographic distribution: N Atlantic; Arctic; Indian Ocean; SW Asia (Cyprus, Turkey); Asia (China, Japan, Korea); NE Pacific (Alaska); Antarctic.

Uses: Used as food in Japan (Johnston 1966).

Heterosaundersella hattoriana Tokida

Common name: Japanese: Karafuto-mozuku (Tokida 1954).

Description: Plants epiphytic on other algae in intertidal zone; erect thallus composed of prostrate part and erect part. The prostrate part is composed of the filaments penetrating inside the host plant and dense assimilatory filaments arising from them. The erect part of thallus filiform, simple, caespitose, up to 5 cm in length and 5 mm in diam. when dried, flaccid, gelatinous, yellowish brown in color.

Geographic distribution: Asia (Japan).

Uses: Used as food in Japan (Tokida 1954).

Ishige okamurae Yendo

Common names: Chinese: Tieding cai (Tseng 1983); Korean: Pae (Song et al. 2013).

Description: This species has leathery branched narrow fronds consisting of cylindrical hairs, uniseriate plurilocular sporangia lacking sterile terminal cells, apical growth, pyrenoid-less discoid plastids and an isomorphic life history (Yendo 1907, Lee et al. 2003).

Geographic distribution: Asia (China, Japan, Korea, Taiwan).

Uses: Used as food in China and Korea; the simplest way involves powdering the dried blades, then mixing with condiments; or the dried thallus may be steamed, then boiled, each about 15 min, mixed with soy sauce, and eaten (Tseng 1983, Bangmei and Abbott 1987); also eaten in Korea (Tseng 1983, Song et al. 2013).

Their extracts have anti-inflammatory (Kim et al. 2009), anti-diabetic (Min et al. 2011), and antioxidant (Zou et al. 2008) action. Some fraction extracts (phlorotannins) exhibited inhibitory activity against acetylcholinesterase that could be used as potential functional food ingredients or nutraceuticals for preventing Alzheimer's disease (Yoon et al. 2009).

Ishige sinicola (Setchell & N.L. Gardner) Chihara

Description: Plants attached on the rocks by discoid holdfast, dark brown to olive green, irregularly dichotomously branched, terete or foliose with short stipe. Profusely branched upright filaments develop from the center of disk which further develops into dichotomously branched, parenchymatous erect thalli.

Geographic distribution: Asia (China, Japan, Korea and Taiwan); W Pacific (California and Mexico).

Uses: Used for foodstuff in Korea (Kang 1968, Tseng 1983). In the region of the E China and S China seas (central and south China) this is a common seaweed prepared in many ways. Perhaps the simplest way involves powdering the dried blades, then mixing with condiments; or the dried thallus may be steamed, then boiled, each about 15 min, mixed with soy sauce, and eaten (Bangmei and Abbott 1987).

Ishige foliacea Okamura

Common name: Korean: Neolppae (Song et al. 2013).

Description: Plants attached on the rocks by discoid holdfast, irregularly dichotomously branched, terete or foliose with short stipe, up to 20 cm in height. Erect thalli composed of densely packed, entangled, colorless, thick-walled, isodiametric medullary filaments and pigmented cubic cortical cells.

Geographic distribution: Asia (Japan, Korea); NE Pacific (Gulf of California).

Uses: Used as food in Korea (Song et al. 2013).

Laminaria yezoensis Miyabe

Description: Thallus is medium to dark brown with a large disk- or suction-cup-like holdfast, a somewhat rigid stipe up to nearly 1 m long although it is often shorter, and a thick blade that can be nearly as wide as long. The blade is usually split and has mucilage ducts, which are visible microscopically (Lindeberg and Lindstrom 2015b).

Geographic distribution: Asia (Japan, Russia); NE Pacific (Alaska, British Columbia).

Uses: Made into Chiimi-kombu in Japan and Russia (Chapman and Chapman 1980); consumed also in China and Korea (Zemke-White and Ohno 1999).

Nemacystus decipiens (Suringar) Kuckuck

Synonyms: *Cladosiphon decipiens*, *Mesogloia decipiens*

Common names: Chinese: Haida, Hai yun (Tseng 1983); Japanese: Mozuku (Smith 1904, Madlener 1977, Chapman and Chapman 1980), Mozuka (Rhoads and Zunic 1978); Pacific name (Tonga Is): Tanga'u (Ostraff 2003).

Description: Slippery seaweed that is golden brown to dark brown in color; fronds have a highly branched structure and can reach up to 30 cm in height. Individual branches usually reach a maximum of 4 to 5 cm in length. This brown seaweed is most commonly observed during the summer.

Geographic distribution: India Ocean; SW Asia (Arabian Gulf, Bahrain, India, Kuwait, Pakistan, Saudi Arabia); Asia (China, Japan, Korea); Australia; Pacific Is (Hawaiian Is, Tonga Is).

Uses: Eaten raw as a salad (Kuda et al. 2005, Hu 2005). This sea vegetable is highly prized in Japan, and consumed with soy sauce and vinegar (Smith 1904, Johnston 1966, Chapman and Chapman 1980, Arasaki and Arasaki 1983, Ohno et al. 1998, Zemke-White and Ohno 1999, Harrison 2013); is considered an edible species also in the Tonga Is (Polynesia) (Ostraff 2003).

Papenfussiella kuromo (Yendo) Inagaki
(formerly *Myriocladia kuromo*)

Common name: Japanese: Kuromo (Kuda et al. 2005).

Description: Plants solitary or caespitose, arising from small discoid base, branched, solid, cord-shaped, slimy, tomentose, dark brown, up to 50 cm in length, haplostichous, composed of multiaxial medullary filaments and cortical layer of long and short assimilatory filaments; sub-cortical layer absent and boundary between medulla and cortex distinct; phaeophycean hairs absent (Parker and Guiry 2014c).

Geographic distribution: Atlantic Is (Azores, Canary Is, Selvage Is); Asia (China, Japan, Korea, Russia).

Uses: Consumed as traditional food in the Noto Peninsula area, Japan (Johnston 1966, Kuda et al. 2005).

Petalonia binghamiae (J. Agardh) K.L. Vinogradova (formerly *Endarachne binghamiae*)

Common names: Japanese: Habonori (Arasaki and Arasaki 1983); Korean: Miyeoksoi (Yang et al. 2010).

Description: Erect thalli collected in the field are dorsiventrally flattened and arise from an encrusting base or small discoid holdfast. The erect blades, up to 13 cm long and 2 cm wide, are lanceolate or obovate, simple, solid and thin, gradually tapering at the base. The plurilocular sporangia lack paraphyses and are positioned in continuous sori covering most of the surface area of the blades (Parente et al. 2003).

Geographic distribution: Atlantic Is (Azores); Tropical and subtropical W and E Atlantic; Asia (China, Japan, Korea, Taiwan); Australia and New Zealand; Pacific Is (Hawaiian Is); NE Pacific (California).

Uses: Used as food in China, Japan, Taiwan and Korea (Johnston 1966, Arasaki and Arasaki 1983, Tseng 1983, Mori et al. 2004, Yang et al. 2010). In coastal Shandong (China), this alga is cut into pieces with pork and used as a filling ("xian") in Chinese dumplings or prepared as a cooked vegetable by the "chao" method (Bangmei and Abbott 1987).

Pseudochorda nagaii (Tokida) Inagaki

Synonym: *Chordaria nagaii*

Common name: Japanese: Nise-tsurumo (Tokida 1954).

Description: Epilithic or epiphytic in intertidal and subtidal zone. Erect thalli attached by a small disk, solitary or caespitose, slightly compressed, solid but later becoming partly hollow, lubricous, up to 1.5 m in height and 2 mm in diam. medium to dark brown in color; fronds simple (Tokida 1954).

Geographic distribution: Asia (Japan, Russia).

Uses: Used as food in Japan (Tokida 1954).

Saccharina angustata (Kjellman) C.E. Lane, C. Mayes, Druehl & G.W. Saunders (formerly *Laminaria angustata*)

Common names: Chinese: Hai Dai, Hai Tai, Kunpu (Madlener 1977); English: Tender Kombu, Shreaded Kombu (Chapman and Chapman 1980); Japanese: Kizami-Kombu (Chapman and Chapman 1980); Mitsuishi-

Kombu, Sopaushi, Shiohoshi-Kombu, Urakawa-Kombu, Shamani Kombu, Tokachi-Kombu, Dashi-Kombu, Mizu-Kombu (Madlener 1977).

Description: Grows as one very long, dark, linear frond arising from a very short stipe; the holdfast is rather small and composed of many branched haptera. The blade itself is over a meter long with wavy edges.

Geographic distribution: Asia (Japan, Russia).

Uses: Used as food in Japan and Russia (Johnston 1966, Chapman and Chapman 1980, Arasaki and Arasaki 1983, Zemke-White and Ohno 1999, Roo et al. 2007, Kim 2011, Harrison 2013).

Saccharina cichorioides (Miyabe) C.E. Lane, C. Mayes, Druehl & G.W. Saunders (formerly *Laminaria cichorioides*)

Common name: Japanese: Chiimi-Kombu (Chapman and Chapman 1980).

Description: Sporophyte differentiated into a basal holdfast, a small but firm, cylindrical stipe slightly flattened at the top and a single, undivided blade (phylloid), light brown to deep chocolate brown, that is highly variable in structure.

Geographic distribution: Asia (Japan, Korea, Russia).

Uses: Used as food in Japan and Russia (Johnston 1966, Chapman and Chapman 1980).

Saccharina cichorioides f. *coriacea* (Miyabe) Selivanova, Zhigadlova & G.I. Hansen (formerly *Laminaria coriacea*)

Description: Diplohaplontic lifecycle with alternation of large sporophyte (thallus divided into haptera or rhizoids, stipe and blade) bearing unilocular meiosporangia with paraphyses (sori) and microscopic dioecious and oogamous, heteromorphous gametophytes.

Geographic distribution: Asia (China, Japan, Russia).

Uses: Used as food in China and Japan (Johnston 1966, Arasaki and Arasaki 1983, Hu 2005, Harrison 2013).

Saccharina diabolica (Miyabe) C.E. Lane, C. Mayes, Druehl & G.W. Saunders (formerly *Laminaria diabolica*)

Common names: English: Thin snow Kombu, Black Kombu, Cloudy Kombu (Kirby 1953), White-pulpy Kombu (Chapman and Chapman 1980); Japanese: Black Kombu, Cloudy Kombu (Kirby 1953), Oni-kombu (Tokida 1954), Kuro-tororo Kombu, Oboro Kombu (Chapman and Chapman 1980).

Description: Thallus of this very common kelp is light to medium brown with a finely branched holdfast (haptera), a cylindrical stipe up to 50 cm long without mucilage ducts, and a blade up to 3.5 m long (Li et al. 2009).

Geographic distribution: Asia (Japan, Russia).

Uses: Used as food in Japan and Russia (Tokida 1954, Chapman and Chapman 1980, Zemke-White and Ohno 1999, Roo et al. 2007, Li et al. 2009, Harrison 2013).

Saccharina gyrata (Kjellman) C.E. Lane, C. Mayes, Druehl & G.W. Saunders (formerly *Kjellmaniella gyrata*)

Common names: English: Sea Banner (Chapman and Chapman 1980); Japanese: Tororo Kombu (Madlener 1977).

Description: Diplohaplontic lifecycle with alternation of large sporophyte (thallus divided into haptera or rhizoids, stipe and blade) bearing unilocular meiosporangia with paraphyses (sori) and microscopic dioecious and oogamous, heteromorphous gametophytes (Pereira 2009).

Geographic distribution: Asia (Japan, Russia).

Uses: Used as food in Japan (Johnston 1966, Chapman and Chapman 1980, Arasaki and Arasaki 1983, Kim 2011).

Saccharina japonica (Areschoug) C.E. Lane, C. Mayes, Druehl & G.W. Saunders

Synonyms: *Laminaria japonica*, *L. ochotensis*

Common names: Chinese: Hai Dai, Hai Tai, Kunpu (Madlener 1977); English: Royal Kombu (Chapman and Chapman 1980); Japanese: Makombu, Shinori-Kombu, Hababiro-Kombu, Oki-Kombu, Uchi Kombu, Moto-Kombu, Minmaya-Kombu, Ebisume, Kombu, Hirome, Umiyama-Kombu, Rishiri-Kombu, Para-Kompo, Dashi Kombu, Menashi-Komou, Birodo-Kombu, Teshio-Kombu, Kuro-Kombu, Koteshio, Hosome-Kombu, Shio-Kombu (Madlener 1977), Hoiro-kombu (Chapman and Chapman 1980); Korean: Hae tae, Tasima (Madlener 1977), Dasima (Sohn 1998).

Description: This species grows as a single blade (reaching 10 m in length) with a short stipe. The holdfast is rather small compared to the overall size of the plant it supports, and composed of irregular haptera. Often multiple plants will grow together from an entangled mass of holdfasts. The blade is entire, tapering towards the tip and rounding out at the base, with the widest portion about 1/3 of the way up the blade. The various subspecies of *S. japonica* vary in blade form specifics (rippled edges or not, etc.), range in color from golden yellow to olive brown, and are even reported to differ in taste (Braune and Guiry 2011).

Geographic distribution: Asia (China, Japan, Korea).

Uses: *S. japonica* is large, edible seaweed used in Japanese, Chinese and Korean cooking. It can be added to beans during the cooking process to help them become more digestible. It's also used as a condiment and to make vegetable broth. Like all sea vegetables, it is rich in vitamins and minerals, including iodine (Tokida 1954, Johnston 1966, Kang 1968, Okazaki 1971, Bonotto 1976, Madlener 1977, Tseng 1983, Bangmei and Abbott 1987, Hotta et al. 1989, Sohn 1998, Zemke-White and Ohno 1999, Roo et al. 2007, Harrison 2013, Zhao et al. 2015). This species is also used as food in Russia (Madelener 1977, Chapman and Chapman 1980, Arasaki and Arasaki 1983).

See recipes with *Sacharina* in Annex I.

Saccharina japonica f. *longipes* (Miyabe) Selivanova, Zhigadlova & G.I. Hansen (formerly *Laminaria diabolica* f. *longipes*)

Common name: Japanese: Enaga-onikombu (Tokida 1954).

Description: Stipe 30–70 cm long; the lamina is broadly lanceolate, gradually tapering toward the base, 20–40 cm in breadth at the broadest portion, with a broad median fascia, 7–11 cm, rarely up to 20 cm (Tokida 1954).

Geographic distribution: Asia (Japan, Russia).

Uses: Used as food in Japan (Tokida 1954).

Saccharina longissima (Miyabe) C.E. Lane, C. Mayes, Druehl & G.W. Saunders (formerly *Laminaria longissima*)

Common names: Chinese: Hai Dai, Hai Tai, Kunpu (Madlener 1977); Japanese: Naga-Kombu, Ma Kombu, Nimotsu-Kombu, Gimberi-Kombu, Kimberi-Kombu, Mizu-Kombu, Shima-kombu, Wakaoi (Madlener 1977); English: Barner Kombu (Chapman and Chapman 1980).

Description: Large, flexible thalli with a rubbery texture; size can range from less than 1 m to several. Color ranges from light brown to dark brown.

Geographic distribution: Asia (China, Japan, Russia).

Uses: Used as food in China, Japan and Russia (Madelener 1977, Zemke-White and Ohno 1999, Kim 2011, Hu 2005).

Saccharina longipedalis (Okamura) C.E. Lane, C. Maves, Druehl & G.W. Saunders (formerly *Laminaria longipedalis*)

Description: Diplohaplontic lifecycle with alternation of large sporophyte (thallus divided into haptera or rhizoids, stipe and blade) bearing unilocular meiosporangia with paraphyses (sori) and microscopic dioecious and oogamous, heteromorphous gametophytes.

Geographic distribution: Asia (Japan).

Uses: Used as food in Japan (Johnston 1966, Harrison 2013).

Saccharina religiosa (Miyabe) C.E. Lane, C. Mayes, Druehl & G.W. Saunders (formerly *Laminaria religiosa*)

Common names: Japanese: Hosame-kombu, Saimatsu-kombu (Chapman and Chapman 1980); Korean: Aegidasima (Sohn 1998).

Description: Yellow brown, to 3 m in length; with a claw-like holdfast, a small, smooth, flexible stipe, and an undivided laminate blade to 3 m (Suzuki 2011).

Geographic distribution: Asia (Japan, Korea, Russia).

Uses: Used for foodstuff in Korea, Japan and Russia (Johnston 1966, Kang 1968, Bonotto 1976, Chapman and Chapman 1980, Arasaki and Arasaki 1983, Hotta et al. 1989, Sohn 1998, Zemke-White and Ohno 1999, Mori et al. 2004, Harrison 2013).

This species is used in treatment of menstrual disorders (Read and How 1927). Extracts have anticoagulant and antimicrobial activity (Nisizawa 1979, Baik and Kang 1986).

Saccharina sculpera C.E. Lane, C. Mayes, Druehl & G.W. Saunders

Synonym: *Kjellmaniella crassifolia*

Common name: Japanese: Gagome (Tokida 1954, Airanthi et al. 2011).

Description: Sporophyte perennial (life-span unknown, at least two years), composed of holdfast, stipe and undivided blade; its peculiar shaped, regular gyrations on the blade broadly linear-Ianceolate in shape, measuring 1–2 m in length and 15–30 cm in breadth (Tokida 1954).

Geographic distribution: Asia (Japan, Korea).

Uses: Used as food in Japan and Korea (Tokida 1954, Johnston 1966, Kang 1968, Arasaki and Arasaki 1983, Airanthi et al. 2011).

Sargassum coreanum J. Agardh

Synonym: *Sargassum ringgoldianum* subsp. *coreanum*

Description: Thallus to 10–20 cm or more in length, with one to a few simple, terete to compressed, stipes 1–20 cm long arising from a discoid-conical holdfast; stipes bearing radially or distichously borne, long primary branches, produced seasonally from the stipe apices and subsequently deciduous, leaving scars or other residues on the stipe.

Geographic distribution: Asia (Japan, Korea).

Uses: Anti-hyperlipidemic, antihypertensive and anti-arteriosclerosis activities of S. *coreanum*, called "ooba-moku" in Japan, on rats when comparing them with various marine algae. Perhaps these beneficial effects are related to the high contents of minerals, dietary fibers and antioxidants shown in this study. Therefore this species can be used as a functional food material (Kuda and Ikemori 2009).

Sargassum fulvellum (Turner) C. Agardh

Synonym: *Sargassum enerve*

Common names: Japanese: Hondawara (Smith 1904, Madlener 1977); Korean: Mojaban (Madlener 1977, Song et al. 2013); English: Gulf weed (Hu 2005).

Description: Yellow-brown, 40–80 cm long, main axes bearing slender, leaf-like retroflex lateral branches, 5–6 cm long, sharply dentate, the primary branches bearing globular vesicles 3–4 mm in diam. (Hu 2005).

Geographic distribution: Asia (China, Japan, Korea and Taiwan).

Uses: Used as food in main dishes, and traditional medicine, in Japan and Korea (Johnston 1966, Kang 1968, Bonotto 1976, Madlener 1977, Chapman and Chapman 1980, Arasaki and Arasaki 1983, Lee 2008, Harrison 2013, Song et al. 2013). When the plant is young it is eaten in soup or with soy-bean sauce (Smith 1904).

Sargassum fusiforme (Harvey) Setchell

Synonyms: *Cystophyllum fusiforme, Hizikia fusiformis, Turbinaria fusiformis*

Common names: Chinese: Hai tso, Chiau tsai, Hai ti tun, Hai toe din, Hai tsao, Hoi tsou (Madlener 1977), Chu-chiau ts'ai (Tseng 1983); Japanese: Hijiki (Subba Rao 1965, Madlener 1977), Hiziki (Yendo 1902, Pereira 2011); Korean: Nongmichae (Madlener 1977), Tot (Sohn 1998, Song et al. 2013).

Description: Many stipes arising from a common holdfast area. These form the main axes of the thalli around which short, terete branchlets of variable size grow in whorls; when wet, *S. fusiforme* has a yellow-brown color, but dries to nearly black when removed from water making the once supple, hydrated tissues thick and tough (SIA 2014).

Geographic distribution: Asia (China, Japan, Korea).

Uses: In Japan and Korea is used as food in salads (Subba Rao 1965, Johnston 1966, Kang 1968, Bonotto 1976, Madlener 1977, Hotta et al. 1989, Sohn 1998, Zemke-White and Ohno 1999, Lee 2008, Harrison 2013, Song et al. 2013). Used as food in Amoy, China (Chapman and Chapman 1980). The Japanese food product from this species, called "Hiziki", is prepared by boiling the seaweed with soy sauce and seasoning. Japanese people consider Hiziki as a special health food (Arasaki and Arasaki 1983, Ohno and Largo 2006). In Japan, the young fronds are collected and dried in the sun, in this condition they are a well-known article of merchandise; cooked in soy it is eaten by the peasantry, but not by the higher classes (Yendo 1902, Smith 1904).

This is a popular seaweed throughout China, with many methods of preparation. In Lushun City in Liaoning, *S. fusiforme* is frequently cooked together with mussels, using the "shao" method. The people of Shandong mix the alga with a little oil and use the "shao" method, or mix it with oil, salt, and green onions as a filling in dumplings. In northeastern Shandong, *Sargassum* is cooked with bean curd in a dish called "Shao dou fu." In a part of southern Fujian, this alga is used as a special food for post-partum women. In another part of the province, *Sargassum* is cooked with fish in a small amount of oil, cooled until it sets, then cut and eaten as a cold dish. In Pintan county (near the city of Fuchou), people boil it in two changes of water to extract the strong flavors, then drain, chop, and mix it with brown sugar to make a sweet filling for steamed buns. These are eaten especially during the New Year and other Chinese festivals (Chapman and Chapman 1980, Tseng 1983, Bangmei and Abbott 1987, Roo et al. 2007).

Sargassum hemiphyllum (Turner) C. Agardh

Common name: Filipino: Aragan (Agngarayngay et al. 2005).

Description: Fronds 80–100 cm high, up to 130 cm. Holdfast composed of irregularly ramifying filamentous rhizoidal, some of which develop as stolon and give rise to a new shoot, these filamentous rhizoidal 1.5–4 cm long, 1–5 mm wide; main axis terete, smooth, 3–5 cm long about 2 mm in diam. Primary branches giving rise from main axis, terete, slightly compressed and narrower toward apices, 1–2 mm in diam. Secondary branches spirally giving rise from primary branches, terete to compressed, sometimes with spiny processes, more slender than that of primary branches, less than 1 mm in diam., alternate, up to 13 cm long, at intervals of 2–4 cm, beset with leaves, vesicles and receptacles. Leaves coarse, lower leaves oblong-elliptical to oblong-obovate or lanceolate, 1–3 cm long, 5–8 cm wide, margin entire or undulate, bases cuneate and asymmetrical, midrib inconspicuous, absent or not apparent, cryptostomata small and inconspicuous, irregularly scattered on both sides of midrib; upper leaves elongated elliptical, lanceolate or bat-shaped, 1.5–3.5 cm long 1.5–4 mm wide, apex obtuse or acute, margin entire or undulate, bases cuneate and asymmetrical, midrib inconspicuous, vanishing near the middle part of leaves, one row of small and inconspicuous cryptostomata scattered on both sides of midrib. Vesicles spherical, obovate or elongated elliptical, 3–10 mm long, 2–7 mm in diam., apex rounded, with sharp tip and (or) auricular tip, margin with or without ridge, pedicel flattened, 3–20 mm long, cryptostomata rarely scattered on the vesicles and pedicels. Plant dioeciously. Female receptacle terete or slightly compressed, 2–3 mm long,

0.8 mm in diam., margin and upper part, solitary or racemose, stalked (Yang 2014a).

Geographic distribution: Asia (China, Japan, Korea, the Philippines, Taiwan).

Uses: Used for foodstuff in China and Korea (Tseng 1983, Gray 2015). In the Philippines, the tender tops are used in salads; coat the pot when boiling fish (Agngarayngay et al. 2005).

Extracts of this species have anti-inflammatory (Hwang et al. 2014), antioxidant and immune-stimulating activity (Hwang et al. 2010).

Sargassum henslowianum C. Agardh

Description: Fronds 40 cm high; holdfasts discoid, 0.6–1 cm in diam., giving rise to one or two main axes. Main axes smooth, terete, often less than 2 cm, up to 2.5 cm long, brown, 1.5–5 mm in diam., giving rise to several cylindrical to sub-cylindrical primary branches, slightly compressed after drying, angled, 1–2 mm in diam.; secondary branches and leaves both giving rise from primary branches, secondary branches, cylindrical, 0.2–1.5 cm long, variable; leaves papyraceous, lanceolate, generally 4–8 cm long, up to 18 cm, 2–7 mm broad, with some shallow teeth at the margins, with oblique cuneate bases and acute apices, distinct percurrent midrib, cryptostomata scattered on both sides of the midrib. Vesicles spherical, sub-spherical to ovate, 4–9 mm long, 3–5 mm in diam., rounded at the apices (Yang 2015).

Geographic distribution: Asia (China, Hong Kong, Japan, Taiwan); SE Asia (Vietnam).

Uses: Used as food in China (Tseng 1983).

Sargassum horneri (Turner) C. Agardh

Common names: English: Devil weed (Bushing 2014); Japanese: Aka-moku (Tokida 1954), Akamoku (Airanthi et al. 2011); Korean: Kwaengsaegi-Mojaban (Sohn 1998).

Description: Immature specimens have flat, symmetrical, fern-like blades with notched tips. As the alga grows, it becomes loosely branched in a zigzag pattern, develops small air bladders, and may reach lengths of more than 6 m. Very thick, nearly impenetrable forests may form (Bushing 2014).

Geographic distribution: Asia (China, Japan, Korea, Taiwan, the Philippines); NE Pacific (California to Mexico).

Uses: Used for foodstuff in China, Japan and Korea (Tokida 1954, Madlener 1977, Tseng 1983, Sohn 1998, Zemke-White and Ohno 1999, Airanthi et al. 2011, Terasaki et al. 2012, Harrison 2013, Gray 2015), treat goiter (Tokuda et al. 1986), animal fodder and production of alginates (Kang 1968, Bonotto

1976, Tseng 1983). Their extracts have antiviral (Hoshino et al. 1998), osteoblastogenesis and osteoclastogenesis (Yamaguchi and Matsumoto 2012) activity.

Sargassum miyabei Yendo

Synonym: *Sargassum kjellmanianum*

Common names: Japanese: Hahakimoku, Miyabe-moku (Tokida 1954).

Description: Thallus to 10–20 cm or more in length, with one to a few simple, terete to compressed, stipes 1–20 cm long arising from a discoid-conical holdfast; stipes bearing radially or distichously borne, long primary branches, produced seasonally from the stipe apices and subsequently deciduous, leaving scars or other residues on the stipe. Primary branches 10 cm to 20 cm or more long, distichously, tristichously or radially branched with terete, angular, compressed or three-side axes; the receptacles are distinctly longer than the female ones, often measuring 10–15 mm, or sometimes up to 30 mm, in length, while the receptacles of the female plant measure 3–8 mm in length (Tokida 1954).

Geographic distribution: Asia (Hong-Kong, Japan, Korea, Russia); SE Asia (the Philippines, Vietnam).

Uses: Used for foodstuff in Japan, especially for the manufacture of potash salts (Tokida 1954).

Sargassum muticum (Yendo) Fensholt

Common names: English: Wireweed (Mondragon and Mondragon 2003, Gutschmidt 2010f); Japanese weed, Jap weed (Hairon 2014), Tamahahakimoku (Oshie-Stark et al. 2003).

Description: *S. muticum* is a species originating in Japan but has become well established on the coasts of Europe and N America. It is regarded as a pest by mariners, fouling boat propellers and fishing nets and disturbing the natural balance of the indigenous marine flora and fauna. It was accidentally introduced to western N America and Europe, and is steadily extending its range in these areas. It out competes other seaweeds and in these regions is regarded as an invasive species; however it remains as a food source. This long, bushy seaweed has numerous side-branches, which have many leaves like fronds up to 10 cm long. The fronds bear small, gas-filled bladders, either singly or in clusters (Pereira 2010a).

Geographic distribution: Native to waters around Japan, Russia, Korea, and China, but has subsequently spread via ballast water and oyster shells to such far distant places as Europe, Mediterranean to the Adriatic, and N America from Alaska to Baja California (Gutschmidt 2010f).

Uses: Though not of commercial use, many coastal populations make use of *Sargassum* as an edible seaweed. It is reported to be more bitter than other seaweed species, but is high in minerals and nutrients and can be made palatable through various preparation techniques. All kelps are also sources of alginate, a compound used in a variety of applications such as thickening and stabilizing agents in the food industry (Gutschmidt 2010, Pereira 2011, Rodrigues et al. 2015a, 2015b).

S. muticum is often gathered from the shore or floating mats and used as nutrient-rich fertilizer or compost, although it is also used as animal feed, fish bait, and insect repellent. Asian medicine makes use of various species of *Sargassum* to treat afflictions such as fever, high cholesterol, and skin ailments (SIA 2015).

Sargassum naozhouense C.K. Tseng & Lu

Common name: English: Black vegetable (Xie et al. 2013).

Description: Perennial brown algae found in the coastal area of Leizhou Peninsula, Guangdong, China. Its rhizoids are perennial, but its sporophytes are annual (Xie et al. 2013).

Geographic distribution: Asia (China).

Uses: In China (Zhanjiang coast, Guangdong province) is commonly consumed as a sea vegetable or crude drugs for treating internal heat, infections, laryngitis and other ailments among the locals (Wang et al. 2010, Peng et al. 2013, Xie et al. 2013, Abdallah 2014).

Sargassum pallidum (Turner) C. Agardh

Common names: Chinese: Hai hao zi, Da hao zi, Hai gen cai, Hai cao, Hai zao (AAMD 2004).

Description: Perennial brown algae, erect, 30–60 cm, up to 1 m high; discoid holdfast or conical blunt, 1–2 cm diam. Main axe cylindrical, mostly solitary, 2–7 mm diam., leaflets alternate, winter litter after residual remnants of the cone-shaped trunk. Leaves simple, alternate, leaf variation very large, primary leaves obovate, lanceolate, 20–70 mm long; secondary leaves relatively small, linear to lanceolate, sometimes shallow plume jagged crack or sparse, thin ribs obvious; spherical airbags, with 2–5 mm diam. (AAMD 2004).

Geographic distribution: Asia (China, Japan, Korea, Russia); SE Asia (Indonesia).

Uses: Used for foodstuff in China (Tseng 1983, Gray 2015).

Extracts of this species have antioxidant (Ye et al. 2009, Zhang et al. 2012), immunity-enhancing (Zhang et al. 2012), antimicrobial and hemolytic (Gerasimenko et al. 2014) activity.

Sargassum sagamianum Yendo

Description: Thallus to 10–200 cm or more in length, with one to a few simple, terete to compressed, stipes 1–20 cm long arising from a discoid-conical holdfast. Stipes bearing radially or distichously borne, long primary branches, produced seasonally from the stipe apices and subsequently deciduous, leaving scars or other residues on the stipe. Primary branches 10 cm to 200 cm or more long.

Geographic distribution: Asia (Japan, Korea, the Philippines).

Uses: Used as food in Korea (Michanek 1975).

Sargassum siliquastrum (Mertens ex Turner) C. Agardh

Description: Thallus to 10–20 cm or more in length, with one to a few simple, terete to compressed, stipes 1–20 cm long arising from a discoid-conical holdfast; stipes bearing radially or distichously borne, long primary branches, produced seasonally from the stipe apices and subsequently deciduous, leaving scars or other residues on the stipe. Primary branches 10 cm to 20 cm or more long, distichously, tristichously or radially branched with a terete, angular, compressed or three-sides axes; basal laterals simple or branched, compressed and relatively narrow (in most species) leaf-like, 3–15 mm broad, entire or with dentate margins; upper laterals usually branched, with slender, compressed to terete, ramuli (Womersley 1987).

Geographic distribution: Asia (Hong Kong, Japan, Korea, and Vietnam); Pacific Islands (New Caledonia).

Uses: In Korea is used for foodstuff (Tseng 1983), and for goiter treatment (Tokuda et al. 1986).

Sargassum swartzii C. Agardh (formerly *Sargassum wightii*)

Common name: Tamil: Kattaikkorai (Kaliaperumal et al. 1995).

Description: The plants have smooth, flat branches, and the leaves of this plant are linear with sparse serrations. The vesicles are oval to elliptical without spines. This seaweed is 1–1.5 m high, and found on rocks in subtidal zones along shorelines with moderate wave activity (JIRCAS 2012g).

Geographic distribution: W Indian Ocean (Bangladesh, India, Kenya, Pakistan, Sri Lanka, Tanzania); Asia (China, Japan, Korea, Taiwan, Vietnam); SW Asia (Indonesia, Malaysia, Singapore); Pacific Is (New Macedonia).

Uses: Used as food in India (Kaliaperumal et al. 1995, Gray 2015); it is used as raw material for the production of sodium alginate; it also contain 8–10% of mannitol, which can be used as a substitute for sugar (Kaliaperumal et al. 1995).

Extracts of this species have antiviral (Subramaniam et al. 2014), larvicidal (Khanavi et al. 2011), antioxidant and anti-cholinesterase (Syad et al. 2013) activity.

Sargassum thunbergii (Mertens ex Roth) Kuntze

Synonym: *Turbinaria thunbergii*

Common names: Chinese: Shuweizhao (= mouse tail algae); Djichungi (= earth worm) (Koh et al. 1993); Japanese: Umitoranoo (Zhuang et al. 1995).

Description: Thallus upright, cartilaginous-firm, fronds with a conical outline, the smooth axes are laterally covered with the genus-characteristic cone-shaped phylloid-complexes; witch may have embedded float bladders. This plant shows a high degree of morphological differentiation, having a perennial holdfast, a stipe, branches, leaves and vesicles. Different from other species belonging to the genus *Sargassum*, *S. thunbergii* shows a distinctive feature in the branch systems. Every primary branch develops to an individual-like main axis (Koh et al. 1993, Braune and Guiry 2011).

Geographic distribution: Atlantic Is (Canary Is); Asia (China, Japan, Korea, Taiwan).

Uses: In Japan and Korea is used for foodstuff, especially for the manufacture of potash salts (Tokida 1954, Kang 1968), as animal fodder and manure (Kang 1968, Bonotto 1976, Arasaki and Arasaki 1983).

Extracts of this species have vermifuge (Bonotto 1976, Tseng 1983), antitumor (Zhuang et al. 1995), and hypocholesterolemic (Wei et al. 2012) activity.

Saundersella simplex (De A. Saunders) Kylin

Common names: English: Golden bottlebrush epiphyte; Japanese: Motsuki-chasômen (Tokida 1954).

Description: The tubular and mucilaginous unbranched thallus is strictly epiphytic on *Analipus japonicus*. This golden brown thallus frequently dominates its host, growing up to 18 cm long (Lindeberg and Lindstrom 2014e).

Geographic distribution: Arctic; Atlantic Is (Greenland); Asia (Japan); NE Pacific (Alaska, British Columbia).

Uses: Used as food in Japan (Tokida 1954).

Silvetia babingtonii (Harvey) E.A. Serrão, T.O. Cho, S.M. Boo & Brawley

Synonym: *Pelvetia wrightii*

Common name: Japanese: Yezo-ishige (Tokida 1954).

Description: Olive green or yellowish brown plant about 30 cm long (up to 90 cm long), composed of thick, narrow, dichotomous fronds that often appear irregularly branched because of axis breakage (Tokida 1954).

Geographic distribution: Asia (Japan, Korea, Russia).

Uses: Used for foodstuff in Japan and Korea, especially for the manufacture of potash salts (Tokida 1954).

Silvetia siliquosa (C.K. Tseng & C.F. Chang) E.A. Serrão, T.O. Cho, S.M. Boo & S.H. Brawley (formerly *Pelvetia siliquosa*)

Common names: Chinese: Lijao kai (Tseng 1983); Korean: Ttumbugi (Sohn 1998).

Description: *Silvetia* genus includes perennial *Pelvetia*-like fucacean algae that lack a midrib, found in the upper intertidal zone (Serrão et al. 1999).

Geographic distribution: Asia (China and Korea).

Uses: In China and Korea is used for foodstuff (Kang 1968, Tseng 1983, Bangmei and Abbott 1987, Sohn 1998, Zemke-White and Ohno 1999, Harrison 2013), and for extraction of alginates (Kang 1968).

Sphaerotrichia divaricata (C. Agardh) Kylin

Synonym: *Sphaerotrichia japonica*

Common name: Kusa-mozuku (Tokida 1954).

Description: Sporophytes are irregularly branched structures growing up to 40 cm high. The axes and branches are worm-like and slippery, and rather cartilaginous, attaining diam. of 0.5–1 mm. The thalli have a uniaxial architecture and grow through the activity of trichothallic meristems (Hoek 1995).

Geographic distribution: N Atlantic; Atlantic Is (Azores); Baltic Sea; Asia (China, Japan); Australia; NE Pacific (British Columbia).

Uses: Used as food in Japan (Tokida 1954).

Sphaerotrichia firma (E.S. Gepp) A.D. Zinova
(formerly *Chordaria firma*)

Common name: Japanese: Ishi-mozuku (Tokida 1954).

Description: Thalli threadlike, gelatinous, 15–20 cm high, single or assembled in clumps, covered by long colorless hairs and fixed to the substratum by a basal disk. The main axis with well distinct irregularly branches and carries branches inserted more or less at right angles in all directions. The branches, tapered near the apex, lead ramuli very short (2–3 cm) often curved (Riouall 1985).

Geographic distribution: Mediterranean; Asia (China, Japan, Korea).

Uses: Used as food in Japan (Tokida 1954, Johnston 1966).

Stephanocystis crassipes (Mertens ex Turner) Draisma, Ballesteros, F. Rousseau & T. Thibaut

Synonym: *Cystoseira crassipes*

Common name: Japanese: Nebuto-moku (Tokida 1954).

Description: Similar to *Stephanocystis geminata* but smaller in the size of the vesicles and receptacles (Tokida 1954).

Geographic distribution: Asia (Japan, Russia).

Uses: Used as food in Japan (Tokida 1954).

Stephanocystis geminata (C. Agardh) Draisma, Ballesteros, F. Rousseau & T. Thibaut

Synonym: *Cystophyllum geminatum*

Common names: English: Chain bladder (McConnaughey 2012), Japanese: Yezo-moku (Tokida 1954).

Description: This richly branched brown seaweed grows up to 5 m in length. A stout, discoidal holdfast anchors it to rocks. From this holdfast arises an erect woody stipe that branches radially to give rise to several main branches, which, in turn, produce long secondary branches and, ultimately, short terminal branchlets. The terminal branchlets near the base of each individual tend to be flattened, while those in the upper half are mainly cylindrical. The whole seaweed, therefore, has a rather dual appearance. The terminal branchlets carry oval, air-filled bladders called pneumatocysts (vesicles). These can be single, paired, or in short chains, and suggested the common name of this species. Elongated, warty receptacles develop on the terminal branchlets of mature individuals. The warts surround the openings to internal conceptacles, which contain the microscopic gamete-producing organs (Klinkenberg 2014c).

Geographic distribution: Asia (Japan); NE Pacific (Alaska, British Columbia, Aleutian Is).

Uses: Used as food in Japan (Tokida 1954).

Stephanocystis hakodatensis (Yendo) Draisma, Ballesteros, F. Rousseau & T. Thibaut

Synonyms: *Cystophyllum hakodatense, Cystoseira hakodatensis*

Common names: Japanese: Uga-no-moku (Tokida 1954), Uganomoku (Airanthi et al. 2011).

Description: This species is of a more robust frond than either of the same genus and bears markedly larger receptacles. The vesicles in the typical specimens are sometimes solitary but more frequently are arranged in a moniliform series generally with shallow constrictions or occasionally deep, stalk-like ones between successive vesicles (Tokida 1954).

Geographic distribution: Asia (Japan, Korea); SE Asia (the Philippines).

Uses: Used as food in Japan (Tokida 1954, Airanthi et al. 2011, Terasaki et al. 2012).

Tinocladia crassa (Suringar) Kylin

Synonyms: *Eudesme crassa, Mesogloia crassa*

Common names: Japanese: Futo-mozuku (Chapman and Chapman 1980), Somen-nori.

Description: Thalli lubricous, 15–25 cm tall, irregularly branched; branches of one or two orders, cylindrical, 2–3 mm in diam. (Abbott and Hollenberg 1976).

Geographic distribution: Asia (China, Japan, Korea, Russia); NE Pacific (California).

Uses: Used as food in Japan (Johnston 1966, Chapman and Chapman 1980, Arasakai and Arasaki 1983).

Undaria peterseniana (Kjellmann) Okamura (formerly *Laminaria peterseniana*)

Common names: Japanese: Wakame (Zanveld 1955), Ao-wakame (Arasaki and Arasaki 1983), Korean: Nolmiyok (Sohn 1998).

Description: Linear-lanceolate blade, smooth lubricious surface, two-edged stipe, and sori as median fascia on the blade (Lee and Yoon 1998).

Geographic distribution: Asia (Japan and Korea).

Uses: In Japan is utilized as various forms of wakame; in Korea is fried in oil, boiled soup, soaked with vinegar, soy sauce and sugar in water (fresh, dried, salted) (Zanveld 1955, Johnston 1966, Kang 1968, Okazaki 1971, Bonotto 1976, Chapman and Chapman 1980, Arasaki and Arasaki 1983, Hotta et al. 1989, Sohn 1998, Zemke-White and Ohno 1999, Harrison 2013).

Undaria pinnatifida (Harvey) Suringar (Fig. 64)

Common names: Chinese: Qun dai cai (Lembi and Waaland 1988); English: Asia kelp, Apron-ribbon vegetable (Santhanam 2015), Sea mustard (Kirby 1953), Precious sea grass (Rhoads and Zunic 1978), Wakame (Bunker et al. 2010); Japanese: Wakame (Suba Rao 1965, Madlener 1977); Korean: Miyok (Madlener 1977, Sohn 1998), Miyeouk (Lembi and Waaland 1988).

Description: The thallus can reach 1 to 3 m in length. The blade is lanceolate and broad with a prominent midrib, and translucent with the color ranging from green to yellowish-brown to dark brown. The blade could also be described as triangular and lobed. The appearance of the blade evolves and changes over time; it is initially simple, flattened, and broad with a

Figure 64. *Undaria pinnatifida* (Harvey) Suringar (Ochrophyta, Phaeophyceae).

pronounced or distinct midrib; older plants have thicker blade tissue which splits horizontally down to the midrib to form fingers or straps, becoming more transversally lobed, and becomes pinnate with age. The margins of the blade can also be described as wavy. The distal portion of the blade and the straps eventually become tattered. The stipe of *U. pinnatifida* is wavy or corrugated above the holdfast. The stipe is also usually short (10–30 cm in length and up to 1 cm in diam.) and in mature plants bears convoluted wing-like reproductive outgrowths or frills (sporophylls) (Pereira 2015a).

Geographic distribution: Kelp indigenous to the NW Pacific Ocean and the cold temperate coastal regions of Japan, China, Korea, and Southeast Russia. *U. pinnatifida* has been spread around the world by international shipping and mariculture, and has extended its range to include four continents since 1980's. It is now growing in the temperate Pacific Ocean, the SE temperate Indian Ocean, the Mediterranean, and the temperate N and S Atlantic. *U. pinnatifida* was grown in the French Bretagne as food, which increased exposure of this seaweed to Europeans (Pereira 2015a).

Uses: Used in soups, baked and raw in China, Japan, Korea and Vietnam (Suba Rao 1965, Chapman and Chapman 1980, Arasaki and Arasaki 1983, Hotta et al. 1989, Simoons 1990, Sohn 1998, Zemke-White and Ohno 1999, Hong et al. 2007, Harrison 2013); blanched and salted wakame is the major wakame product. Fresh wakame is plunged into water at 80°C for one min and cooled quickly in cold water. About 30 Kg of salt per 100 Kg of seaweed are mixed and stored for 24 hours. This dehydrates the wakame; excess water is removed and the seaweed stored at –10°C. When ready for packaging, it is taken from storage, the midribs are removed and the pieces placed in plastic bags for sale. It is a fresh green color and can be preserved for long periods when stored at low temperatures (McHugh 2003). Cut wakame is a very convenient form, used for various instant foods such as noodles and soups. It is one of the most popular dried wakame products. It is made from blanched and salted wakame which is washed with freshwater to remove salt, cut into small pieces, dried in a flow-through dryer and passed through sieves to sort the different sized pieces. It has a long storage life and is a fresh green color when rehydrated (McHugh 2003).

Undaria is traditionally served as a luxury food among Japanese and Korean people. This seaweed, called Wakame in Japanese, is sold as boiled or dried and is especially appreciated as an ingredient for soya-bean soup (Misoshiru) and seaweed salad (Ohno and Largo 2006).

This species is used for nicotine poisoning cure, as anti-hypertensives (Takagi 1975), and treatment of stomach ailments, hemorrhoids, anal fistulas, leucorrhea, nocturnal enuresis, urinary diseases, and dropsy (Tseng and Zhang 1984).

Undaria undarioides (Yendo) Okamura

Common name: Japanese: Wakame (Chapman and Chapman 1980).

Description: It has a broadly ovate lamina whose margins become undulated and sometimes entirely but usually pinnated with short blunt segments with broad sinuses. The presence of a distinct midrib, of rugose-bullations, of cryptostomata and of glandular dots is common to the two plants (Okamura 1915).

Geographic distribution: Asia (Japan, Korea).

Uses: Used in Japan as various forms of wakame (Chapman and Chapman 1980, Hu 2005).

2.3.3 Rhodophyta

Acanthopeltis japonica Okamura

Common names: Japanese: Toriashi, Yuikiri (Subba Rao 1965, Arasaki and Arasaki 1983).

Description: Thallus sub-cylindrical sympodially branched with erect axes attached to the substratum by several elongated haptera. Erect axes up to 15 cm high, 4 mm diam., branched in an alternate or somewhat dichotomous manner. All the segments, except the basal portions, completely invested with numerous spirally arranged disk-shaped, sub-orbicular, leaf-like structures. After the formation of the leaflet terminally on a branch, a new lateral branch is initiated as a protuberance at the base of the leaflet. This new branch gradually enlarges and elongates forming a short cylindrical axis which terminally soon becomes broad and leaf-like. A new branch is then initiated on the upper surface, at the base of the leaflet. Protuberances may be formed on the upper surface of the same leaflet, each developing into an axis, thus forming a dichotomy. The margins and surface of the leaflets may have many bristle-like projections. Tetraspores formed in dilated processes near the margin of ramuli. Cystocarps roundish-oval, produced in the marginal setae.

Geographic distribution: Asia (Japan and Korea).

Uses: Used for foodstuff and for agar production, in Japan and Korea (Subba Rao 1965, Kang 1968, Okazaki 1971, Bonotto 1976, Chapman and Chapman 1980, Arasaki and Arasaki 1983, Hu 2005).

Ahnfeltiopsis flabelliformis (Harvey) Masuda
(formerly *Gymnogongrus flabelliformis*)

Common names: Japanese: Okitsu-nori (Smith 1904, Subba Rao 1965, Chapman and Chapman 1980); English: Fan seaweed (Stender and Stender 2014a).

Description: Dark olive green with tough fan-shaped branches (Stender and Stender 2014a).

Geographic distribution: Asia (China, Japan, Korea, Russia, Taiwan, Vietnam); Pacific Is (Hawaiian Is).

Uses: Used for foodstuff and agar production in Japan and Korea (Smith 1904, Subba Rao 1965, Bonotto 1976, Arasaki and Arasaki 1983, Tseng 1983, Bangmei and Abbott 1987).

Ahnfeltiopsis paradoxa (Suringar) Masuda

Common name: Japanese: Harigane (Arasaki and Arasaki 1983).

Description: The thalli of reproductively mature female gametophyte are erect, stiff-cartilaginous, multiaxial fronds. The erect fronds grow from well developed basal crusts, strongly attached to the substratum; tetrasporangial plants crustose (Braune and Guiry 2011).

Geographic distribution: Asia (Japan, Taiwan).

Uses: Used as food in Japan (Arasaki and Arasaki 1983).

Betaphycus gelatinus (Esper) Doty ex P.C. Silva

Synonyms: *Betaphycus gelatinum*, *Eucheuma gelatinae*

Common names: English: agar-agar, Celebes agar agar, Macassar agar agar; Japanese: Kirinsai (Ohno et al. 1998).

Description: Fronds are apically flat, pliable and arising from marginal cylindrical teeth. Tolerates direct wave action (SuriaLink 2014).

Geographic distribution: Indian Ocean; Asia (China, Japan, Taiwan); SE Asia (Indonesia, the Philippines, Vietnam); Australia; Pacific Is (New Caledonia).

Uses: Considered and edible species in Vietnam and Japan (Zemke-White and Ohno 1999, Lideman et al. 2011), and used as food for seaweed salads in China (Arasaki and Arasaki 1983, Tseng 1983, Bangmei and Abbott 1987, Simoons 1990, Ohno et al. 1998, Sidik et al. 2012, Kılınç et al. 2013). In the Philippines and Malaysia, fresh collected seaweeds are washed, dried, and then boiled with sugar (Zaneveld 1959, Sidik et al. 2012, Harrison 2013). *B. gelatinus* is used for making jellies and cakes in Vietnam (Huynh and

Nguyen 1998, Roo et al. 2007); is utilized in Indonesia for making vegetable soup with coconut milk (Istini et al. 1998).

Harvested from the wild but also cultivated to some extent in China; tends to grow slowly. The generic name "Betaphycus" may be on slightly shaky grounds among taxonomists but the carrageenan synthesized by this genus are distinct from a commercial standpoint and the plants deserve a separate classification from that viewpoint (SuriaLink 2014). This species produce beta carrageenan (Pereira 2004, Pereira et al. 2009a).

Campylaephora hypnaeoides J. Agardh

Common names: Japanese: Ego (Yendo 1902, Tokida 1954); Ego nori (Chapman and Chapman 1980), Egonori (Tokida 1954, Arasaki and Arasaki 1983), Igoneri (Schwarzfluss 2014), Yego-nori (Subba Rao 1965).

Description: Thallus dark red to yellowish, erect at first, later like matted wool; main axes irregularly dichotomously branched, in the upper parts slightly unilateral; the branch tips dichotomously pincer-like incurved, sometimes sickle-shaped; axes and branches with cortical filaments; conical-discoid holdfast consisting of rhizoidal cells (Braune and Guiry 2011).

Geographic distribution: Asia (China, Korea, Japan and Russia).

Uses: This species is used as food and for Agar extraction in Japan and Korea (Tokida 1954, Subba Rao 1965, Kang 1968, Okazaki 1971, Bonotto 1976, Chapman and Chapman 1980, Arasaki and Arasaki 1983, Tseng 1983, Hu 2005, Braune and Guiry 2011); usually used as the raw material of "Kanten" or seaweed jelly (Nakamura 1965); "Igoneri" is also a local dish in Niigata, Honshu Is, Japan, but eaten in some other regions as well; this dish is made with the red algae Egonori (*C. hypnaeoides*); the seaweed is boiled, melted, and then solidified in a fridge; usually is eaten with soy sauce and mustard. Good for health and usually consumed in warmer season (Schwarzfluss 2014).

Ceramium kondoi Yendo

Common name: Japanese: Igisu (Tokida 1954).

Description: Thallus erect, to 30 cm, axes monosiphonous pseudo-dichotomously or irregularly branched, cells ovoid or cylindrical, corticated by filaments of limited growth arranged in bands and forming partial or complete investment. Carpogonial branches four-celled borne on first-formed pericentral (supporting) cells successively along branches. Subsequent to fertilization connection occurs between carpogonium and auxiliary cell, cut off from supporting cell, leading to development of carposporophyte with successive groups of carposporangia finally enveloped by an involucre of filaments formed from cells of the segment below; spermatangia from apical cells of cortical filaments of limited growth. Tetrasporangia developed in cortical bands from filaments of limited growth, singly, in whorls, or irregularly arranged; spores tetrahedrally or occasionally cruciately arranged; cortex thin, but not as above; main branches bearing a branchlet at each axil in opposite directions by turn (Tokida 1954).

Geographic distribution: Asia (China, Japan, Korea, Russia), NE Pacific (Alaska).

Uses: Used for foodstuff in Japan, China and Korea (Tokida 1954, Tseng 1983, Bangmei and Abbott 1987).

Ceramium virgatum Roth (Fig. 65)

Synonym: *Ceramium rubrum*

Common names: Japanese: Fa-tsai; Swedish: Grovsläke (Tolstoy and Österlund 2003).

Description: Small red seaweed growing up to 30 cm tall. It has a filamentous frond that is irregularly and dichotomously branched, with the branches narrowing towards pincer-like tips. The holdfast is a minute conical disk that extends into a dense mass of rhizoidal filaments. The plant is reddish-brown to purple in color and has a banded appearance when viewed closely (Hiscock and Pizzolla 2007).

Geographic distribution: Arctic (Canada); Ne Atlantic; Adriatic Sea; NE Pacific (Alaska); Caribbean Sea; SW Atlantic; SW Asia (India, Turkey); Asia (Japan, Korea); Australia and New Zealand; Antarctic Is.

Uses: Uses as food in Japan (Hu 2005).

C. virgatum is the source of polysaccharides, agaroids, agarose, proteins, nitrogen, iodine, lectins, beta-carotene and a number of unique xanthophylls; antimicrobial, antiviral, anthelminthic, fungicidal, antioxidant, antialgal, antiprotozoal, antimycobacterial and cytotoxic, and antitumor BASs; contain pigments which use for textile and fragrance industries (Milchakova 2011, Pereira 2015a).

Chondracanthus intermedius (Suringar) Hommersand (formerly *Gigartina intermedia*)

Description: Thalli 2–3 cm high, cartilaginous, purplish red to black with a bluish iridescence, forming low, pulvinate, densely overlapping masses widely spread over rocks, firmly attached to the substratum by tips of branches that act as holdfasts, irregularly branched; axes 1–2 mm in width; branches 0.5–1.5 mm in diam., 8–12 mm long, often dilated to 2 mm or more forming sub-lanceolate segments ending in sharp points, strongly recurved;

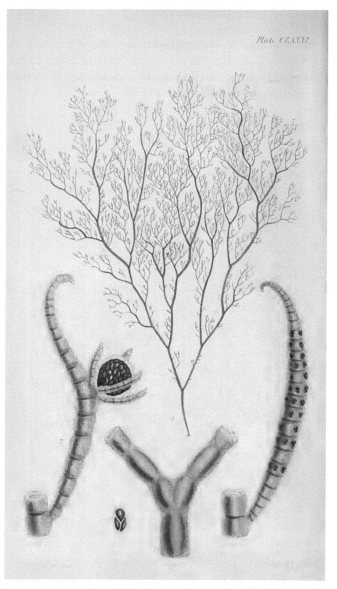

Figure 65. *Ceramium virgatum* Roth (Rhodophyta). Illustration from William H. Harvey (Plate CLXXXI, Phycologia Britannica, 1846–1951).

cystocarps about 1 mm in diam., mostly marginal on the branches of all orders (Hommersand et al. 1993, Guiry and Guiry 2014c).

Geographic distribution: W Pacific (El Salvador, Chile, Mexico); Australia (New South Wales, Norfolk Is, Queensland); Asia (China, Japan, Korea, Taiwan, Vietnam).

Uses: Used for foodstuff in China and Korea (Lee 1965, Tseng 1983, Bangmei and Abbott 1987), and carrageenan production (Bonotto 1976). This alga is eaten in the north of Taiwan. It is cooked in the "shao" method with pork. A jellied salad is made by boiling to extract the gel (the colloid carrageenan), adding the desired flavorings, allowing to cool and set, then cutting it into bite-size pieces (Bangmei and Abbott 1987).

Chondracanthus teedei (Mertens ex Roth) Kützing (formerly *Gigartina teedii*) (Fig. 66)

Common names: Japanese: Cata-nori (Smith 1904), Shikin-nori (Chapman and Chapman 1980); Portuguese: Musgos (Pereira and Correia 2015).

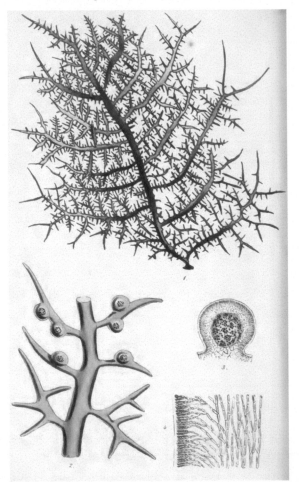

Figure 66. *Chondracanthus teedei* (Mertens ex Roth) Kützing (Rhodophyta). Illustration from William H. Harvey (Plate CLXXXI, Phycologia Britannica, 1846–1951).

Description: The fronds of this alga are cartilaginous-membranous, with purple-violet color that darkens by desiccation, becoming greenish-yellow by decay. The fronds of this alga are cartilaginous-membranous, with purple-violet color that darkens by desiccation, becoming greenish-yellow with decay. The main axes of the fronds, as its ramifications are wide, reaching 1 cm in the older portions (Pereira 2010a).

Geographic distribution: NE Atlantic; Tropical and subtropical W Atlantic; SE Atlantic (Brazil); Atlantic Is (Azores, Canary Is, Cape Verde Is); Adriatic Sea; Mediterranean Sea; Indian Ocean, Asia (Japan, Korea).

Uses: Used as food in Japan (Subba Rao 1965, Chapman and Chapman 1980).

Chondracanthus tenellus (Harvey) Hommersand (formerly *Gigartina tenella*)

Common names: Japanese: Matsuba-gusa, Suginori (Naoko et al. 2007).

Description: Thalli composed of one or more erect axes from a discoid or crustose holdfast, the erect axes initially cylindrical and either remaining cylindrical or becoming compressed or flattened; axes repeatedly pinnately branched or foliaceous and either simple or basally or marginally proliferous, often bearing numerous vegetative or reproductive branchlets, or sparsely to thickly covered by vegetative or reproductive papillae. Growth multiaxial by means of apical and marginal meristems and consisting of files of outwardly directed branched cortical filaments overlying an internal network of medullary filaments.

Geographic distribution: Asia (China, Japan, Korea, Taiwan, the Philippines); Pacific Is (Samoa, Polynesia, Micronesia, Hawaiian Is).

Uses: Used as food in Japan and Korea (Jung et al. 2006, Naoko et al. 2007).

Chondria crassicaulis Harvey

Common name: Korean: Seoshil (Sohn 1998).

Description: Thalli highly branched, cylindrical, with thickened primary axes (up to approximately 5 mm diam.); branching radial, sometimes secondarily distichous, smaller branches usually with proximal constriction; apices either pointed and attenuate or sunken within a terminal depression.

Geographic distribution: Asia (China, Japan, Korea, Taiwan, the Philippines).

Uses: Used for foodstuff in Korea (Kang 1968, Sohn 1998, Zemke-White and Ohno 1999, Roo et al. 2007, Harrison 2013).

Chondria dasyphylla (Woodward) C. Agardh (Fig. 67)

Common name: English: Diamond cartilage weed (Bunker et al. 2010).

Description: Thalli of cylindrical erect axes or decumbent tufts, brownish-red or yellowish, 8–15 cm high, to 1 mm diam., tubular, coarse-fibrous, attached by basal holdfast; main axes distinct, richly branched, branching sparsely at irregular intervals; branchlets multiple, 3–20 mm long, 0.5–1 mm diam., simple or brachiated, often growing in small clusters (Milchakova 2011).

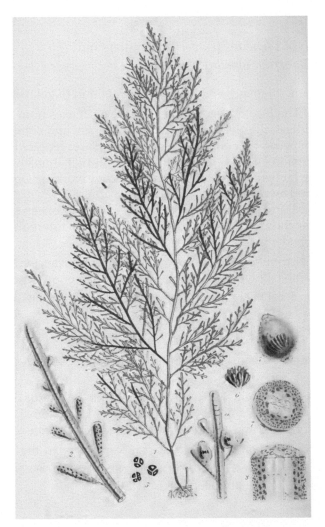

Figure 67. *Chondria dasyphylla* (Woodward) C. Agardh (Rhodophyta). Illustration from William H. Harvey (Phycologia Britannica, 1846–1951).

Geographic distribution: NE Atlantic; Atlantic Is (Azores, Bermuda, Canary Is, Madeira); Adriatic Sea; Black Sea; Gulf of Mexico Caribbean Sea; Tropical and subtropical W Atlantic; Arabian Gulf; Indian Ocean; Asia (China, Japan, Korea, Russia, Taiwan); SE Asia (Indonesia, the Philippines); Australia; Pacific Is (Polynesia, Fiji).

Uses: Used as food in Korea (Milchakova 2011).

This species produces sarganin and glycolipids that inhibit bacterial and fungal growth; *C. dasyphylla* contain phytosterols (fucosterol) with cytotoxicity against breast and colon cancer cell line, and is also a source of angiocardial substances; extracts of this seaweed have larvicidal, antiprotozoal, antiviral, antifertility, and hypoglycemic (Milchakova 2011, Pereira 2015a).

Chondrus armatus (Harvey) Okamura

Common name: Japanese: Togetsunomata (Tokida 1954).

Description: Frond terete-compressed, usually very densely and irregularly branched in divaricate-dichotomous and pinnate manner; branchlets sub-cylindrical, spine-like or filiform, with a spinose apex (Tokida 1954).

Geographic distribution: Asia (China, Japan, Korea, Russia).

Uses: Used as food in Japan (Tokida 1954).

Chondrus ocellatus Holmes

Common names: English: Japanese moss (Chapman and Chapman 1980), Carrageenan; Japanese: Tsuno-mata, Hosokeno-mimi (Smith 1904), Makuri-nori (Subba Rao 1965), Hosokenomimi (Chapman and Chapman 1980), Tsunomata (Arasaki and Arasaki 1983).

Description: This species is about 5–8 cm high, twice or thrice-forked from near the base and the segments are divergent at a wide angle, varying from 0.6–0.8 cm in diam., with rounded axils. At the base the frond tapers to a point, and the branches are slightly constricted at intervals. The cystocarps are surrounded with a raised ring, giving them an ocellate appearance (Holmes 1895).

Geographic distribution: NE Pacific (Alaska); Asia (China, Japan, Korea, Taiwan); Pacific Is (Hawaiian Is).

Uses: Used for foodstuff in Japan, Korea and China, as a jelly (Subba Rao 1965, Arasaki and Arasaki 1983, Tseng 1983, Bangmei and Abbott 1987, Chapman and Chapman 1980, Zemke-White and Ohno 1999, Hu 2005, Roo et al. 2007, Harrison 2013), and for production of carrageenan (Kang 1968, Okazaki 1971, Bonotto 1976), food tranquilizer, homogenizer (Hotta et al. 1989), and to treat intestinal disorders (Dawes 1981).

Extracts of this species have antitumor, immunomodulation (Zhou et al. 2004), antioxidant (Zhou et al. 2014), and antimicrobial (Baik and Kang 1986) activity.

Chondrus pinnulatus (Harvey) Okamura

Common names: Japanese: Kotojitsunomata, Hirakotiji (Chapman and Chapman 1980), Hirasaimi (Tokida 1954).

Description: Frond compressed, regularly, sometimes more or less irregularly, branched in dichotomous manner, often furnished with short pinnules on the margins; branchlets compressed, with a subulate or blunt apex (Tokida 1954).

Geographic distribution: Asia (Japan, Korea, Russia).

Uses: Used as food in Japan (Tokida 1954).

Chondrus yendoi Yamada & Mikami

Common name: Japanese: Ezo-tsunomata, Kuroha-ginnanso (Arasaki and Arasaki 1983).

Description: Relatively small red algae reaching up to a little over than 20 cm in length. It grows from a discoid holdfast and branches four or five times in a dichotomous, fan-like manner. The morphology is highly variable, especially the broadness of the thalli.

Geographic distribution: Asia (Japan, Korea, Russia).

Uses: Used as food in Japan (Arasaki and Arasaki 1983).

Digenea simplex (Wulfen) C. Agardh

Common names: Chinese: hai jen ts'ai, Tser koo ts'ai; Japanese: Makuri (Smith 1904), Makuri-nori (Subba Rao 1965), Kaijinso (Arasaki and Arasaki 1983).

Description: Thallus tufted, purple to dark brownish-red, stiff and wiry at the base, cartilaginous at the apices, axes with loose repeated irregular dichotomous branching, densely covered with bristly stiff multicellular unbranched short shoots on all sides except on the often denuded base; crustose holdfast.

Geographic distribution: Asia (Japan, Korea, Russia, Taiwan).

Uses: Used as food in Japan and SE Asia (Subba Rao 1965, Arasaki and Arasaki 1983).

Eucheuma amakusaense Okamura

Description: Frond terete or compressed, branched pinnately from the margins, or from the surface, often becoming verticillate; frond decumbent at first and later erect; projections from the surface trigonal in shape, branched pinnately, cartilaginous and fleshy color (Yoshida 1992).

Geographic distribution: Asia (China, Japan, the Philippines).

Uses: Used as food in Japan (Yoshida 1992).

Eucheuma arnoldii Weber-van Bosse

Common name: Filipino: Kanutkanot (Agngarayngay et al. 2005).

Description: *E. arnoldii* is characterized by a wide range of colors and shapes; bushy, cartilaginous, and fleshy thalli that are branched in all planes; the branches are, for the most part, vertically aligned (Kraft 1972).

Geographic distribution: Asia (China, Japan, Taiwan, Indonesia, Malaysia, the Philippines, Singapore, Vietnam); Australia.

Uses: Used as food in salads, soups, and mixed with meat (Agngarayngay et al. 2005).

Eucheuma cartilagineum

Description: Plants of the largest species are bushy, reach over 50 cm in length, and can weigh over a Kg. Thalli are often very fleshy and rigidly cartilaginous when fresh, and are erect from an encrusting base or prostrate to entangled and anchored at many points by haptera. Most bear whorled to scattered, simple to compound spines.

Geographic distribution: Asia (Japan).

Uses: In Japan this species is used as food (Zemke-White and Ohno 1999, Roo et al. 2007, Harrison 2013).

Gelidiophycus divaricatus (G. Martens) G.H. Boo, J.K. Park & S.M. Boo (formerly *Gelidium divaricatum*)

Common names: Chinese: Shihua, Tanmae (Santelices 1988); Japanese: Hime tengusa (Santelices 1988).

Description: Thallus purple-red (0.2–2.5 cm), repent, forming entangled mats on upper intertidal rocks; holdfast attached to the substratum by brush-like rhizoidal haptera; uprights mostly absent; decumbent axes terete to slightly compressed, irregularly to sub-distichously branched; lateral branches arising at almost right angles. Uniaxial structure with a more or less conspicuous dome-shaped apical cell; cortex consisting of 24 layers of ovoid pigmented cells; medulla consisting of both longitudinal filaments of elongated cells and unicellular rhizoidal filaments with thickened walls, secondarily formed from the innermost cortical and medullary cells. Life history triphasic with isomorphic generations; gametophytes dioecious; spermatangia in compressed with roundish apices; cystocarps spherical to ovoid on the apical parts of axes and branches, bilocular, with one ostiole on each surface of the branch; carposporangia elongated, terminal. Tetrasporangia irregularly arranged in apical sori and/or in stichidium-like compressed branchlets; cruciately divided (Guiry and Guiry 2014).

Geographic distribution: India Ocean (Maldives); Asia (China, Japan, Korea, Taiwan); SE Asia (the Philippines, Singapore); Pacific Is (Micronesia, Mariana Is).

Uses: Used for foodstuff and for agar production in China, Japan and Korea (Kang 1968, Okazaki 1971, Bonotto 1976, Chapman and Chapman 1980, Tseng 1983, Bangmei and Abbott 1987).

Used also for the treatment of dysentery, blood platelet diseases, stomach ailments, hemorrhoids and anal fistulas (Tseng and Zhang 1984, Tokuda et al. 1986).

Gelidium elegans Kützing

Common name: Japanese: Makusa (Yoshie-Stark et al. 2003).

Description: Thallus cartilaginous, somewhat crispate, 2 to 40 cm tall, composed by one or several erect axes, terete or compressed, distichously, plumously or irregularly branched, red to deep purple, although in some species it can be blackish. Erect axes arise from cylindrical or compressed, branched or unbranched creeping axes with numerous short haptera extending as individual axes or forming massive disk-like holdfasts. Plants sometimes occur in mats of algal turf with extensive basal parts or in more discrete clamps.

Geographic distribution: Asia (Japan, Korea, Russia, Taiwan).

Uses: Considered an edible species in Asia (Jeon et al. 2014).

Gelidium japonicum (Harvey) Okamura

Common names: Japanese: Onigusa, Oyakusa (Chapman and Chapman 1980), Onikusa (Arasaki and Arasaki 1983).

Description: Thallus large, more than 2 cm tall, with median line of main branches thickened like a midrib. Some branches are incurved, and others lack the character (Akatsuka 1981).

Geographic distribution: Asia (China, Japan, Korea, Taiwan).

Uses: Used as food in China (Bangmei and Abbott 1987).

Gelidium pacificum Okamura

Common name: Japanese: Obusa (Arasaki and Arasaki 1983).

Description: Thalli 5–35 cm long; the erect axes are flattened throughout and are 2–3 mm wide. Branching can be relatively sparse or abundant. When the branching is abundant, it can be up to four orders, with branches alternate or opposite along the axis. The main branches are elongated, flexuous, and slightly narrower than the erect axes but otherwise similar in appearance and branching pattern. Often branches of different lengths and different degrees of branching are intermixed along the same axis. However, all branches are flattened and broad. Even terminal pinnules can be up to 1 mm wide. Although the branching angle may vary from one specimen to another and from one order of branching to another, the branches of *G. pacificum* are directed upward, sometimes with the base of the branch slightly incurved (Santelices and Miyata 1995).

Geographic distribution: Asia (China, Japan, Korea, Russia, Taiwan).

Uses: In Korea is used for foodstuff and for agar production (Okazaki 1971, Bonotto 1976, Tseng 1983, Bangmei and Abbott 1987).

Gelidium vagum Okamura

Common name: Japanese: Yore-kusa (Tokida 1954).

Description: The present species is characterized by the branches not uniform in breadth, which are usually broader in primary segments (2–3 mm) and gradually or abruptly tapering upward to filiform segments (Tokida 1954).

Geographic distribution: Asia (China, Japan, Korea, Russia); NE Pacific (British Columbia to California).

Uses: Used as food in Japan (Tokida 1954); used in Japan, Korea and China for agar extraction (Chapman and Chapman 1980).

Gloiopeltis complanata (Harvey) Yamada

Common name: Japanese: Hana-fu-nori.

Description: Rusty red to golden yellow red alga that grows up to 5 cm tall. Several smooth, narrow cylindrical branches arise from a common basal crust. This crust remains throughout the year and produces a new plant every spring. The thalli are spineless and rubbery in texture, with infrequent forking branchlets (Yamada 1932).

Geographic distribution: Asia (China, Japan, Korea, Taiwan, the Philippines).

Uses: Used as food in Japan and Korea (Kang 1968, Bonotto 1976, Toma 1994).

Gloiopeltis furcata (Postels & Ruprecht) J. Agardh

Synonym: *Gloiopeltis coliformis*

Common names: Chinese: Chi tsai, Chiao tsai, Kaau tsoi, Hung tsai, Lu kio tsai, Hung tsoi (Madlener 1977); English: Cretan sea plant (Rhoads and Zunic 1978); Japanese: Kita-funori, Fukuro-funori, Funori (Tokida 1954); Korean: Bultunggasari (Sohn 1998).

Description: Rusty red to golden yellow red alga that grows up to 5 cm tall; several smooth, narrow cylindrical branches arise from a common basal crust. This crust remains throughout the year and produces a new plant every spring. The thalli are spineless and rubbery in texture, with infrequent forking branchlets (Lee et al. 1996).

Geographic distribution: Asia (China, Taiwan, Korea, Japan, Russia); Bering Sea and Aleutian Is; NE Pacific (Alaska to Baja California, Mexico).

Uses: The genus *Gloiopeltis* has a wide distribution throughout the north China coast. People in this region often individually collect these species as food. There are many methods of preparation. In Shandong, *Gloiopeltis* is fried with noodles and eaten hot. In southern Fujian, it is frequently steamed ("zheng"), and the resulting gelatinous cake is then fried in a little oil. These seaweeds are also cooked with fish (Chapman and Chapman 1980, Bangmei and Abbott 1987). In Korea this species is used in mixed soup, and seasoning (Kang 1968, Bonotto 1976, Madlener 1977, Sohn 1998).

G. furcata has long been utilized as a food source in Asia (Japan, Vietnam) where it is also used as a sizing material in silk and other textile industries (Tokida 1954, Arasaki and Arasaki 1983, Zemke-White and Ohno 1999, Harrison 2013). *G. furcata* is also a raw material for textile binding. Studies now show that extracts of *G. furcata* inhibit the growth of several human cancer cell lines, and are able to significantly lower blood glucose levels (Kang 1968, Okazaki 1971, Bonotto 1976, Madlener 1977, Tseng 1983, Hotta et al. 1989).

Gloiopeltis tenax (Turner) Decaisne (formerly *Dumontia tenax*)

Common names: Chinese: Hai lo; English: Glueweed, Jelly moss; Japanese: Funori, Yanagi-funori (Chapman and Chapman 1980), Kobu-nori; Korean: Chamgasari (Sohn 1998).

Description: *G. tenax* is a rusty red to golden yellow red alga that grows up to 5 cm tall. Several smooth, narrow cylindrical branches arise from a common basal crust. This crust remains throughout the year and produces a new plant every spring. The thalli are spineless and rubbery in texture, with infrequent forking branchlets. *G. tenax* is very similar to two other species: *G. furcata* and *G. complanata*. The three species are grown, harvested, and often used interchangeably. *G. tenax* inhabits the mid to high intertidal zone of rocky coasts where it grows abundantly, limited only by available space (Chiang 1969).

Geographic distribution: Asia (China, Taiwan, Korea, Japan, Russia); Bering Sea and Aleutian Is; NE Pacific (Alaska to Baja California, Mexico).

Uses: In Korea this species is used in mixed soup, and seasoning (Tseng 1983, Sohn 1998, Zemke-White and Ohno 1999). *G. tenax* has long been utilized as a food source in Asia, namely in China and Taiwan (Chapman and Chapman 1980, Bangmei and Abbott 1987, Harrison 2013), where it is also used as a sizing material in silk and other textile industries. *G. tenax* is also a raw material for textile binding.

Studies now show that extracts of *G. tenax* inhibit the growth of several human cancer cell lines, and are able to significantly lower blood glucose levels (Tseng 1983, Tseng and Zhang 1984, Tokuda et al. 1986).

Gracilaria salicornia (C. Agardh) E.Y. Dawson

Synonym: *Corallopsis salicornia*

Common names: Filipino: Susueldot-baybay (Velasquez 1972), Samo sa lawod (Leyte 2015); Indonesian: Bulung-buka (Zaneveld 1955), Boeloeng (Chapman and Chapman 1980); Arab: Canot-canot; Ilo: Caocaoayan (Guiry and Guiry 2014); Filipino: Kinkintal (Domingo and Corrales 2001), Lunglonggangan (Agngarayngay et al. 2005); Gorilla ogo, Ogo, Robusta (Hart et al. 2014).

Description: Thalli semi-erect to erect bright yellow to dark green in color. Branching dichotomous to trichotomous, divaricately arranged. Branches distinctly divided into terete or cylindrical, sub-clavate to clavate segments which are swollen at the distal end and constricted at the base (FAO 1996).

Geographic distribution: Indian Ocean; Asia (China, Japan, Malaysia, the Philippines, Taiwan, Thailand, Singapore, Vietnam); Australia (Papua New Guinea); Pacific Is (Mariana Is, Micronesia, Fiji, Guam, Hawaiian Is, Palau, Solomon Is).

Uses: Used as food in salads, soups, and mixed with meat, in Hawaii, Thailand, Indonesia, the Philippines, Vietnam, and Japan (Zaneveld 1955, Subba Rao 1965, Velasquez 1972, Lewmanomont 1978, Chapman and Chapman 1980, Lewmanomont 1998, Zemke-White and Ohno 1999, Agngarayngay et al. 2005, Nang 2006, Roo et al. 2007, Lee 2008, Harrison 2013, Kılınç et al. 2013, Hart et al. 2014, Irianto and Syamdidi 2015, Leyte 2015).

Grateloupia asiatica S. Kawaguchi & H.W. Wang

Description: Spores germinate to form crusts, filaments and spherical structures that subsequently differentiate and develop into ramified upright red thalli (Adharini and Kim 2014).

Geographic distribution: Asia (China, Japan, Korea); SE Asia (Vietnam).

Uses: Considered an edible species in Japan and Korea (Terasaki et al. 2012, Adharini and Kim 2014).

Grateloupia divaricata Okamura

Synonym: *Grateloupia incurvata*

Common name: Japanese: Katanori (Tokida 1954, Chapman and Chapman 1980).

Description: Compressed, tufted, dark purplish brown or yellow brown fronds, to 120 mm high, main axis 1–4 mm broad. Axes and branchlets tapered at base and apex.

Geographic distribution: Asia (China, Japan, Korea, and Russia); SE Asia (the Philippines, Vietnam); Australia.

Uses: Used for foodstuff in Japan and Korea (Tokida 1954, Kang 1968, Okazaki 1971, Bonotto 1976, Hotta et al. 1989).

Grateloupia elliptica Holmes (formerly *Pachymeniopsis elliptica*)

Description: Thallus flat, membranous, with short stipe, elliptical fronds, dark red or brown.

Geographic distribution: Asia (Japan, Korea).

Uses: Used for foodstuff in Korea (Kang 1968, Okazaki 1971, Bonotto 1976).

Extracts of this species have anti-inflammatory (Khan et al. 2008), and antimicrobial (Baik and Kang 1986) activity.

Grateloupia lanceolata (Okamura) Kawaguchi

Common name: Fudaraku

Description: Thallus flat, membranous, with short stipe, fronds lanceolate, red, yellow, or green shades.

Geographic distribution: NE Atlantic (France); NE Pacific (California); NW Pacific (China, Japan).

Uses: Used for human consumption (Abowei and Ezekiel 2013, Seo et al. 2013).

Grateloupia turuturu Yamada

Synonym: *Halymenia sinensis*

Common names: English: Devil's tongue weed (Bunker et al. 2010); Korean: Jinuari (Sohn 1998); Portuguese: Ratanho (Pereira 2009b).

Description: Thallus flat, membranous, with short stipe, the single fronds linear to broad-lanceolate, undivided or irregularly dividing from the base, narrowing towards the base as well as the stipe; sometimes proliferating on the margins and the surface; consistency gelatinous-slippery but firm; discoid holdfast; violet to crimson-red, often greenish at the top thallus (Braune and Guiry 2011, Pereira 2015a).

Geographic distribution: NE Atlantic; Atlantic Is (Canary Is); Tropical and subtropical W and SE Atlantic; Mediterranean; Asia (China, Japan, Korea, Russia); Australia and New Zealand.

Uses: Used as food in China (Bangmei and Abbott 1987); in Japan, is commonly used as a sea vegetable (Fujiwara-Arasaki et al. 1984, Munier et al. 2013); and in Korea is used in salads (Sohn 1998); also considered an edible species in Portugal (Rodrigues et al. 2015b).

Commercially used for carrageenan-production (Braune and Guiry 2011). Carrageenan of *G. turuturu* also showed anticoagulant activity (Efimov et al. 1983). *G. turuturu* is also characterized by its richness in dietary fiber (nearly 60% DW) and therefore appears to be a good source of food fiber for human consumption (Denis et al. 2010, Rodrigues et al. 2015b). This is of interest as the beneficial effect of fiber on health is already well-known. This seaweed is also rich in proteins, like *Palmaria palmata,* another red alga now authorized in France as a sea vegetable. Its lipid content is low, like all red seaweeds used in human nutrition, and its eicosapentaenoic acid content is similar to those reported for edible red seaweeds such as *Chondrus cripus* or *Gracilaria verrucosa* (Fleurence et al. 1994, Pereira 2015a).

Hydropuntia fisheri (B.M. Xia & I.A. Abbott) M.J. Wynne
(formerly *Gracilaria fisheri*)

Description: Thalli are terete to slightly compressed, bushy or prostrate arising from discoid holdfast or creeping multicellular haptera. Cortex and medulla and growth pattern are the same as described for the genus *Gracilaria* (Gurgel and Fredericq 2004).

Geographic distribution: Asia (China, Malaysia, Myanmar, the Philippines, Singapore, Thailand, Vietnam).

Uses: Used as food and for agar extraction in Thailand (Lewmanomont 1998, Zemke-White and Ohno 1999, Roo et al. 2007, Benjama and Masniyom 2012, Harrison 2013).

Iridophycus subdichotomus Nagai

Common name: Japanese: Chishima-ginnan (Tokida 1954).

Description: Frond three-four times repeatedly sub-dichotomo-palmate (Tokida 1954).

Geographic distribution: Asia (Japan, Russia).

Uses: Used as food in Japan (Tokida 1954).

Laurencia tropica Yamada (formerly *Laurencia flexilis* var. *tropica*)

Description: Erect thalli, up to 40 cm high, attached by discoidal to rhizoidal to encrusting holdfasts; thalli sparingly to highly branched in all directions or bilateral; branches cylindrical to flat; ultimate branchlets blunt to truncate, often claviform, not much narrower than other branches.

Geographic distribution: Asia (China, Japan, Korea, Taiwan, the Philippines, Vietnam); Pacific Is (Micronesia, Mariana Is).

Uses: Used as food in China (Bangmei and Abbott 1987).

Lomentaria catenata Harvey

Description: Thallus erect, tufted, repeated irregular opposite branching, tubular-hollow, filled with watery mucus (Braune and Guiry 2011).

Geographic distribution: Asia (China, Japan, Korea); Australia.

Uses: Used as food (Gray 2015), and produces an agarose-carrageenan hybrid polysaccharide (Takano et al. 1994); extracts have antioxidant (Kim et al. 2008), anticoagulant (Pushpamali et al. 2008), antimicrobial (Kanagasabhapathy et al. 2008) activity.

Mastocarpus pacificus (Kjellman) L.P. Perestenko

Synonyms: *Gigartina ochotensis, G. pacifica*

Common name: Japanese: Hoso-ibonori (Tokida 1954).

Description: Fronds narrow linear throughout or narrowly cuneate above, marginal papillae usually rather sparse or almost lacking while the plant is sterile (Tokida 1954).

Geographic distribution: Asia (Japan, Russia, Taiwan); NE Pacific (Alaska).

Uses: Used as food in Japan (Tokida 1954).

Mazzaella laminarioides (Bory de Saint-Vincent) Fredericq

Synonyms: *Iridophycus cornucopiae, Iridaea laminarioides*

Common names: Japanese: Kuroba-ginnansô, Atsuba-ginnansô (Tokida 1954); Spanish: Luga, Luga corta, Luga cuchara.

Description: Numerous thick, golden-brown, erect, fleshy, petal-shaped blades arise from a common basal sheet which is often woven tightly into the substratum it occupies (Tokida 1954).

Geographic distribution: Atlantic Is (Gough Is); Mediterranean; Asia (Japan, Russia); NE Pacific (Alaska to California); SE Pacific (Chile); Antarctic Is.

Uses: Used as food in Japan (Tokida 1954); is collected as a raw ingredient of carrageenan production in Chile (Gomez and Westermeier 1991, Buschmann et al. 2005).

Nemalion multifidum (Lyngbye) Chauvin

Common name: Japanese: Tsukomo-nori (Chapman and Chapman 1980).

Description: The thalli is golden brown, cylindrical with axes that can reach one meter long and 3.5 mm in diam. They are tuned to the apex, simple to some branching dichotomous; are formed by a core of many fine filaments, colorless, parallel to each other, and an outer shell of pigmented filaments; chloroplasts starry, red violated and with a pyrenoid.

Geographic distribution: N Atlantic; Baltic Sea; Asia (Japan).

Uses: Used as food in Japan (Chapman and Chapman 1980).

Nemalion vermiculare Suringar

Synonym: *Nemalion elminthoides* var. *vermiculare*

Common names: English: Crop of threads, Sea noodles, Threadweed; Japanese: Somen-nori (Smith 1904), Tsukomo-nori, Umi-somen, Umisomen (Chapman and Chapman 1980); Korean: Guksunamul (Sohn 1998).

Description: Consists of multiple, tubular fronds, or thalli, attached at a discoid holdfast. Each frond, or thallus, is roughly 2 mm in diam. and up to 30 cm in length. Occasionally the thalli will be branched dichotomously in one or two places. *N. vermiculare* often feels slippery to the touch and is red to dark brownish purple in color (Masuda and Horiguchi 1988).

Geographic distribution: Asia (China, Japan, Korea, Russia).

Uses: Used as food in seaweed salad (fresh) in China, Japan and Korea (Zemke-White and Ohno 1999); preserved and used in soups in Japan (Smith 1904, Tokida 1954, Chapman and Chapman 1980, Tseng 1983, Bangmei and Abbott 1987, Sohn 1998, Hug 2005, Roo et al. 2007, Harrison 2013).

Neodilsea yendoana Tokida

Common names: Japanese: Akaba, Akahata (Tokida 1954).

Description: Foliose, linear-lanceolate to orbiculate, with distinct stipe and discoid holdfast; medulla of multinucleate, secondarily pit-connected, stellate cells with long arms, mixed with rhizoidal filaments; inner cortical cells oblong to stellate; outer cortical cells uninucleate, progressively smaller toward cuticle; markedly thinner frond; dioecious (Tokida 1954).

Geographic distribution: Asia (Japan, Russia).

Uses: Used as food in Japan (Tokida 1954). Provides raw material for "Ginnansô" which belongs to the group of "algal slimes" similar to funoran and has similar uses in E Asia on account of its adhesive properties (Hoppe 1979).

Neohypophyllum middendorffii (Ruprecht) M.J. Wynne

Synonym: *Hypophyllum middendorffii*

Common names: Japanese: Chikaputsuro, Setakemaa (Tokida 1954).

Description: Similar to *Delesseria* spp.

Geographic distribution: Asia (Japan, Russia); NE Pacific (Alaska).

Uses: Used as food in Japan (Tokida 1954).

Neorhodomela larix (Turner) Masuda

Common names: English: Black pine (Lindeberg and Lindstrom 2014b), Japanese: Fujimatsumo (Tokida 1954).

Description: Thallus is stout, black to brownish black, reaching 30 cm tall. The main axis has stout, unbranched determinate branches whorled around it like a bottle brush; indeterminate branches are rare (Lindeberg and Lindstrom 2014b).

Geographic distribution: Asia (Bearing Sea; Japan, Korea, Russia); NE Pacific (Alaska to California).

Uses: Used as food in Japan (Tokida 1954).

Odonthalia corymbifera (S.G. Gmelin) Greville

Common name: Japanese: Hakesaki-nokogirihiba (Tokida 1954).

Description: Frond flat, midrib absent; reproductive organs on minute marginal branchlets (Tokida 1954).

Geographic distribution: Asia (Japan, Korea, Russia).

Uses: Used as food in Japan (Tokida 1954).

Odonthalia kamtschatica (Ruprecht) J. Agardh

Common name: Japanese: Kamchakka-nokogirihiba (Tokida 1954).

Description: This species has an expansive thalli, conspicuous midribs and cystocarps arranged in a flexuose-racemose manner (Tokida 1954, Masuda 1981).

Geographic distribution: Asia (Japan, Russia); NE Pacific (Alaska, British Columbia).

Uses: Used as food in Japan (Tokida 1954).

Odonthalia floccosa (Esper) Falkenberg

Common names: English: Bewildering brush, Sea brush (Lindeberg and Lindstrom 2014c); Japanese: Fusa-nokogirihiba (Tokida 1954).

Description: Thallus is dark red to brownish black (tips sometimes bleaching blond), growing to 40 cm tall. Branching is profuse and alternate, with branches cylindrical to somewhat flattened basally, arranged in two rows on opposite sides of the axis (distichous branching). Tertiary branches bear clusters of short, pointed branchlets. Holdfast is discoidal (Lindeberg and Lindstrom 2014c).

Geographic distribution: Asia (Bearing Sea, Japan); NE Pacific (Alaska to California).

Uses: Used as food in Japan (Tokida 1954).

Polyopes affinis (Harvey) Kawaguchi & Wang

Synonyms: *Carpopeltis affinis, Grateloupia affinis*

Common names: Hawaiian: Limu pala-wai, Limu li-pala wai (MacCaughey 1918); Japanese: Come-nori (Yendo 1902, Smith 1904), Kome-nori (Chapman and Chapman 1980); Matsunori (Chapman and Chapman 1980); Korean: Chamkkama (Song et al. 2013).

Description: Thallus stiped, branching repeated-alternating or also fan-shaped in one plane, branches compressed-flat, with or without distinct midrib in the lower axis; texture membranous to fleshy-firm; discoid holdfast (Braune and Guiry 2011).

Geographic distribution: Asia (China, Japan, Korea, Indonesia, the Philippines).

Uses: Used as food in Japan and Korea (Chapman and Chapman 1980, Song et al. 2013). These plants are dried, and after dipping in fresh water are eaten with vinegar or soy; their use is not very general (Yendo 1902). This alga is also consumed in Hawaii (MacCaughey 1918).

Polyopes lancifolius (Harvey) Kawaguchi & Wang
(formerly *Grateloupia okamurae*)

Description: Thalli purple-red, erect, cartilaginous, compressed, 10–30 cm tall, dichotomously branched, main branches 3–5 cm broad, margins entire or with bifurcate branchlets, having cylindrical short stalked, thallus becomes crispy on drying. In cross section, cortical region consists of about 15 layers of cells; the innermost cortical cells are stellate, other cortical cells are elliptical to fusiform, progressively smaller toward cuticle. The medulla is loosely constructed, with numerous branched medullary filaments. Cystocarp compact, with many small carposporangia, on the upper parts of branches. Tetrasporangia are cruciately divided, in a nemathecium on the upper parts of branchlets (Huang 2014g).

Geographic distribution: N Atlantic (Channel Is); Asia (China, Japan, Korea, Taiwan).

Uses: Used for foodstuff (Kang 1968, Okazaki 1971, Bonotto 1976).

Polyopes prolifer (Hariot) Kawaguchi & Wang

Synonyms: *Carpopeltis flabellata*, *Grateloupia flabellata*

Common name: Japanese: Kome-nori (Subba Rao 1965, Chapman and Chapman 1980).

Description: Small dark red fan-shaped seaweed, 5–6 cm long, thalli compressed filiform, repeatedly dichotomously branched; growing on intertidal rocks (Hu 2005).

Geographic distribution: Asia (China, Japan, Korea, the Philippines, Taiwan).

Uses: Used as food in Japan (Subba Rao 1965, Chapman and Chapman 1980, Hu 2005).

Polysiphonia morrowii Harvey

Common name: Japanese: Moroitogusa (Tokida 1954).

Description: Thallus developing a primary erect system of cylindrical polysiphonous branches, mostly having indeterminate growth; prostrate branches often developed secondarily by attaching decumbent branches or determinate attaching branches; rhizoids, mostly unicellular, formed by pericentral cells, from which they are usually separated by a cell wall and pit connection, on ventral sides of decumbent and attaching branches in most species, sometimes epiphytic species forming a discoid cellular attachment derived from germling; tetrasporangia on stichidial branchlets arising tufts in branchlet axils (Tokida 1954).

Geographic distribution: Mediterranean; Asia (China, Japan, Korea, Russia); New Zealand; SE Pacific (Chile).

Uses: Used as food in Japan (Tokida 1954).

Porphyra akasakae (formerly *P. akasakai*)

Common name: Japanese: Murone-amanori (Tokida and Hirose 1975).

Description: This species grows in the tidal belt along the coast of Kesen-numa Bay and Shizugawa Bay, Miyagi Prefecture (Japan), and in waters of high salinity. Cultivation of this species is only practiced in these bays. The frond is variously shaped, sub-orbiculate, elliptical, lanceolate, oblanceolate or linear, and deeply undulates on the margins. Fertile fronds are dioecious. Female fronds are generally wide, ranging in size from 9 cm long by 9 cm wide to 12.5 cm long by 1.5 cm wide, whereas male frond are generally narrow, ranging in size from 10 cm long by 8 cm wide to 24 cm long by 1 cm wide. Monospores are formed in young fronds, 0.1–1 mm long, at their early stages of growth. The carposporangial areas are dark reddish brown in color, being formed uniformly over the whole surface of the upper part of the female frond, whereas the antheridial area is pale yellowish in color (Tokida and Hirose 1975).

Geographic distribution: Asia (Japan, Korea).

Uses: Used as food in Japan (Harrison 2013).

Porphyra angusta Okamura & Ueda

Common name: Japanese: Kosuji-nori (Chapman and Chapman 1980).

Description: Thallus a foliose blade, one or two cells in thickness and ranging in size from a few mm to several cm in length. Blades arise singly from a small discoid holdfast; stipe absent or minute; basal cells have rhizoids; blades ranging in morphology from orbicular to linear, with margins ruffled.

Geographic distribution: Asia (Japan, Korea, Taiwan).

Uses: Used as food in Japan (Chapman and Chapman 1980).

Pyropia dentata (Kjellman) N. Kikuchi & M. Miyata

Synonym: *Porphyra dentata*

Common name: Japanese: Oni-amonari (Arasaki and Arasaki 1983).

Description: Thallus a foliose blade, one or two cells in thickness and ranging in size from a few mm to over 3 m in length. Blades arise singly from a small discoid holdfast; stipe absent or minute. Basal cells have rhizoids; linear Blades, with margins smooth or dentate.

Geographic distribution: Asia (China, Japan, Taiwan).

Uses: Used for foodstuff in Asia (Kang 1968, Okazaki 1971, Bonotto 1976, Tseng 1983, Bangmei and Abbott 1987).

Used for treatment of goiter, cough, bronchitis, edema and measles (Oh et al. 1990), and for prevention of scurvy (Dawes 1981).

Porphyra marginata C.K. Tseng & T.J. Chang

Description: Thalli generally 12–24 cm, sometimes reaching 40 cm in height, light or yellowish brownish-purple, with shortly stiped blade arising from a discoid holdfast; the blades orbiculate or sub-orbiculate, perforated, with cordate or umbilicate base and undulate to much folded marginal portions, membranaceous and monostromatic (Tseng and Chang 1958).

Geographic distribution: Asia (China).

Uses: Used as food in China (Bangmei and Abbott 1987).

Porphyra okamurae Ueda

Common name: Kumo-nori.

Description: Fronds typically grow up to 20 cm long and are irregularly lobed with a central holdfast. Fronds have a double layer of tissue, but range from being thin to broad, and are usually divided into lobes.

Geographic distribution: Asia (Japan, Korea).

Uses: Used as food in Japan (Chapman and Chapman 1980).

Porphyra ochotensis Nagai

Common name: Japanese: Ana-amanori (Tokida 1954).

Description: Plants monostromatic with vegetative blades more than 60 μ thick (Tokida 1954).

Geographic distribution: Asia (Japan, Russia); NE Pacific (Alaska).

Uses: Used as food in Japan (Tokida 1954).

Pterosiphonia bipinnata (Postels & Ruprecht) Falkenberg

Common names: English: Black tassel (Lindeberg and Lindstrom 2014d); Japanese: Itoyanagi (Tokida 1954).

Description: This species is light to dark red in color, and grows to at least 12 cm tall. The branches are cylindrical and near the tips of the branches, the terminal orders of branching are all in one plane (rather than coming off all around the axis as in Polysiphonia). You should be able to distinguish this branching pattern by spreading a branch tip out on your wet finger and then examining it carefully with a 10x field lens. Many branches are nearly the size of the axis bearing them, and thus branching can appear to be pseudodichotomous (Lindeberg and Lindstrom 2014d, Klinkenberg 2016).

Geographic distribution: Asia (Japan, Russia); NE Pacific (Alaska to California).

Uses: Used as food in Japan (Tokida 1954).

Pyropia onoi (Ueda) N. Kikuchi & M. Miyata (formerly *Porphyra onoi*)

Description: Thalli monostromatic, linear, ovate, and orbicular or funnel shaped margins entire or dentate, planar, undulate, or ruffled.

Geographic distribution: Asia (Japan, Russia).

Uses: Used as food in Japan (Chapman and Chapman 1980).

Pterocladiella capillacea (S.G. Gmelin) Santelices & Hommersand (formerly *Pterocladia capillacea*)

Common names: Chinese: Yimaocai (Santelices 1988); Hawaiian: Limu loloa (Reed 1906, MacCaughey 1918); Japanese: Obakusa (Arasaki and Arasaki 1983), Japanese: Kata-obakusa (Shimada and Masuda 2002); English: Branched Wing Weed (Bunker et al. 2010); Korean: Kaeumu (Sohn 1998); Portuguese: Musgo (Neto et al. 2005, Pereira and Correia 2015).

Description: Cartilaginous, dark purplish-red, flattened fronds, 2 mm wide, 4–20 cm tall. Pinnate or bi-pinnate, often bare at base, branches opposite or alternate, often tapering at both ends; frequently give a "Christmas-tree" appearance; fronds in loose tuft from a rhizoidal base (Neto et al. 2005).

Geographic distribution: NE Atlantic; Atlantic Is (Azores, Canary Is, Madeira); Adriatic Sea; SE and SW Atlantic; Asia (China, Japan, Korea, Taiwan); Australia and New Zealand; Pacific Is (Hawaiian Is).

Uses: Used as food in China, Korea, and Japan (Bangmei and Abbott 1987, Sohn 1998, Zemke-White and Ohno 1999, Roo et al. 2007, Harrison 2013), and for agar extraction (Hu 2005, Neto et al. 2005, Pereira et al. 2013); used also by the Hawaiian people (Reed 1906, MacCaughey 1918).

Pyropia haitanensis (T.J. Chang & B.F. Zheng) N. Kikuchi & M. Miyata (formerly *Porphyra haitanensis*)

Common name: Chinese: Zicai (Lembi and Waaland 1988).

Description: *P. haitanensis* as a typical warm temperate zone species originally found in Pingtan, Putian, and Huian of Fujian province (China).

Because of its high nutritional and economic value among seaweeds, the cultivation of *P. haitanensis* has come to be one of the most important fisheries industries, and its artificial cultivation is now primarily conducted in Zhejiang, Fujian, and Guangdong provinces of China. However, till now, the cultivation of *P. haitanensis* still relies on natural populations, with very limited germplasm development and no genetic improvement. Serious germplasm degeneration and the frequent disease problems have restrained the development of the cultivation industry. Growing evidence from land plant studies has showed that the germplasm resource determines the quality and output of one product. Therefore, it is important to further investigate the genetics of *Porphyra*. Furthermore, it is urgent to carry out breeding studies and to cultivate elite species to raise the industrial economic efficiency and expand the scale of *P. haitanensis* cultivation (Yan et al. 2007).

Geographic distribution: Asia (China).

Uses: Used as food in China (Bangmei and Abbott 1987, Zemke-White and Ohno 1999, Roo et al. 2007, Harrison 2013).

Pyropia kuniedae (Kurogi) M.S. Hwang & H.G. Choi (formerly *Porphyra kuniedae*)

Common name: Japanese: Maruba-nori (Chapman and Chapman 1980).

Description: Leafy plants of *P. kuniedae* are rounding, ovate or reniform with very thin blade of about 20–30 µ thickness, and spermatangia and zygote-sporangia on the same blade.

Geographic distribution: Asia (Japan, Korea).

Uses: Used as food in Japan and Korea (Chapman and Chapman 1980, Zemke-White and Ohno 1999, Sohn 2006, Roo et al. 2007, Harrison 2013).

Pyropia pseudolinearis (Ueda) N. Kikuchi & M. Miyata, M.S. Hwang & H.G. Choi (formerly *Porphyra pseudolinearis*)

Common name: Japanese: Uppurui-nori (Tokida 1954, Arasaki and Arasaki 1983).

Description: Thallus is a linear blade, one cell layer thick, reaching about 40 cm long and 2 cm wide, usually reddish pink or reddish brown.

Geographic distribution: NE Pacific (Alaska); Asia (Japan).

Uses: Used as food in Japan (Tokida 1954, Arasaki and Arasaki 1983).

Pyropia seriata (Kjellman) N. Kikuchi & M. Miyata (formerly *Porphyra seriata*)

Common name: Japanese: Ichimatsu-nori (Chapman and Chapman 1980).

Description: Membranous olive to brown-purple fronds up to 20 cm long, irregularly lobed and split from central holdfast. The fronds have no definite shape, they can be narrow or broad, divided by lobes, thin, but always have a double layer of tissue.

Geographic distribution: Asia (China, Japan, Korea, Russia).

Uses: Used as food in Japan and Korea (Chapman and Chapman 1980, Zemke-White and Ohno 1999, Sohn 2006, Roo et al. 2007, Harrison 2013).

Pyropia suborbiculata (Kjellman) J.E. Sutherland, H.G. Choi, M.S. Hwang & W.A. Nelson (formerly *Porphyra suborbiculata*)

Common names: Chinese: Tsz Tsai, Tsu Tsoi, Tsu Tsai, Chi Tsai, Hung Tsoi, Hung Tsai, Chi Choy, Hai Tsai, Hai Tso (Madlener 1977), Zi-cai (Arasaki and Arasaki 1983); English: Red Laver (Madlener 1977); Filipino: Gamet (Agngarayngay et al. 2005); Japanese: Mambiama, Maruba-amanori (Madlener 1977), Iwanori (Arasaki and Arasaki 1983); Korean: Kim (Madlener 1977).

Description: Thallus membranous, orbicular, 2–6 cm high, to 10 cm broad, sessile, commonly aggregate, often overlapping each other, with smooth or slightly undulate margins, purple-red to dark brownish-red, often with lighter marginal stripe, cordate or sometimes cuneate at base with small discoid attachment, Blade monostromatic, thin, 40–70 μ; margins inrolled, edges dentate. In transverse section, cells 28–30 μ high and 15–25 μ broad, with single, stellate chloroplast. Monoecious, spermatangial and carposporangial patches develop along margins. Growing during autumn-winter season on the uppermost (splash zone) to middle intertidal rocks, exposed to strong wave action (Milstein et al. 2015).

Geographic distribution: Tropical and subtropical W Atlantic; Asia (China, Japan, Korea, Taiwan, the Philippines, Vietnam); Australia and New Zealand; NE Pacific.

Uses: Used as food in China, Japan, Korea, Vietnam and US, in salads, soups, and mixed with meat (Kang 1968, Okazaki 1971, Bonotto 1976, Madlener 1977, Arasaki and Arasaki 1983, Tseng 1983, Simoons 1990, Nang and Dinh 1998, Zemke-White and Ohno 1999, Agngarayngay et al. 2005, Roo et al. 2007).

Used for clearing lungs, and treatment in relieving tension and anxiety, pulmonary and lymphatic tuberculosis, goiter, toothache, high blood pressure and kidney-urinary problems (Oh et al. 1990).

Pyropia tenera (Kjellman) N. Kikuchi & M. Miyata, M.S. Hwang & H.G. Choi (formerly *Porphyra tenera*)

Common names: Chinese: Tsz Tsai, Tsu Tsoi, Tsu Tsai, Chi Tsai, Hung Tsoi, Hung Tsai, Chi Choy, Hai Tsai, Tai Tso (Madlener 1977), Zicai (Lembi and Waaland 1988); Japanese: Asakusa nori, Amanori, Hoshi-nori, Kuro-nori, Sushi nori, Chishima kuro-nori, Tisima (Madlener 1977); Korean: Kim (Madlener 1977); Russian: Nuru (Madlener 1977).

Description: A small red alga with thallus small, irregularly shaped, foliaceous (leaf-like) and membranaceous but tough attached to the substratum (e.g., rocks) by small discoid hold-fast; frond more or less divided, crinkled and undulate at edges; color green in its earlier stages, becoming brownish purple, or purplish-red (Levring et al. 1969).

Geographic distribution: Indian Ocean; NW Pacific (China, Japan, Korea).

Uses: Used for foodstuff in China, Japan and Korea (Kang 1968, Bonotto 1976, Madlener 1977, Arasaki and Arasaki 1983, Tseng 1983, Zemke-White and Ohno 1999, Hu 2005, Roo et al. 2007, Harrison 2013).

Extracts of this species have hypocholesterolemic (Tsuchiya 1969), antioxidant (Fujimoto and Kaneda 1980), and antiulcer (Sakagami et al. 1982) activity.

See recipes with *Porphyra* in Annex I.

Pyropia yezoensis (Ueda) M.S. Hwang & H.G. Choi (formerly *Porphyra yezoensis*)

Common names: Chinese: Zicai (Lembi and Waaland 1988); English: Open sea nori (Chapman and Chapman 1980); Japanese: Susab-nori (Chapman and Chapman 1980), Susabi-nori (Arasaki and Arasaki 1983), Amanori (Lembi and Waaland 1988); Korean: Kim (Lembi and Waaland 1988).

Description: It can be a deep red or purple with an elliptical or lanceolate shape. Individual plants may reach 15 to 35 cm in length and 20 to 30 cm in width. Most naturally growing *P. yezoensis* propagates on the surface of rocks around high tide and a little below.

Geographic distribution: Tropical and subtropical W Atlantic; Asia (China, Japan, Korea, Russia); NE Pacific (Alaska).

Uses: Used as food in China, Japan, Korea, Hawaii and Israel (Arasaki and Arasaki 1983, Hu 2005, Lipkin and Friedlander 2006, Roo et al. 2007, Harrison 2013, Hart et al. 2014).

See recipes with *Porphyra* in Annex I.

Rhodoglossum pulchrum (Kützing) Setchell & N.L. Gardner

Common names: Japanese: Akaba-ginnansô, Usuba-ginnansô (Tokida 1954).

Description: The frond tissue is composed of three layers, cortical, intermediate and medullary. The intermediate layer lies between the cortex of small sub-globular cells and the central medulla of filamentous cells mostly vertically arranged. It is composed of beautiful networks of fibrous cells (Tokida 1954).

Geographic distribution: Asia (Japan); NE Pacific (Alaska).

Uses: Used as food in Japan (Tokida 1954); provides raw material for Ginnansô which belongs to the group of "algal slimes" similar to funoran and has related uses in E Asia on account of its adhesive properties (Tokida 1954, Hoppe 1979).

Rhodomela sachalinensis Masuda

Synonym: *R. macracantha*

Common name: Japanese: Niretsu-fujimatsu (Tokida 1954).

Description: Frond gracile, black to brownish black, reaching 30 cm tall, with slender, rather sparsely arising branchlets (Tokida 1954).

Geographic distribution: Asia (Japan, Russia).

Uses: Used as food in Japan (Tokida 1954).

Sarcodia dentata (Suhr) R.E. Norris

Synonym: *Kallymenia dentata*

Common names: Tosaka, Tosaka nori (Smith 1904).

Description: Toothed, usually with sharp teeth pointing outwards.

Geographic distribution: Indian Ocean; Asia (Japan); SE Asia (Indonesia); Auckland Is.

Uses: Is preserved by drying, and is eaten as a condiment or mixed with soy-bean sauce (Smith 1904).

Turnerella mertensiana (Postels & Ruprecht) F. Schmitz

Common names: English: Red Sea-cabbage (Lindeberg and Lindstrom 2014); Japanese: Yezo-nameshi, Oba-sô (Tokida 1954).

Description: Thallus is a dark red blade. It lacks a stipe and attaches to the substratum by a small discoidal holdfast. Young blades are frequently undivided (entire) and nearly round, reaching at least 30 cm in diam., but older blades split and become tattered, irregular in shape, and very thick. The large, white "gland cells" in the cortex can be seen with the aid of a hand lens (Lindeberg and Lindstrom 2014f).

Geographic distribution: Asia (Japan, Korea, Russia); NE Pacific (Alaska, British Columbia).

Uses: Used as food in Japan (Tokida 1954).

Provides raw material for "Ginnansô" which belongs to the group of "algal slimes" similar to funoran and has related uses in E Asia on account of its adhesive properties (Hoppe 1979).

2.4 Indo-Pacific (India, Indochina, the Philippines, Indonesia, Australia, New Zealand)

2.4.1 Chlorophyta

Acetabularia major G. Martens

Description: Thalli light green, calcified, about 4–7 cm in height, stipes slender, rather rigid; disk cup-shaped, about 10–15 mm diam.; the mature cap composed of 60–85 joined, terminally tapered or rounded elongate rays, which are associated basally with whorls of enlargements, the basipetal inferior corona and the apical superior corona from which project sterile laterals (Huang 2014a).

Geographic distribution: Asia (China, Japan, Taiwan); SE Asia (Indonesia, Malaysia, the Philippines, Vietnam); Australia.

Uses: Used as food in Malaysia (Sidik et al. 2012).

Aegagropila linnaei Kützing

Synonym: *Cladophora aegagropila*

Common names: English: Lake Balls; Japanese: Marimo (Yoshii et al. 2004, Zimmerman 2012); Swedish: Getraggsalg (Tolstoy and Österlund 2003); Thai: Kai (Thiamdao et al. 2012).

Description: Lake Balls, or Marimo, as the Japanese refer to them, are a rare and unique growth form of the filamentous green algae species *A. linnaei*. They occur in only a few isolated habitats worldwide because, unlike most algae, the species lacks a desiccation (dryness) resistant life stage which would allow it to be carried to distant bodies of water. The balls are formed from a densely-packed clump of algal strands which grow outward in all directions, and can reach up to 25 cm in diam. New balls can be produced from the free-floating form of the same species, or from the breakup and re-growth of an old ball. Found in shallow lakes with sandy bottoms, gentle wave action rolls the clump around, forming a near-perfect sphere and allowing all sides of the ball to receive light for photosynthesis. Seen rolling lazily around the lake bottom, and even rising and falling on columns of warm water, the Marimo can almost seem sentient (Zimmerman 2012).

Geographic distribution: NE Atlantic; Asia (Japan); SE Asia (Indonesia); SW Asia (Sri Lanka, Turkey); Australia.

Uses: Edible species, known as "Kai" in northern Thailand, is an economically and ecologically important green alga (Thiamdao et al. 2012).

Boodlea composita (Harvey) F. Brand

Description: Thalli bright green, spongiose, composed of branched uniseriate filaments forming three-dimensional reticulate masses. Branching on the main axes is generally pinnate, alternate to irregular in certain portions with the branches anastomosing with one another by means of special attachment cells, somewhat matted, forming amorphous structures; primary axes measure 100–350 µ in diam. Well-developed rhizoid prominent in juvenile plants, less so in mature (Huang 2015a).

Geographic distribution: NE Atlantic; Atlantic Is (Bermuda, Cape Verde Is); Tropical and subtropical Atlantic; Caribbean; Indian Ocean; Asia (China, Japan, Taiwan); SE Asia (Indonesia, Malaysia, Thailand, the Philippines, Singapore, Vietnam); Australia and New Zealand; Pacific Is (Easter Is, Fiji, Mariana Is, Marshall Is, Micronesia, Hawaiian Is, Polynesia, Samoa; Solomon Is).

Uses: Considered an edible species (Rao 1979).

Caulerpa corynephora Montagne

Synonym: *Caulerpa racemosa* var. *corynephora*

Description: Thallus slender, stolons only rarely branched, erect fronds up to 14 cm high, rachis terete or slightly compressed, unbranched or only rarely so, with a naked basal part (0.5–1 cm) and a series of more or less distant (sub-) opposite branchlets; these ramelli gradually enlarge from base to apex (clavate), with regularly rounded apices (Coppejans and Prud'Homme Van Reine 1992).

Geographic distribution: Tropical and subtropical W Atlantic; Indian Ocean; Asia (Japan, Taiwan, Thailand, Indonesia, the Philippines, Singapore, Vietnam); Pacific Is (Marshall Is, Micronesia, Fiji, Hawaiian Is, Solomon Is).

Uses: Is commonly used in Thailand as a salad vegetable (Lewmanomont 1978, Lewmanomont 1998).

Caulerpa hodkinsoniae J. Agardh

Synonym: *Caulerpa annulata*

Description: Stolon course, 2–3.5 mm in diam., naked, cartilaginous, epilithic. Erect fronds medium to dark green, with simple or occasionally branched axes, 3–7 (25) cm high and 1.5–2 cm across, with the axis 1.5–3 mm in diam. and constricted just above the position of the opposite pairs of ramuli, moniliform from the base up with the upper segment. Ramuli oppositely and distichously arranged, elongate-ovoid to clavate, 4–10 mm long and 2–3.5 mm in diam., thick walled (often shiny when dried) (Womersley 1984).

Geographic distribution: Australia, Tasmania.

Uses: Edible species from Tasmania, Australia (Sanderson and Benedetto 1988).

Caulerpa racemosa var. *macrophysa* (Sonder ex Kützing) W.R. Taylor

Description: Thalli glass green, 3–5 cm tall, forming dense clusters consisting of the prostrate terete, naked branched stolon and erect, terete branches. The erect branches are simple or branched bearing crowded stalked spherical ramuli, 2–5 mm in diam. Thalli attached to sandy-muddy substratum by colorless rhizoidal holdfast. Thalli are coenocytic, strengthened by an internal system of branching cylindrical in growths of the wall (trabeculae) (Huang 2014b).

Geographic distribution: Tropical and subtropical W Atlantic; Indian Ocean; Asia (Japan, Taiwan, Thailand, Indonesia, the Philippines, Singapore, Vietnam); Pacific Is (Marshall Is, Micronesia, Fiji, Hawaiian Is, Solomon Is).

Uses: Is commonly used in Thailand as a salad vegetable (Lewmanomont 1978, Lewmanomont 1998).

Caulerpa brachypus Harvey

Common name: English: Mini caulerpa.

Description: Green leafy blade-like fronds attached by stem-like rhizoids to horizontal runner-like stolons. The thallus (leaf-like portion) is green sometimes with a yellowish margin, and is ribbon-like or tongue-like in appearance (Masterson 2007).

Geographic distribution: Indo-Pacific tropics and subtropics, including East and Southeast Asia, the Pacific Is, East Africa, the Indian Subcontinent and Indian Ocean, Australia and New Zealand.

Uses: Commonly eaten species (Deane 2014a).

Caulerpa brownii (C. Agardh) Endlicher

Common name: Sea rimu (Wassilieff 2012b).

Description: Stolon usually robust, 1–3 mm in diam., moderately densely covered with simple ramuli 0.5–2.5 mm long and 150–500 µ in diam., tapering abruptly to a spinous tip, epilithic or on jetty piles. Erect fronds medium to dark green, with simple or several times irregularly branched axes, usually 3–40 cm high and 3–8 mm across; axes terete, 0.5–1.5 mm in diam., densely covered throughout with irregularly placed ramuli. Ramuli simple on basal part of axes and 1.5–4 mm long, becoming basally furcate and often bifurcate over most of the axes, 3–4 mm long and 100–350 µ in diam., terete, upwardly curved, tapering close to their apices to a spinous tip (Womersley 1984).

Geographic distribution: Tasmania, Australia and New Zealand.

Uses: Sea rimu (*C. brownii*), is an edible species from Tasmania (Australia) and New Zealand, and looking very much like the foliage of a large tree (Sanderson and Benedetto 1988, Wassilieff 2012).

Caulerpa cactoides (Turner) C. Agardh

Description: Stolon very coarse, 2–5 mm in diam., naked, epilithic or in a sandy mud substratum. Erect fronds medium to dark green, with simple or occasionally branched axes usually 10–40 cm high and 1–5 cm across, with the axis 2–7 mm in diam. bearing opposite (occasionally alternate) distichously arranged vesiculate ramuli; lower part of axes usually bare and constricted in rough-water plants but not in calm-water plants, upper axis usually constricted above the pairs of ramuli. Ramuli usually clavate, sometimes sub-pyriform, 0.7–3 cm long and 3–10 mm in diam., thick walled (especially in rough-water plants) (Womersley 1984).

Geographic distribution: SW Asia (Bangladesh); SE Asia (Indonesia); Australia, Tasmania.

Uses: Edible species from Tasmania (Sanderson and Benedetto 1988).

Caulerpa chemnitzia (Esper) J.V. Lamouroux

Synonym: *Caulerpa peltata*

Common names: English: Big parasol green seaweed; Filipino: Ar-arusip, Butbutones, Saluysoy (Domingo and Corrales 2001); Pacific name (Tonga Is): Fuofua (Ostraff 2003).

Description: Thalli with stolons producing descending rhizoidal branches and ascending branches bearing several ramuli consisting of a short pedicel ending in a disk-like head. The disk-like head differentiates this species from the spherical ramuli of *C. racemosa* (Domingo and Corrales 2001).

Geographic distribution: Widely distributed in tropical seas (Braune and Guiry 2011).

Uses: Used as food in the preparation of salads and main dishes, in Indonesia, Malaysia, the Philippines and Tonga (Pacific Is) (Zaneveld 1955, 1959, Chapman and Chapman 1980, Arasaki and Arasaki 1983, Trono 1998, Zemke-White and Ohno 1999, Ostraff 2003, Roo et al. 2007, Lee 2008, Harrison 2013, Irianto and Syamdidi 2015).

Caulerpa fergusonii G. Murray

Description: Thallus epilithic, forming dense mats. Smooth rhizomes entangled, frequently branched, closely appressed to the substratum, 1 mm in diam. Rhizoidal branches very irregularly spaced from only a few mm to several cm distant, 1–4 mm long. Assimilators dark to medium green, rather densely set: 0.5–2.5 cm apart, slightly recurved alternately.

Geographic distribution: Indian Ocean; Asia (Japan, Indonesia, Malaysia, the Philippines); Australia (Papua New Guinea); Pacific Is (Fiji).

Uses: Used as food in Malaysia (Levring et al. 1969, Sidik et al. 2012).

Caulerpa flexilis J.V. Lamouroux ex C. Agardh

Description: Plants dark green, 50–400 mm tall, upright branches (axes) arise from a coarse, runner, covered with forked spines, short side branches in two rows along axes are covered with numerous, cylindrical ultimate branches (ramuli) forked at their bases (Edgar 2008).

Geographic distribution: Australia (Tasmania, Victoria).

Uses: Edible species from Tasmania (Sanderson and Benedetto 1988).

Caulerpa geminata Harvey

Description: Stolon slender to moderately robust, 0.7–2 mm in diam., naked, usually epilithic. Erect fronds light to medium green, with simple to several times branched axes, usually 2–15 cm high and 4–10 mm across, axes 0.7–1.5 mm in diam., basally usually bare and above loosely covered with vesiculate ramuli on all sides or sometimes distichous or almost so, not or slightly constricted between ramuli. Ramuli sub-spherical or ovoid to elongate-ovoid to clavate, rounded apically and basally constricted to a short pedicel on a knot or slightly raised area of the axis, 1.5–7 mm long and 1–3 mm in diam., thin walled (Womersley 1984).

Geographic distribution: Asia (Korea, Indonesia); Australia (Tasmania, Victoria); New Zealand; Pacific Is (Fiji, Solomon Is).

Uses: Edible species from Tasmania (Sanderson and Benedetto 1988).

Caulerpa lamourouxii (Turner) C. Agardh
(formerly *Caulerpa racemosa* var. *lamourouxii*)

Description: Thalli 5–8 cm tall, erect axes compressed, strap-shaped, simple or branching, with sharp edges, to 4 mm wide, issuing distichous ramuli from margins, either clavate, slender, slightly inflated distally and pedicillate or spherical and inconspicuously pedicillate, to 8 mm long, ramuli usually sparsely and irregularly distributed (especially the spherical forms), becoming more regularly distributed (especially the clavate forms); stolons slender, to 2 mm in diam., issuing long descending branches, to 5 mm long, with prominently branched rhizoids at the ends (Belleza and Liao 2007).

Geographic distribution: Mediterranean (Greece); Atlantic Is (Bermuda); Tropical and subtropical W and SW Atlantic; SW Asia (Arabian Gulf, Cyprus, Lebanon, Turkey); Asia (Japan); SE Asia (Indonesia, the Philippines, Vietnam); Australia, Pacific Is (Fiji).

Uses: Used as food in salads (Lee 2008).

Caulerpa lentillifera J. Agardh

Common names: English: Green caviar, Sea grapes; Filipino: Ararosip, Ararusip, Arurusip, Lato, Lelato (Agngarayngay et al. 2005); French: Raisins de la mer; Japanese: Kumejima, Umibudô, Umi-budō, and Umibudou; Spanish: Uva de mar, Caviar verde (Radulovich et al. 2013).

Description: The seaweed resembles bunches of little grapes. Each "grape" is tiny (0.1–0.2 cm) usually spherical on a stalk. The "grapes" are usually tightly packed on a vertical "stem", often forming a sausage-like shape (2–10 cm long). This species is distinguished by the distinct constriction where the "grape" attaches to the stalk. This bunch of "grapes" emerges from a long horizontal "stem" that creeps over the surface. Colors range from bright green to bluish and olive green (Tan 2008a).

Geographic distribution: Native to tropical areas of the Indian and Pacific Oceans, though it is found as an invasive species in other parts of the Pacific such as the California coast and Hawaii.

Uses: Edible seaweed, used in the preparation of salads in Asia (Trono and Toma 1997, Trono 1998, McHugh 2003, Agngarayngay et al. 2005, Roo et al. 2007, Rao 2010); used as food in Pakistan (Rahman 2002), the

Philippines (Zemke-White and Ohno 1999, Leyte 2015), and Malaysia (Matanjun et al. 2009, Ahmad et al. 2012, Harrison 2013).

Caulerpa longifolia C. Agardh

Description: Stolon course, 1–4 mm in diam., cartilaginous, naked, epilithic. Erect fronds medium to dark green, with simple or rarely branched axes, usually 15–65 cm high and 1–3 cm across; axes terete, naked below and 1–3 mm in diam. above, bearing normally five regular rows (occasionally four or six rows) of slender, terete, ramuli. Ramuli separated by about their basal width, 0.5–1.5 cm long and 300–600 µ in diam., linear, directed and curved upwards, tapering near their apices to a short, blunt or spinous tip (Womersley 1984).

Geographic distribution: Australia (Tasmania, Victoria).

Uses: Edible species from Tasmania (Sanderson and Benedetto 1988).

Caulerpa macrodisca Decaisne (formerly *Caulerpa peltata* var. *macrodisca*) (Fig. 68)

Description: The rhizoids give rise to a great number of green shoots, which are simple and phylloid, or exhibit axes bearing ramuli of various shapes (Zaneveld 1969).

Geographic distribution: SE Asia (Indonesia, the Philippines, Singapore, Thailand, Vietnam); SW Asia (Sri-Lanka); Pacific Is (Samona Is).

Uses: This variety is eaten in Bangka and Thailand (Lewmanomont 1978); in Luzon (Ilocos and Cagayan Provinces) it is used raw as a salad (Zaneveld 1969).

Caulerpa microphysa (Weber-van Bosse) Feldmann

Common name: Filipino: Ar-arusip (Domingo and Corrales 2001).

Description: Thalli forming dense clusters consisting of prostrate, naked, branched stolons bearing crowded, stalked, minute spherical ramuli. This species has the tiniest ramuli of all *Caulerpa* species (Domingo and Corrales 2001).

Geographic distribution: Tropical and subtropical W Atlantic; Atlantic Is (Bermuda); Indian Ocean, Asia (Japan, Thailand, Indonesia, the Philippines, Vietnam); Australia (Papua New Guinea); Pacific Is (Micronesia, Fiji, Hawaiian Is).

Uses: Used as food in the Philippines (Domingo and Corrales 2001).

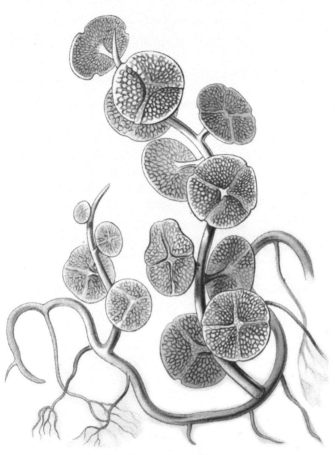

Figure 68. *Caulerpa macrodisca* Decaisne (Chlorophyta). Illustration from Ernst Haeckel (Plate 64, Siphoneae, Kunstformen der Natur, 1904).

Caulerpa obscura Sonder

Description: Stolon robust, 1–4 mm in diam., cartilaginous, bearing a sparse to moderate covering of simple or basally furcate, terete to tapering, spinous or rounded-tipped ramuli 1–4 mm long, and 200–300 μ in diam., directed forward towards the stolon apex, epilithic. Erect primary fronds dark green, with usually simple, occasionally branched axes, usually 10–30 cm high and 2–6 cm across; axes terete, 0.7–2 mm in diam., bearing on all sides dense second-order laterals 1–4 cm long which bear numerous ramuli irregularly arranged or sometimes tending to lie more or less in two rows; second-order laterals 350–450 μ in diam.; ramuli 2–10 mm long and 250–350 μ in diam., terete, simple or furcate just below their apices, tapering abruptly to the tip which bears 1–3 single or divided spines (Womersley 1984).

Geographic distribution: Australia (Tasmania, Victoria).

Uses: Edible species from Tasmania (Sanderson and Benedetto 1988).

Caulerpa racemosa (Forsskål) J. Agardh (Fig. 69)

Synonyms: *Caulerpa clavifera*, *Caulerpa racemosa* var. *clavifera*

Common names: English: Sea Grapes, Mouse plant (Madlener 1977), Green sea feathers (Novaczek 2001); Fijian: Fuofua, Nama, Nama levulevu (South 1995, Novaczek 2001); Japanese: Surikogidzuta (Madlener 1977); Filipino: Ararucip lai-lai (Galutira and Velasquez 1963), Ararusip (Agngarayngay et al. 2005); Pacific name (Tonga Is): Fuofua (Ostraff 2003); Hawaiian: Limu fuafua (Chapman and Chapman 1980).

Description: This plant has basal runners (like a root running along the seabed) up to 3 mm wide, anchored by roots. Upright branches are 3–10 cm tall and bear grape-like branchlets. *Caulerpa* is bright green and when seen underwater, may appear iridescent. Branchlets are variable in shape: they may be round or oval or shaped like the end of a trumpet, or be crowded together, or spaced apart. Regardless of the shape or size, all sea grapes are quite delicious. Some are more peppery tasting than others (Novaczek 2001).

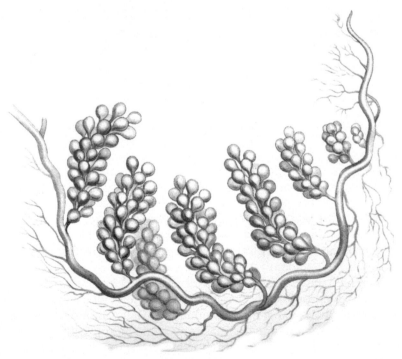

Figure 69. *Caulerpa racemosa* (Forsskål) J. Agardh (Chlorophyta). Illustration from Ernst Haeckel (Plate 64, Siphoneae, Kunstformen der Natur, 1904).

Geographic distribution: Common in tropical seas, Caribbean; NW Pacific (Japan); NE Pacific (Mexico); W Pacific and Indo-Pacific (the Philippines, Indonesia, Vietnam); Pacific Is (Micronesia); Indian Ocean; Australia, New Zealand; W Africa.

Uses: *Caulerpa* species are eaten raw or cooked in the Caribbean and Central America, Hawaii, Tonga and Cook Is (Chapman and Chapman 1980, N'Yeurt 1999, Ostraff 2003, Harrison 2013). Eat fresh on bread, dalo, yam or breadfruit. Use in salads or cook in coconut milk (Novaczek 2001). The aboriginal Yemei, tribe on Taiwan's Hungt'ou Hsii (Lanyu Is), use this alga as a delicacy by frying it in pork fat or peanut oil and then mincing and eating it (Bangmei and Abbott 1987). Considered a delicacy in the Philippines, is consumed in salads (Zaneveld 1955, Trono 1998, Agngarayngay et al. 2005, Roo et al. 2007). This species is also used as food in Pakistan, Bangladesh, Japan, S Pacific Is, Vietnam and Malaysia (Galutira and Velasquez 1963, Subba Rao 1965, Zemke-White and Ohno 1999, Rahman 2002, Roo et al. 2007, Ahmad et al. 2012, Harrison 2013, Pereira 2015a). This species is quite often eaten as a dessert after a rice meal (Johnston 1966).

Caulerpa is consumed as a salad in Fiji and Malaysia, the most popular way to eat it is to wash in fresh water and drain it immediately before serving, place it in a small bowl, pour several spoonfuls of diluted fresh coconut cream (miti) containing finely-chopped onion, sprinkle some kora (which is essentially a coconut "cheese") and add chopped fresh chillie (Sidik et al. 2012). This is then served to accompany fish and a root crop like boiled taro or cassava. Another popular way of preparing "nama" in Fiji includes marinating with lemon juice and then adding coconut cream (lolo) with some finely chopped chillie and canned fish, usually herring or pilchard (South 1993c). Fijians consider this preparation as a delicacy, and it is traditionally consumed on Sundays with the special meal served after church services (IMR 2011). In Bangladesh this species is used as food (Islam 1998); and is also eaten in New Zealand (Ostraff 2003).

They contain compounds that help reduce high blood pressure. They are also rich in folic acid and vitamins A, B_1 and C. In the Philippines these plants are used as a folk remedy for rheumatism (Novaczek 2001).

Caulerpa scalpelliformis (R. Brown ex Turner) C. Agardh (Fig. 70)

Description: Stolon slender, 0.5–1 mm in diam., in small forms of young plants, robust, 1.5–3 mm in diam., in large, rough-water plants, cartilaginous, naked, epilithic. Erect fronds medium to dark green, simple to occasionally branched, from 4–10 cm high and 3–6 mm broad in slender forms, to 20 cm high and 2–3 cm broad in robust plants, terete for the basal 1–3 cm, then strongly compressed with an axis 2–3 mm broad in slender plants to 4–8 (10) mm broad in robust plants, bearing alternately distichous, closely

Figure 70. *Caulerpa scalpelliformis* (R. Brown ex Turner) C. Agardh (Chlorophyta). Illustration from William H. Harvey (Plate XVII, Phycologia Australica, 1858–1863).

adjacent ramuli. Ramuli 2–4 mm long and 1–2 mm broad in slender plants, to 1–1.5 cm long and 3–5 mm broad in robust plants, strongly compressed, scalpelliform, basally broadest but often slightly constricted, usually slightly upwardly curved and with an acute angle between adjacent ramuli, tapering gradually over their lower half to three quarters, then abruptly to a distinct spinous tip (Womersley 1984).

Geographic distribution: Atlantic Is (Canary Is); Caribbean Is (Barbados); Tropical and subtropical W and E Atlantic; Indian Ocean; Asia (Japan, Indonesia, Singapore, Vietnam); Australia (Tasmania, Victoria).

Uses: Considered an edible species in Tasmania (Sanderson and Benedetto 1988), and potential food use in India (Kumar et al. 2011a).

Caulerpa serrulata (Forsskål) J. Agardh

Common names: English: Serrated green seaweed (Tan 2008), Toothed spiral; Filipino: Gal-galacgac (Domingo and Corrales 2001); Japanese: Yorezuta (Oshie-Stark et al. 2003); Pacific name (Tonga Is): Kaka (Ostraff 2003).

Description: Thalli made up of stolons giving rise to ascending branches with flattened, slightly curved or spirally twisted upper portion. Rhizoidal holdfasts attached to sandy or rocky substratum (Domingo and Corrales 2001).

Geographic distribution: Widely distributed in tropical seas (Braune and Guiry 2011).

Uses: In the Pacific Is (Tonga), Thailand, Malaysia, Singapore, and the Philippines, this seaweed is reported to be edible and used in soups, pickles and salads (Zaneveld 1955, 1959, Subba Rao 1965, Lewmanomont 1978, Ostraff 2003, Sidik et al. 2012, Irianto and Syamdidi 2015) and used in medicines as an antibacterial and antifungal agent as well as to lower blood pressure (Tan 2008b).

Caulerpa sertularioides (S.G. Gmelin) M.A. Howe

Common names: English: Green feather alga, Delicate feathery green seaweed; French: Caulerpe plume; Filipino: Salsalamagi (Domingo and Corrales 2001); Pacific name (Tonga Is): Louniu (Ostraff 2003); Spanish: Pluma de mar (Radulovich et al. 2013).

Description: Thallus to 6 cm tall, with terete stolons 0.25–1.0 mm in diam., bearing sparse erect branches which are simple or occasionally dichotomously divided, naked or branched at the base and bearing plumose, pinnate, undivided branchlets; branchlets cylindrical throughout, not constricted at the base, to 8 mm long and 200 µ in diam., up-curved, with pointed tips (N'Yeurt 1999).

Geographic distribution: Widely distributed in tropical seas (Braune and Guiry 2011).

Uses: *C. sertularioides* is an edible crop consumed by humans in some parts of the world, namely in the Caribbean and Central America (Zaneveld 1969, Sidik et al. 2012, Harrison 2013). This alga is consumed in a raw state and as salad (Levring et al. 1969); in the Philippines, the alga is used for dietary and medicinal purposes (Trono 1998, Zemke-White and Ohno 1999, Roo et al. 2007, Harrison 2013). In the Persian Gulf, this species is considered a potential edible alga (Mohammadi et al. 2013).

As a food source, *C. sertularioides* may provide important anti-oxidants. In fact, studies on the extracts from populations in Palau showed the highest anti-oxidant activity of any alga examined (Santoso et al. 2004). The green feather alga has also been sold, along with other macroalgae, to aquarium hobbyists in retail shops and via internet sites throughout the US and overseas (Stam et al. 2006).

In India, Bangladesh, Indonesia, Thailand and Tonga (Pacific Is), this species is used as food (Rao 1970, Lewmanomont 1978, Kaliaperumal et al. 1995, Islam 1998, Ostraff 2003, Irianto and Syamdidi 2015); the slightly sour and pungent nature of the young and fresh thalli gives it a spicy taste (Kaliaperumal et al. 1995).

Caulerpa taxifolia (M. Vahl) C. Agardh (Fig. 71)

Common name: Filipino: Lukay-lukay.

Description: Thalli with upright leaf-like fronds arising from creeping stolons. The fronds are compressed laterally and the small side branchlets

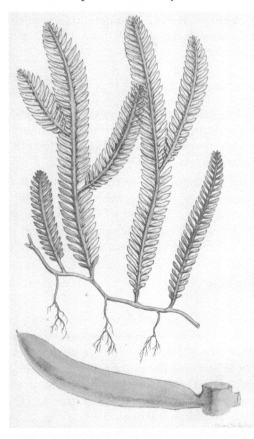

Figure 71. *Caulerpa taxifolia* (M. Vahl) C. Agardh (Chlorophyta). Illustration from William H. Harvey (CLXXVIII, Phycologia Australica, 1858–1863).

are constricted at the base (where they attach to the midrib of each frond), are opposite in their attachment to the midrib and curve upwards and narrow towards the tip. Frond diam. is 6–8 mm and frond length is usually 3–15 cm in the shallows, 40–60 cm in deeper waters (Braune and Guiry 2011, Pereira 2015a).

Geographic distribution: Caribbean Is, Gulf of Guinea, Red Sea, E African coast, Maldives, Seychelles, northern Indian Ocean coasts, southern China Sea, Japan, Hawaii, Fiji, New Caledonia and tropical/sub-tropical Australia; introduced in the Mediterranean Sea in 1984 from the Oceanographic Museum of Monaco (Galil 2006).

Uses: This green seaweed is reported to be edible, consumed in the Philippines in salads (Trono 1998, Zemke-White and Ohno 1999, Roo et al. 2007, Harrison 2013), to have antibacterial and antifungal properties, and used to treat tuberculosis and high blood pressure (Tan 2013).

Caulerpa trifaria Harvey

Description: Stolon usually moderately slender, 0.7–1.5 mm in diam., bearing scattered, short spines about 0.5 mm long, epilithic, on jetty piles or in sandy mud substratum. Erect fronds medium green, with simple or occasionally branched axes, usually 5–25 cm high and 4–12 mm across; axes terete, 400–700 μ in diam., with short spinous ramuli near their base, above bearing three (two when juvenile or rarely on mature fronds) regular rows of slender, terete ramuli. Ramuli separated by less than their basal width, 3–9 mm long and 200–300 μ in diam., terete, with a slight to moderate upward curve, tapering close to their apices to a short spinous tip (Womersley 1984).

Geographic distribution: Australia (Tasmania, Victoria).

Uses: Considered an edible species in Tasmania (Sanderson and Benedetto 1988).

Caulerpa veravalensis Thivy & V.D. Chauhan

Description: Plants stoloniferous, the branching stolon rather stout, 1.5–2.0 mm diam., with thick descending rhizoid-bearing branches at intervals of a few mm and erect branches at intervals of 1.0–2.5 cm; stalk of foliar branches 1.0–2.5 mm long, cylindrical, bulbous at the base; blade flat, linear to lanceolate, 4.2–21.5 mm wide, simple or occasionally forked, pinnately divided with a flat mid-rib of 1.0–2.0 mm width; ramuli flat, opposite to alternate, ascending, oblong, slightly arcuate, 1.0–1.5 mm wide, 5–7 mm long, with round apex, with sides parallel throughout or with base a little narrowed, 1.0–2.0 mm apart, occasionally having bifurcate apex (Thivy and Chauhan 1963).

Geographic distribution: SW Asia (India, Pakistan).

Uses: This species is reported as edible (Kumari et al. 2010, Gomez-Gutierrez et al. 2011, Kumar et al. 2011a).

Caulerpa vesiculifera (Harvey) Harvey

Description: Stolon moderately robust, 1–3 mm in diam., naked, epilithic. Erect fronds medium to dark green with simple or often with several branched axes, usually 8–35 cm high and 3–5 mm across, with axes 1–2 mm in diam. densely covered to their bases with vesiculate ramuli, 7–9 around the axes; ramuli with a slight to conspicuous, broadly conical, pedicel from the axis, ovoid above the constriction, 1–2.5 mm long and 0.7–1.5 mm in diam. (Womersley 1984).

Geographic distribution: Asia (the Philippines); Australia (Tasmania, Victoria).

Uses: Considered an edible species in Tasmania (Sanderson and Benedetto 1988).

Chaetomorpha aerea (Dillwyn) Kützing

Common name: Japanese: Tarugata-juzumo (Tokida 1954).

Description: Plants occur as numerous fines, single, dark green unbranched threads attached to rock or seagrass at their bases (Baldock 2007).

Geographic distribution: NE Atlantic; Adriatic Sea; Baltic Sea; Black Sea; Mediterranean Sea; Tropical and subtropical W Atlantic; Caribbean Sea; Indian Ocean; Asia (China, Japan, Korea, Taiwan, Indonesia, the Philippines, Singapore, Vietnam); Australia (Tasmania, Victoria); New Zealand; Pacific Is (Easter Is, Hawaiian Is).

Uses: Considered an edible species in the Caribbean Is and Tasmania (Radulovich et al. 2013, Sanderson and Benedetto 1988).

C. aerea is a source of lectin, a specific protein important for regulation of immunity system and cancer diagnosis (Milchakova 2011).

Chaetomorpha antennina (Bory de Saint-Vincent)

Synonym: *Chaetomorpha media*

Common names: Filipino: Riprippiis (Agngarayngay et al. 2005); Hawaiian: Limu hulu-ilio, Limu ilio, Limu manu (Reed 1906, MacCaughey 1918); Malay: Lumut-laut (Zaneveld 1950).

Description: Green plants growing in dense shaped brush tufts 3–5 cm, but may reach 10–16 cm or more, formed by numerous densely juxtaposed uniseriate filaments. Non-branched filaments comprised of large multinucleated cells and a single cross-linked chloroplast and many pyrenoids.

Geographic distribution: Atlantic Is; Tropical and Subtropical W Atlantic; SE Atlantic; Caribbean Sea; Indian Ocean; Asia (China, Taiwan); SE Asia (Indonesia, Malaysia, the Philippines, Thailand, Vietnam); Australia; Pacific Is (Samoa, Polynesia, Micronesia, Hawaiian Is, Mariana Is, Solomon Is).

Uses: Used as food in salads in Asia (Zaneveld 1950, 1959, Lewmanomont 1978, Kalesh 2003, Agngarayngay et al. 2005, Sidik et al. 2012); is also eaten in Hawaii (Reed 1906, MacCaughey 1918, Levring et al. 1969).

Chaetomorpha billardierii Kützing

Description: Light green, becoming yellow-green, loose-lying in extensive entangled masses (up to several meters across) of fairly straight filaments, in sheltered lower eulittoral to upper sublittoral habitats; filaments without attachment cells, of similar diam. throughout; cells with chloroplasts openly reticulate and provided of numerous pyrenoids (Womersley 1984).

Geographic distribution: Australia (Tasmania).

Uses: Considered an edible species in Tasmania (Sanderson and Benedetto 1988).

Chaetomorpha coliformis (Montagne) Kützing

Common name: English: Mermaids necklace (Kai Ho 2015).

Geographic distribution: SW Asia (India); Australia (Tasmania, Victoria); New Zealand; SE Pacific (Chile).

Uses: Considered an edible species in Tasmania (Sanderson and Benedetto 1988); this alga releases a burst of flavor when bitten with a taste that is similar to cucumbers, and used primarily as a garnish (Kai Ho 2015a).

Chaetomorpha linoides Kützing

Common name: Tamil: Nool pasi (Kaliaperumal et al. 1995).

Description: Plants bright green to yellowish-green; composed of unbranched filaments; plants twist together to form clumps or tangles; tangles remain quite rigid when removed from water (Kaliaperumal et al. 1995).

Geographic distribution: Atlantic Is (St. Helena); Tropical and subtropical W Atlantic; Indian Ocean; SE Pacific (Chile).

Uses: In India is eaten as salad or cooked with fish, meat, etc. (Kaliaperumal et al. 1995).

Chaetomorpha linum (O.F. Müller) Kützing (Fig. 72)

Common names: English: Flax brick weed (Bunker et al. 2010), Spaghetti algae (Barnes 2008); Japanese: Warakuzumo (Tokida 1954); Swedish: Krullig borsttråd (Tolstoy and Österlund 2003).

Description: Delicate green seaweed; also known as spaghetti algae, it grows as a filamentous loosely entangled mass. Usually free-floating, it may also

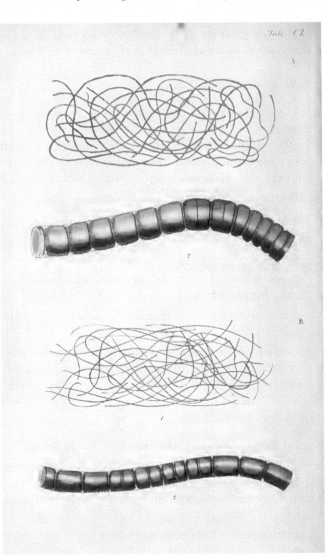

Figure 72. *Chaetomorpha linum* (O.F. Müller) Kützing (Chlorophyta). Illustration from William H. Harvey (Plate CL, Phycologia Britannica, 1846–1951).

be attached to rocks and shells. The filaments themselves are unbranched and usually between 5 and 30 cm in length. The unattached filaments are wiry, stiff and curled in appearance. It is bright light to dark green in color (Barnes 2008).

Geographic distribution: NE Atlantic; Atlantic Is (Azores, Bermuda, Canary Is, Madeira, Selvage Is); Adriatic Sea; Mediterranean Sea; Caribbean Is; Tropical and subtropical W and E Atlantic; Indian Ocean; Asia (China, Japan, Korea, Russia, Taiwan, Indonesia, the Philippines, Singapore, Thailand, Vietnam); Australia and New Zealand; Pacific Is (Micronesia).

Uses: Considered an edible species in India (Rao 1970), and the Philippines (Davidson 2004).

C. linum has antipyretics and antiseptic properties; used in treatment of asthma and cough and as parts of traditional cosmetics (refreshing liquid, skin powder, pulp form for skin sunlight protection) in the Pacific Is (Milchakova 2011).

Cladophora glomerata (Linnaeus) Kützing

Common names: Swedish: Grönslick (Tolstoy and Österlund 2003); Thai: Kai (Thiamdao et al. 2012).

Description: Plants are up to 20 cm high and light to dark green in color. The texture of thallus is soft and slightly mucilage. Plants usually formed in dense tufts well branched. Rhizoids are primary and adventitious and descend from the bases of thallus or from the lower segments of the fronds. Primary branches fused together or not and branched in dichotomous manner (Pereira 2015a).

Geographic distribution: Adriatic Sea; NE Atlantic; Atlantic Is (Iceland, Madeira); Mediterranean; W and SW Atlantic; SW Asia (Arabian Gulf, Cyprus, India, Iraq, Pakistan, Turkey); Asia (Japan, Taiwan); SE Asia (Thailand, Vietnam); Australia; Pacific Is (Samoa, Hawaiian Is).

Uses: Edible *Cladophora*, known as "Kai" in northern Thailand, is an economically and ecologically important green alga (Thiamdao et al. 2012, Pereira 2015a).

Cladophora vagabunda (Linnaeus) Hoek

Synonyms: *Cladophora fascicularis, C. monumentalis*

Description: Thallus filamentous, spongy, soft tufts, anywhere from 5–50 cm in length. Branches mostly on one side, at times, strongly re-branched and claw-like, maximum number of branches at joints one to four, rarely five; rhizoids fine, often connecting to adjacent filaments by hapteroid-like rhizoids. *C. vagabunda* grows from 4 cm diam. on wave-swept habitats to 30 cm high in protected habitats; pale green to grass green (Russel and Balazs 2000).

Geographic distribution: NE Atlantic; Atlantic Is (Bermuda, Canary Is, Cape Verde Is, Madeira, Selvage Is); Tropical and subtropical W Atlantic; Caribbean; Adriatic Sea; Indian Ocean; Asia (China, Japan, Korea, Taiwan); SE Asia (Malaysia, Thailand, the Philippines, Singapore, Vietnam); Australia and New Zealand; Pacific Is (Easter Is, Fiji, Mariana Is, Marshall Is, Micronesia, Hawaiian Is, Polynesia, Samoa; Solomon Is).

Uses: Considered an edible species in Peru (Rao 1979, Browman 1980, Manivannan et al. 2008, Radulovich et al. 2013) and in India is eaten with shrimp and salt (Kaliaperumal et al. 1995).

C. vagabunda is the source of valuable protein; used in pharmaceutical medicines; 5% aquatic extract improves blood formula under adaptation to stress (Milchakova 2011).

Chaetomorpha valida (J.D. Hooker & Harvey) Kützing

Description: Thallus dark green, loose-lying and forming entangled masses to 15 cm across and 1–5 cm thick, of curved, often crisped, rigid filaments, in the uppermost sublittoral of sheltered habitats; filaments without attachment cells and of similar diam. throughout (Womersley 1984).

Geographic distribution: Australia (Tasmania, Victoria); New Zealand; Pacific Is (Fiji).

Uses: Considered an edible species in Tasmania (Sanderson and Benedetto 1988).

Cladophora rugulosa G. Martens

Common name: Filipino: Kulkulasisi (Agngarayngay et al. 2005).

Description: Dark green filaments some of which are branched.

Geographic distribution: S Africa; Asia (China, Japan, Taiwan, the Philippines, Vietnam); Australia.

Uses: Used as food in salads in the Philippines (Agngarayngay et al. 2005).

Cladophora rupestris (Linnaeus) Kützing (Fig. 73)

Common names: English: Slobán, Common green branched Weed (Bunker et al. 2010); Filipino: Kulkulasisi (Agngarayngay et al. 2005); Swedish: Bergborsting (Tolstoy and Österlund 2003).

Description: The bundles, dark green, made up of rough, straight threads have a coarse and rather rigid feel (Braune and Guiry 2011).

Figure 73. *Cladophora rupestris* (Linnaeus) Kützing (Chlorophyta). Illustration from William H. Harvey (Phycologia Britannica, 1846–1951).

Geographic distribution: NE Atlantic; Arctic (White Sea); Adriatic Sea; W Mediterranean; SW Atlantic (Brazil); Asia (China, Japan, Korea, the Philippines); Australia; Antarctic Is.

Uses: Potential use as food in Norway (Mæhre et al. 2014); used as food in salads in the Philippines (Agngarayngay et al. 2005).

Codium australicum P.C. Silva

Description: Firm, dark green, erect, repeatedly dichotomous, to 50 cm high, branches terete, 4–10 mm in diam. below, tapering gradually to 2–3 mm in diam. near apices. Utricles cylindrical to slightly clavate, (100) 130–210 (300) in diam., 460–800 μ long; apices rounded to truncate, wall thin or slightly thickened (15 μ, rarely to 35 μ then lamellate); hairs (or scars) common, borne shortly below apices; medullary filaments 20–35 μ in diam. with the plug 65–200 μ distant from utricle base; gametangia elongate-ovoid, 80–160 μ in diam. and 230–390 μ long, 1–3 per utricle, borne on a protuberance at or below middle of utricle (Womersley 1984).

Geographic distribution: Australia (Tasmania, Victoria); New Zealand.

Uses: Considered an edible species in Tasmania (Sanderson and Benedetto 1988).

Codium dimorphum Svedelius

Description: Medium to dark green, smooth and firm, applanate and strongly adherent, rounded with marginal lobes, 2.5–10 mm thick, to 10 cm across. Utricles in small to large clusters (45–100 μ in diam., 0.5–1.5 mm long, usually slightly constricted just below apices then slightly broader below, apices rounded to sub-truncate with slightly thickened wall (4–7 μ thick), often more so peripherally; utricles from thallus margin often broader (to 125 μ in diam.) below their apices, which are thickened (to 30 μ); hairs absent; medullary filaments extending from the evenly tapering lower part of utricles, wall undulate, 15–35 μ in diam., without plugs. Gametangia narrowly ellipsoid to cylindrical, 44–75 μ in diam., 240–380 μ long (Womersley 1984).

Geographic distribution: Asia (Japan); Australia (Tasmania); New Zealand; SE Pacific (Chile).

Uses: Considered an edible species in Tasmania (Sanderson and Benedetto 1988).

Codium duthieae P.C. Silva

Description: Thallus light to medium green, erect, sub-dichotomously to laterally branched, to 60 cm high, with branches terete and 3–10 mm in diam. but usually compressed (to 2 cm broad) at and near points of branching, branches moderately soft. Utricles cylindrical to clavate, 130–700 μ in diam. near apices, 450–2500 μ long; apices broadly rounded, wall usually remaining thin, occasionally slightly thickened; hairs (or scars) occasional to numerous. Medullary filaments usually 40–75 μ in diam., with plugs usually adjacent to utricle base (Womersley 1984).

Geographic distribution: Indian Ocean; SE Asia (Vietnam); Australia (Tasmania, Victoria); New Zealand.

Uses: Considered an edible species in Tasmania (Sanderson and Benedetto 1988).

Codium indicum S.C. Dixit

Synonym: *C. iyengarii*

Description: Holdfast discoid 0.6–2 cm diam.; stipe erect up to 2 cm high, terete 5–7 mm diam., distally flattened, expanded 1–4 cm broad, bifurcate. Thallus erect up to 40 cm high, throughout flat, expanded up to 5 cm broad, dichotomously divided (to five orders); dichotomies flattened up to 4 cm broad, at 3 cm distant; axils broadly rounded; distally or laterally strongly proliferous, forming coxcomb-like habit. Proliferations simple or dichotomous (two–five orders) of variable sizes, flat up to 3 cm broad and 20 cm long, dichotomies flattened; thallus dissecting out into individual utricles.

Geographic distribution: Africa (Somalia); Indian Ocean; Australia.

Uses: Considered an edible species in Pakistan (Rahman 2002).

Codium galeatum J. Agardh

Description: Medium green, firm, erect, regularly and frequently dichotomously branched, to 1 m high, with branches uniformly terete, not tapering, 4–7 mm in diam. Utricles stout, irregularly cylindrical to clavate, 200–500 µ in maximum diam. (below apex), 650–1500 µ long; utricle apical wall rounded to truncate, moderately to (usually) markedly thickened forming a laminate galeate cap up to 125 µ thick; hairs (or scars) occasional, one to few per utricle; medullary filaments usually 30–80 µ in diam., with a prominent plug adjacent to origin from utricle. Gametangia elongate-ovoid to cylindrical, 55–185 µ in diam., 275–530 µ long (Womersley 1984).

Geographic distribution: Australia (Tasmania, Victoria); Antarctic Is.

Uses: Considered an edible species in Tasmania (Sanderson and Benedetto 1988).

Codium geppiorum O.C. Schmidt

Synonym: *C. bulbopilum*

Common names: Fijian: Sagati, Totoyava (Ostraff 2003); Filipino: Pocpoclo (Domingo and Corrales 2001).

Description: Thalli procumbent, branching irregularly to trichotomously. Fronds deep green forming extensive mats on coral reefs or rocks; differs from *C. edule* in its darker green and shorter fronds.

Geographic distribution: Tropical and subtropical Atlantic; Indian Ocean; Asia (China, Taiwan, Indonesia, Malaysia, the Philippines, Singapore, Thailand, Vietnam); Australia and New Zealand; Pacific Is (Samoa, Polynesia, Micronesia, Fiji, Mariana Is, Marshall Is, Solomon Is).

Uses: Commonly used as food, eaten raw as a seasoned salad in Fiji, Cook Is, Polynesia and the Philippines (South 1993a, N'Yeurt 1999, Domingo and Corrales 2001, Ostraff 2003, Conte and Payri 2006); *C. geppiorum* being favorite dish eaten raw or with fish (Payri et al. 2000).

Codium harveyi P.C. Silva

Description: Medium green, erect, repeatedly dichotomous, to 30 cm high, branches terete, 3–5 mm in diam. below, tapering to 1–2 mm in diam. at apices. Utricles short and irregularly swollen, 170–600 in diam., 340–850 µ long, apices broadly rounded with wall scarcely to slightly thickening to 24–50 µ; hairs (or scars) frequent. Medullary filaments 30–60 µ in diam., with the plug usually closely adjacent to utricle, in some plants more distant on older utricles. Gametangia elongate-ovoid, often tapering above, 70–140 µ in diam. and 150–240 µ long, 1–4 per utricle, borne on a short protuberance on lower half of utricle (Womersley 1984).

Geographic distribution: SE Asia (Indonesia); Australia (Tasmania, Victoria); New Zealand.

Uses: Considered an edible species in Tasmania (Sanderson and Benedetto 1988).

Codium tenue (Kützing) Kützing

Common names: Filipino: Papu-lo, Puk-puklo, Pupulo (Zaneveld 1969).

Description: Thallus spongy, dark green, anchored to rocks or shells by a weft of rhizoids, varying in size from 4 to 5 cm long. Dichotomously branched; erect or repent.

Geographic distribution: Indian Ocean (India, Mauritius, Sri Lanka, S Africa); Asia (Japan, Korea, Indonesia, Malaysia, Taiwan, Vietnam); Pacific Is (Micronesia, Marshall Is).

Uses: Eaten raw as a salad in the Philippines (Zaneveld 1969, Levring et al. 1998), Indonesia (Istini et al. 1998, Zemke-White and Ohno 1999, Roo et al. 2007), Thailand (Lewmanomont 1978, Harrison 2013) and Korea (Bonotto 1976).

Cold water extract of *C. tenue* containing comparatively lower uronic acid and higher sulfate showed stronger blood anticoagulant and antihelminthic activity (Shanmugam et al. 2001, Levring et al. 1998).

Codium papillatum C.K. Tseng & W.J. Gilbert

Common name: Filipino: Popoklo (Velasquez 1972).

Description: Thalli green, spongy-like, large and robust, to about 8 cm. Thalli sub-dichotomously branched, often unilateral. Utricles 650–950 µ in length, 150–300 µ in diam., sub-cylindrical or ovoid, some with few conical papillae at vertical view, apices truncate or sub-truncate. Apical cell wall thickened, about 15–30 µ thick, clearly stratified and foveolate, and hairs or hair scars present (Chang 2015c).

Geographic distribution: Arabian Gulf; Indian Ocean; Asia (China, Taiwan); SE Asia (the Philippines).

Uses: Used as food in the Philippines (Velasquez 1972).

Codium tomentosum Stackhouse (Fig. 74)

Common names: Hawaiian: Limu aala-ula (Reed 1906, MacCaughey 1918); Indonesian: Soesoe lopek (Chapman and Chapman 1980); Japanese: Miru (Smith 1904); Malay: Susu-lopek, Laur-laur (Subba Rao 1965); Portuguese: Chorão, Chorão-do-mar, Pingarelhos (Oliveira 1990, Pereira 2009b, Pereira 2010a).

Description: A small green alga (up to 30–50 cm long), with a dichotomously branched, cylindrical frond; the frond is solid and spongy with a felt-like touch and has many colorless hairs which can be seen when the plant is immersed in water. The holdfast is disk-like and formed from many fine threads (Pereira 2010a).

Geographic distribution: Native to the NE Atlantic Ocean from the British Is southwards to the Azores and Cape Verde Is; it has also been recorded around the coasts of Africa and in various other parts of the world.

Uses: This species is sold in the markets of Makassar and is used as food in several parts of Malaysia (Subba Rao 1965). It is eaten raw as a salad (Zaneveld 1969). The alga is used as food in India (Rao 1970), Indonesia, Thailand, Japan and on the Hawaiian Is consumed in soups or with soya sauce or vinegar (Reed 1906, Levring et al. 1969, Lewmanomont 1978, Arasaki and Arasaki 1983, Istini et al. 1998, Zemke-White and Ohno 1999, Roo et al. 2007, Harrison 2013, Irianto and Syamdidi 2015). In Japan, after drying they are preserved in salt; they are prepared for food by boiling or baking in water, and are put in soups; or, after washing, by mixing with soy-bean sauce and vinegar (Smith 1904). In India is consumed as food in the form of salad or adding to soup and dried plants are prepared as tea (Kaliaperumal et al. 1995). This species is also considered an edible species in Galicia (Spain) and Portugal (Lodeiro et al. 2012, Rodrigues et al. 2015a, 2015b).

Figure 74. *Codium tomentosum* Stackhouse (Chlorophyta). Illustration from William H. Harvey (Phycologia Britannica, 1846–1951).

Halimeda discoidea Decaisne

Common name: Money plant (GCE 2010a).

Description: A heavily calcified algae that is abundant in both shallow and deep water habitats to depths of 30 m or more. It features large, calcified segments with irregular formations. It has the largest individual segments of all *Halimeda* species. *H. discoidea* has a single holdfast that typically attaches to rocks, shells and hard bottom (GCE 2010a).

Geographic distribution: Atlantic Is (Bermuda, Canary Is, Cape Verde Is); Tropical and subtropical W Atlantic; India Ocean; Asia (China, Japan, Taiwan, Indonesia, the Philippines, Singapore, Thailand, Vietnam); Australia (Papua New Guinea); Pacific Is (Samoa, Polynesia, Micronesia, Fiji, Hawaiian Is, Mariana Is, Marshall Is, Palau, Solomon Is).

Uses: In Bangladesh this species is used as food (Islam 1998, Zemke-White and Ohno 1999, Roo et al. 2007, Harrison 2013).

Monostroma angicava Kjellman

Common name: Japanese: Yezo-hitoegusa (Tokida 1954).

Description: The genus name *Monostroma* means "single layer", which is apt due to the single layer of cells that make up the blades of this alga. Blades can reach 5 to 10 cm in length.

Geographic distribution: Asia (China, Japan, Korea).

Uses: Flat bread is made with corn meal and *M. angicava* in coastal northern China (Bangmei and Abbott 1987); is also used as food in Japan (Tokida 1954).

Monostroma nitidum Wittrock (formerly *Porphyra crispata*)

Common names: Chinese: Zi-cai (Arasaki and Arasaki 1983); Japanese: Aonori, Aonoriko (McConnaughey 1985), Tsukushi-amanori (Arasaki and Arasaki 1983), Hitoegusa (Mouritsen 2013).

Description: Bright green in color, with single cell layer blades (in contrast to *Ulva*); slender at the holdfast and growing wider toward the apex, often with a slight funnel shape that has splits down the side.

Geographic distribution: Asia (China, Japan, Korea, Taiwan, Thailand, the Philippines, Vietnam); New Zealand; Pacific Is (Fiji).

Uses: Used for preparation of miso soup, nori-jam, salads, and seaweed powder for various foods in the Japanese, Korean, Thai and Vietnamese cuisine (Kang 1968, Bonotto 1976, Lewmanomont 1978, Abbott and Cheney 1982, Ohno and Largo 2006, Hong et al. 2007, Roo et al. 2007, Lee 2008).

Besides consumption in China, this species is also exported to Indonesia and the Philippines. Among the green algae, *M. nitidum* is the most popular, as the plant is tender and flavorful. It is also used as a condiment (the dried blades are crushed into a dish, and other ingredients such as cooked vegetables or pieces of cooked fish are dipped into the fine green mixture that results). S Fujian people often fry dry *M. nitidum* and eat it with "Chun Bing", flat bread (Bangmei and Abbott 1987, Harrison 2013); consumed also in Vietnam and Thailand (Zemke-White and Ohno 1999).

Extracts of this species have hypocholesterolemic activity (Tsuchiya 1969).

Udotea indica A. Gepp & E.S. Gepp

Description: The thalli are up to 4 cm long, are very broad, and are slightly calcified. The root-mass forms a small tuft. The terete stipe is up to 1.2 cm long and 1 mm thick. The fronds are green, somewhat rounded, flabellate, orbicular, and sometimes broadly proliferated above; the base is cuneate, distinctly zonate. The blade margins are entire, lobed, or lacerated (Nizamuddin 1963).

Geographic distribution: Indian Ocean; SE Asia (the Philippines); Pacific Is (Micronesia, Marshall Is).

Uses: Considered an edible species (Rao 1979, Devi et al. 2009).

Ulva australis Areschoug (formerly *Ulva pertusa*)

Common names: English: Lacy Sea Lettuce (Chapman and Chapman 1980); Japanese: Ana-awosa, Awosa (Tokida 1954).

Description: Large green, leafy seaweed composed of a bright green, glossy, distromatic (two cell layers thick) blade. This species in particular grows up to 20 cm long and begins as an entire, round blade, but develops basal perforations as it grows.

Geographic distribution: Indo-Pacific (Indonesia through Australia and New Zealand), but has been widely introduced and is now found throughout Asia, East Africa, Mediterranean, and both shores of North America.

Uses: Used for foodstuff in the Mediterranean Basin and Korea (Kang 1968, Bonotto 1976, Tseng 1983, Hotta et al. 1989, Harrison 2013).

Used to treat fever (Hoppe 1979), heat stroke, urinary problems, lymphatic swellings, antipyretics, goiter, high blood pressure, dropsy (Bonotto 1976, Tseng and Zhang 1984, Tokuda et al. 1986), wounds (Mshigeni 1983), burn treatment (Waaland 1981); also used as antimicrobial (Baik and Kang 1986), animal fodder (Bonotto 1976), anticoagulant (Shiomi 1983), and antihelminthic (Sidik et al. 2012).

Ulva conglobata Kjellman

Description: Bright green thalli with its frilled fronds clustered into a flower-like formation (Morton and Morton 1983).

Geographic distribution: SW Asia (India); Asia (China, Japan, Korea, and Taiwan); SE Asia (Malaysia, Vietnam).

Uses: In China is used for making a cooling tea as well as a vegetable (Tseng 1983, Bangmei and Abbott 1987); used as food in Korea (Tseng 1983); used also as antipyretic and for sunstroke treatment (Tseng and Zhang 1984, Tokuda et al. 1986).

Ulothrix flacca (Dillwyn) Thuret

Common name: Japanese: Hoso-hibimidoro (Tokida 1954).

Description: This very slender filamentous alga, when growing, looks very much like a clump of fine green hair (Bangmei and Abbott 1987).

Geographic distribution: NE Atlantic; Adriatic Sea; Baltic Sea; Arctic (Canada); Atlantic Is (Azores, Canary Is, Greenland, Iceland); Tropical and subtropical W Atlantic; Mediterranean; Asia (China, Japan, Korea, Russia, Taiwan, Vietnam); New Zealand; Antarctic Is; NE Pacific (Alaska to California).

Uses: Used as food in China; in Fujian, it is eaten as a vegetable and considered a delicacy; and is cooked by cutting it into small pieces and frying in sesame oil (Bangmei and Abbott 1987).

Ulva reticulata Forsskål

Common names: Filipino: Seda-seda (Leyte 2015); Tamil: Pattu pasi (Kaliaperumal et al. 1995).

Description: This alga is light to dark green and the blades are filled with holes of various shapes and sizes. They are net-like or reticulate (Teo and Wee 1983).

Geographic distribution: Tropical and subtropical W Atlantic; Indian Ocean Is; SW Asia (Arabian Gulf, Bahrain, India, Kuwait, Pakistan, Persian Gulf, Saudi Arabia, Sri Lanka, Yemen); Asia (China, Japan, Korea, Taiwan); SE Asia (Indonesia, Malaysia, the Philippines, Singapore, Thailand, Singapore, Vietnam); Australia; Pacific Is (Hawaiian Is); Antarctic Is.

Uses: Used as food in salads in the Philippines and Vietnam (Zemke-White and Ohno 1999, Hong et al. 2007, Roo et al. 2007, Lee 2008, Harrison 2013, Leyte 2015).

Ulva stenophylla Setchell & N.L. Gardner

Description: Thalli 50–80 cm tall; blades linear-lanceolate, without incisions but with broadly ruffled margins, tapering abruptly at base to a short flattened cuneate stipe; blade 5–10 cm, broad and 60–110 µ in thickness.

Geographic distribution: Indian Ocean; Australia and New Zealand; NE Pacific (British Columbia to California).

Uses: Used as food in India (Kaliaperumal et al. 1995), and New Zealand (Smith et al. 2010).

2.4.2 Ochrophyta – Phaeophyceae

Adenocystis utricularis (Bory de Saint-Vincent) Skottsberg (Fig. 75)

Description: Thallus medium to dark brown, with a small discoid holdfast 1–3 mm across bearing one to several clavate to pyriform, clustered, saccate, mucoid vesicles, each with a short solid stipe, 4–6 cm long and

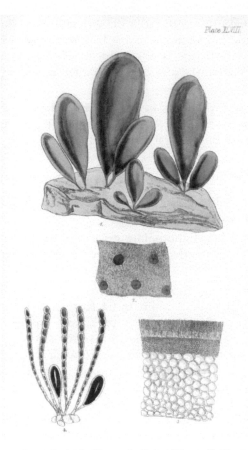

Figure 75. *Adenocystis utricularis* (Bory de Saint-Vincent) Skottsberg (Ochrophyta, Phaeophyceae). Illustration from William H. Harvey (Plate XLVIII, Phycologia Britannica, 1846–1951).

0.5–2 cm in diam., epilithic or on crustose coralline algae; structure multiaxial and haplostichous, with an apical pit containing a tuft of phaeophycean hairs surrounded by laterally adjacent filaments with periclinal cell divisions, differentiating into a pseudo-parenchymatous cortex (Womersley 1987).

Geographic distribution: Antarctic and the sub-Antarctic Is (Antarctic Peninsula, Anvers Is, King George Is, Macquarie Is, S Shetland Is, Trinity Is); Australia and New Zealand (Antipodes Is, Auckland Is, Bounty Is, Campbell Is, Chatham Is, Tasmania); S America (Argentina, Chile).

Uses: Used for foodstuff (Gupta and Abu-Ghannam 2011); extracts have antiviral (Ponce et al. 2003) activity.

Chnoospora minima (Hering) Papenfuss

Synonyms: *Chnoospora pacifica, C. fastigita* var. *pacifica*

Common names: English: Hornwort, Coontail (Santhanam 2015); Hawaiian: Limu wawahi-wa'a, Limukaw-pau (Reed 1906, MacCaughey 1918); Rau ngoai (Indo-China) (Zaneveld 1955, 1959); Polynesian: Imu keikei aoa, Imu makamaka (tree branch algae) (Conte and Payri 2006).

Description: Brown thalli, grows in clumps consisting of upright axes cylindrical, compressed or flattened more commonly in dichotomies; dichotomously branched, not always a regular basis. Fixed to the substratum by structure extended disciform, from which emerge numerous erect branches; thalli 5–6 cm of height (but reaches 10 cm) and a width of 0.5–1.2 mm, with 1.7 to 3 mm in dichotomies (MacCaughey 1918, Santhanam 2015).

Geographic distribution: Tropical and subtropical W and E Atlantic; Indian Ocean; Indo-China; Indonesia and Vietnam; Pacific Ocean (Japan, Australia, Hawaiian Is, Mexico, Polynesia).

Uses: The species is eaten raw, like a salad or relish, in French Polynesia, Vietnam, India, and Indochina (Zaneveld 1955, 1959, Rao 1970, Chapman and Chapman 1980, Conte and Payri 2006). Used as food also in Hawaii and Guam (Reed 1906, MacCaughey 1918).

Cystoseira abies-marina (S.G. Gmelin) (Fig. 76)

Description: At the base a branching, gnarled stem with teeth-like appendages. Upper thallus tufted, forking or lateral branching at more or less the same length, single branches thin, covered with scattered dark dots, and conspicuous bilateral saw-teeth-like; conceptacles as wart-like swellings in the upper parts of the thallus; without swim bladders (Braune and Guiry 2011).

Geographic distribution: Atlantic Is (Azores, Madeira, Canary Is, Cape Verde Is); Mediterranean (Spain, Sardinia, Libya, Italy); Indian Ocean.

Uses: This species is used for food in Gulf of Mannar (Manivannan et al. 2011, Pereira 2015a).

Dictyota dichotoma (Hudson) J.V. Lamouroux (Fig. 77)

Synonym: *Dictyota apiculata*

Common names: English: Divided net weed, Brown fan weed (Bunker et al. 2010); German: Gemeine gabelzunge (Braune 2008); Hawaiian: Limu alani (Reed 1906).

Description: Thallus upright, very polymorphic; flattened, ribbon-like fronds repeatedly forking at the same length in a very regular manner,

Figure 76. *Cystoseira abies-marina* (S.G. Gmelin) (Ochrophyta, Phaeophyceae).

sometimes spiralled; frond width constant from base to tip, or only slightly narrowing; frond-tips blut, rounded or slightly pointed (Braune and Guiry 2011).

Geographic distribution: Common and globally distributed.

Uses: Used as food in India, Malaysia, Thailand and Indonesia (Lewmanomont 1978, Milchakova 2011, Sidik et al. 2012, Mole and Sabale 2013, Irianto and Syamdidi 2015); is also used as food in Hawaii (Reed 1906, Zaneveld 1955).

D. dichotoma is the source of sulfophenols and BASs antitumor and antibiotic activity; is also used in treatments for goiter and scrofula; a preventive medicine for heart disease and stroke (Milchakova 2011, Pereira 2015a).

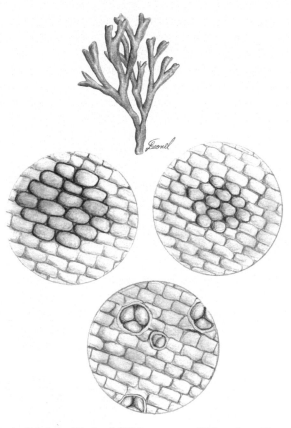

Figure 77. *Dictyota dichotoma* (Hudson) J.V. Lamouroux (Ochrophyta, Phaeophyceae); from top to bottom, female gametophyte - oogonial sori, male gametophyte - antheridial sori, sporophyte - sporangial sori.

Dictyota dichotoma var. *intricata* (C. Agardh) Greville

Synonym: *Dictyota linearis*

Common name: English: Common forked tongue (Santhanam 2015).

Description: Thalli medium-brown of flat blades branching regularly into two's (dichotomous), a narrow variety, var. *intricata*, with relatively long distances between branching, is common in sheltered rocky substrata (Pereira 2014e, Santhanam 2015).

Geographic distribution: NE Atlantic (France to Morocco); Atlantic Is; Mediterranean; SW Atlantic (Brazil); SE Atlantic (S Africa); Indian Ocean; SW Pacific; Australia.

Uses: Used as food in salads (Sidik et al. 2012).

Durvillaea potatorum (Labillardière) Areschoug (Fig. 78)

Common name: English: Bull kelp (Braune 2008).

Description: *D. potatorum* is one of the larger seaweeds, able to exceed 10 m in length and weigh more than 75 Kg (wet weight). Commonly known in Australia as bull kelp (a name used for several large kelp species in different parts of the world), this brown algae has a robust, flat thallus that arises from a large, strong holdfast. Leathery fronds are divided laterally along the length of the stipe into long, wide leafy blades. *D. potatorum* makes a very productive crop. It is long lived, exceeding 14 years, and can grow at a rate of 10 to 14 cm a day during summer months (Guiry and Guiry 2014d).

Geographic distribution: S America (Chile); Australia.

Uses: Used for foodstuff; was dried, roasted, then soaked in fresh water for 12 hours by aborigines in Australia (Chapman and Chapman 1980).

Figure 78. *Durvillaea potatorum* (Labillardière) Areschoug (Ochrophyta, Phaeophyceae). Illustration from William H. Harvey (Plate CCC, Phycologia Australica, 1858–1863).

In Australia, *D. potatorum* is distributed from Robe in S Australia to Bermagui in New S Wales. On the west coast of King Is cast plants are harvested for processing and supply to Kelco Alginates, Girvan, Scotland. Plants can grow up to 6 m in length and can reach 200 Kg in weight (although not typical). Average annual harvest is 20–25.000 tonnes wet weight (Braune and Guiry 2011).

Ecklonia radiata (C. Agardh) J. Agardh

Common name: English: Brown kelp.

Description: Dark brown, 0.3–2 m high, with a single stipe bearing a complanate blade with distinct laterals; holdfast more or less conical, usually 5–10 cm across and high, with sub-dichotomous haptera 2–4 mm in diam.; epilithic; stipe simple, terete, solid, 2–100 cm long, and 2–12 mm in diam. Blade differentiating at its base, with laterals usually developing rapidly from meristems at their bases, more or less alternately distichous; central blade usually smooth, occasionally with surface spines, 1.5–10 cm broad. Laterals 5–40 cm long, 1–10 cm broad, basally narrowed, simple to lobed, smooth to corrugate and often spiny (especially in rough-water forms), margin usually spiny; spines broad-based, mostly 2–10 mm long; structure of a central filamentous medulla and outer cellular cortex with a surface meristoderm; outer cells with numerous discoid phaeoplasts (Womersley 1987).

Geographic distribution: Atlantic Is (Canary Is, Cape Verde Is); Indian Ocean; Australia and New Zealand.

Uses: Used as food in New Zealand (Smith et al. 2010).

Hormophysa cuneiformis (J.F. Gmelin) P.C. Silva

Synonym: *H. triquetra*

Common names: Arabic: Tahalib (FAO 1997); English: Wedgeshaped chainweed (FAO 1997); Filipino: Aragan (Agngarayngay et al. 2005); Tamil: Irrakkai pasi (Kaliaperumal et al. 1995).

Description: Plants arc bushy. Fifteen–20 cm or more in height and brownish in color; attached to the substratum by disk-shaped hold fast from which branches arise; sparsely branched; branches articulated and triangular; margins membranous and dentate: this species is commonly associated with *Cystoseira* and *Sargassum* (Kaliaperumal et al. 1995, JIRCAS 2012f).

Geographic distribution: Indian Ocean; Asia (Japan, Malaysia, Taiwan, Thailand, the Philippines, Singapore, Vietnam); Australia; Pacific Is (Polynesia, Micronesia, Mariana Is, Palau, Fiji, New Caledonia, Solomon Is).

Uses: In the Philippines, the tender tops are used in salads; coat the pot when boiling fish (Agngarayngay et al. 2005); eaten also in India (Rao 1970), and Malaysia (Ahmad et al. 2012).

Hormosira banksii (Turner) Decaisne

Common name: Neptune's necklace (Wassilieff 2012).

Description: These medium to giant-sized seaweeds typically grow at depths below the greens and above the reds. Neptune's necklace is well known to most people who have visited rocky shores. Its branching chains of water-filled bladders help it withstand periods of exposure when the tide goes out. Many seaweeds produce mucilage or slime to protect against drying out (Wassilieff 2012).

Geographic distribution: Australia and New Zealand.

Uses: Used as food in New Zealand (Smith et al. 2010, Wassilieff 2012).

Hydroclathrus clathratus (C. Agardh) M.A. Howe (Fig. 79)

Synonym: *Hydroclathrus cancellatus*

Common names: English: South sea colander (Novaczek 2001), Sponge seaweed; Filipino: Balbalulang (Zaneveld 1959, Velasquez 1972, Domingo and Corrales 2001, Agngarayngay et al. 2005); Japanese: Japanese: Kagomenori (Oshie-Stark et al. 2003); Pacific name: Rimu oma (Maori, New Zealand) (Novaczek 2001, Ostraff 2003); Tamil: Idiappam pasi (Kaliaperumal et al. 1995).

Description: Brownish yellow thalli that are initially sub-spherical and hollow but become extensive and grow to 2–12 cm in diam.; thallus flattened and creeping, perforated by numerous holes, and ultimately irregular in shape (Neto et al. 2005).

Geographic distribution: Widely distributed in warm, subtropical and tropical seas.

Uses: *H. clathratus* has been used for centuries in traditional cuisine and medicine of island cultures, such as Hawaii. This sea plant is rich in calcium and iron, and contains iodine, mannitol, algin, fucoidan, laminarin and folic acid (naturally occurring substances with a range of health benefits) (Levring et al. 1969, Lee 2008); this species is also eaten in New Zealand and Thailand (Lewmanomont 1978, Ostraff 2003, Wang et al. 2010); in India is eaten as salad (Kaliaperumal et al. 1995); consumed in Bangladesh and the Philippines (Velasquez 1972, Zemke-White and Ohno 1999).

Cover in batter and make seaweed fritters, eat fresh or blanch for 2 min and use in a salad, in Philippines and Indonesia (Zaneveld 1959, Trono 1998, Arasaki and Arasaki 1983, Agngarayngay et al. 2005, Irianto and Syamdidi

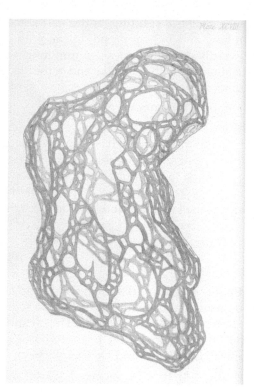

Figure 79. *Hydroclathrus clathratus* (C. Agardh) M.A. Howe (Ochrophyta, Phaeophyceae). Illustration from William H. Harvey (Plate XCVIII, Phycologia Australica, 1858–1863).

2015). In India and Bangladesh this species is used as food (Rao 1970, Islam 1998, Roo et al. 2007, Harrison 2013).

Used as preventative medicine for heart disease; is a good food for pregnant women (Trono 1998, Novaczek 2001). This species is known to possess anti-cancer, anti-herpetic, anti-inflammatory, and anti-coagulant properties and is now used as a mineral supplement in cosmetics and as a soil-additive (fertilizer) for its high concentration of micro-nutrients (Levring et al. 1969, Wang et al. 2010).

Lessonia corrugata Lucas

Common names: Tasmanian Kombu, Strap weed (Kai Ho 2014).

Description: Thallus dark brown, 0.5–2 m high, with numerous blades from the holdfast; epilithic. Holdfast entangled, of compressed prostrate branches 3–10 mm broad, often anastomosing, and producing sub-terete haptera mainly from their margins; stipes becoming erect from the holdfast branches, 2–20 cm long, 1–5 times sub-dichotomous, compressed, 4–15 mm broad. Blades arising by basal splitting, entire and smooth when young, becoming 0.5–2 m long and 1–3 cm broad, linear, with coarse, marginal, basally broad, upwardly curved, spines 2–6 mm long, surface with deep longitudinal corrugations, with the terminal end of the blades disintegrating (adapted from Womersley 1987).

Geographic distribution: Tasmania.

Uses: Tasmanian Kombu, Strap weed (*L. corrugata*) is an endemic seaweed, i.e., the alga is only found in Tasmania. It is found in the high energy upper subtidal zone. If the trials of marketing the alga are successful, an ongoing strategy will be formulated to minimize impacts or even augment current Tasmanian *Lessonia* stocks. This may be achieved through culture. This alga is a good substitute for the Japanese Kombu (Kai Ho 2014).

Padina antillarum (Kützing) Piccone

Synonym: *Zonaria antillarum*

Description: The thallus measures 10.5 cm in height and an overall breadth of 11.5 cm, although the thallus is deeply split into segments 0.5–2.0 cm in width. The blades showed no calcification. Cross- and sagittal sections of the blades, both in mid-region and in more basal portions, show a four-layered organization (Wynne and De Clerck 1999).

Geographic distribution: Atlantic Is (Selvage Is); Tropical and subtropical W and E Atlantic; SW Asia (India, Sri Lanka); SE Asia (Indonesia, the Philippines, Singapore).

Uses: *P. antillarum* is a brown species which proliferates in tropical waters. It is used as seasoning in dried flake form and as a replacement for table salt for high blood pressure patients (Novaczek and Athy 2001); is used as food in India (Rao 1970, Mole and Sabale 2013).

Padina australis Hauck

Common names: English: Fan-leaf seaweed (Santhanam 2015); Filipino: Dunggan-dunggan (Leyte 2015); Indonesian: Agar-agar daun besar (Zaneveld 1959).

Description: Thallus is yellow-brown color, and does not change the original color when dried. Thallus is upright with rhizoid holdfast, fan shape, membrane-like texture, 6–18 cm in height and 5–15 cm in width. Concentric hair bands are too narrow which make it difficult to recognize, and then fertile (1.0–2.0 mm) in width, has reproductive organs and sterility zones (2.0–2.3 mm) in width, lacks reproductive organs alternating arranged on the outer surface of thallus. Proliferations are formed on the hair bands at the older part of thallus. Slight calcification on the inner rather than outer surface of the thallus, and without the *Vaughaniella* stage. White hairs are concentric arranged on two surfaces of thallus, and brown fluffs are covered over the base to middle part of thallus (Wang 2014, Santhanam 2015).

Geographic distribution: SE Atlantic; Indian Ocean; SW Asia (Arabian Gulf; Bangladesh, India, Iran, Kuwait); Asia (China, Japan, Korea, Taiwan); SE Asia (Indonesia, Myanmar, the Philippines, Singapore, Thailand, Vietnam); Australia; Pacific Is (Polynesia, Easter Is, Micronesia, Fiji, Hawaiian Is, Palau, Solomon Is).

Uses: In SE Asia this species is sometimes eaten as a salad, but frequently it is collected for preparing a gelatin-like sweetmeat (Zaneveld 1959, Johnston 1966, Rao 1970, Lewmanomont 1978, Chapman and Chapman 1980, Sidik et al. 2012, Irianto and Syamdidi 2015, Leyte 2015).

Padina boryana Thivy

Synonym: *Padina commersonii*

Common names: English: Ear-like seaweed; Samoan: Limu lautaliga (Skelton 2003).

Description: Blades spreading, lightly to moderately calcified, whitish with a bluish tinge above, tan below, with undulate margins, entire or variously split, to 9 cm long and 8.5 cm wide, to 14 cm long and 18 cm wide; fronds either arising at the ends of creeping or tangled *Vaughaniella* stages, or stalked or sessile from a bulbous fibrous holdfast; thalli bi-layered, becoming tri-layered in some specimens and 80–110 µ thick toward the base; cells of the ventral layer often slightly smaller (30–50 µ long, 30–35 µ wide and 35–42 µ tall) than those of dorsal layer (40–60 µ long, 25–25 µ wide and 45–60 µ tall) (Huisman and Parker 2011b).

Geographic distribution: Adriatic Sea, Mediterranean; Equatorial E Atlantic (S. Tomé & Príncipe); Indian Ocean, Red Sea; Asia (China, Japan, Taiwan, Indonesia, Malaysia, Thailand, Singapore, Vietnam); Australia; Pacific Is (Samoa, Micronesia, Mariana Is, Polynesia, Marshall Is, Fiji, Hawaiian Is, Solomon Is).

Uses: Used as food in India (Rao 1970), Thailand and Samoa (Lewmanomont 1978, Skelton 2003).

Padina distromatica Hauc

Description: Thalli erect or prostrate, attached by a rhizoidal holdfast, up to 20 cm long, complanate, flabellate or becoming lacerate often to the base, segments 1–5 cm broad, calcified on one or both surfaces. Growth initiated by an entire marginal row of apical cells in an involute apical fold, directed towards the upper thallus surface. Thallus 2 cells thick; whitish hairs in concentric lines on one of both thallus surfaces; outermost layer of cortical cells with many discoid chloroplasts; sporangial sori forming concentric rows or isolated patches between the hair lines on one or both thallus surfaces.

Geographic distribution: Indian Ocean; SE Asia (the Philippines, Singapore, Thailand).

Uses: Used as food in Thailand (Lewmanomont 1978).

Padina minor Yamada

Common name: Japanese: Usuyukiuchiwa (Oshie-Stark et al. 2003).

Description: Thallus is light- to gray-brown color, and becoming slightly dark-brown when dried. Thallus is upright with rhizoid holdfast, fan shape, membrane-like texture, 3–12 cm in height and 1–10 cm in width. Concentric hair bands are obvious, but very narrow in width, whereas the zone between hair bands is 1.2–2.0 mm in width. Slight calcification found on the inner than outer surface of the thallus with the *Vaughaniella* stage. White hairs are concentric arranged on the two surfaces of thallus, and brown fluffs are covered over the base to middle part of thallus (Wang 2015).

Geographic distribution: Indian Ocean; Asia (China, Japan, Korea, Taiwan); SE Asia (Indonesia, the Philippines, Thailand, Vietnam); Pacific Is (Polynesia, Micronesia, Fiji, Mariana Is).

Uses: Potentially edible seaweed (Amornlerdpison et al. 2009).

Padina pavonica (Linnaeus) Thivy (Fig. 80)

Common name: English: Peacock's tail (Dickinson 1963).

Description: The fronds are thin and leafy, flattish and entire when young, but often concave, or almost funnel shaped in mature specimens, with a laciniate or irregularly lobed margin. The inner (or upper) surface is covered in a thin coating of slime, and the outer (or lower) surface is banded with zones of light brown, dark brown and olive green. Small, fine hairs form concentric lines, 3–5 mm apart, from the outer margin continuing down the outer (colored) surface of the fronds (Bleach 2007).

Geographic distribution: NE Atlantic; Adriatic Sea; Atlantic Is (Azores, Canary Is; Cape Verde Is, Madeira, Selvage Is); Tropical and subtropical Atlantic; Caribbean Sea; Mediterranean; Indian Ocean; Asia (Japan, Taiwan); SE Asia (Indonesia, Myanmar, the Philippines, Singapore); Australia; Pacific Is (Polynesia, Micronesia, Fiji, Palau).

Uses: *P. pavonica* collected from the Buleji coast (Pakistan) can be utilized as food, medicines and fodder (Fareeha et al. 2013); is also considered an edible species in NE Atlantic (Green 2014).

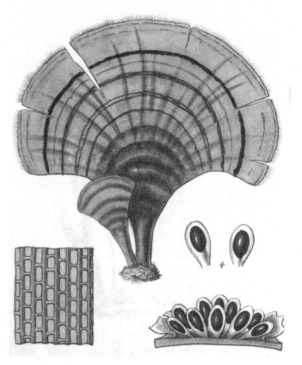

Figure 80. *Padina pavonica* (Linnaeus) Thivy (Ochrophyta, Phaeophyceae). Illustration from Robert K. Greville (Algae Britannicae, 1830).

Padina tetrastromatica Hauck

Description: Thalli flabelliform, usually divided into several small lobes, regularly and distinctly concentrically zonate due to the regular rowan of fructiferous organs; easily recognized due to dark double lines of sporangia; enclosing a line of colorless hairs in between; blades composed of two layers of cells (Paula 2014d).

Geographic distribution: SE Atlantic; Indian Ocean; SW Asia (Arabian Gulf, India, Iran, Oman, Sri Lanka); Asia (China); SE Asia (Indonesia, Malaysia, the Philippines, Thailand); Australia.

Uses: Used as food in Malaysia and Sri Lanka (Michanek 1975, Sethi 2012, Sidik et al. 2014).

Sargassum cinctum J. Agardh

Common name: Filipino: Aragan (Agngarayngay et al. 2005), Pinong samo (Leyte 2015).

Description: Thalli dark brown in color, up to 50 cm long, arising from disk-like holdfast; primary branches terete but compressed near the nodes; leaves linear-lanceolate or ovate; 0.3–2.5 cm long, 3–7 mm broad, serrate with obtuse apices and asymmetrical bases, costate, vesicles oval with flat winged petioles and long leaf like apicules, receptacles 5–10 mm long in racemose branches serrate, spiny, ancipate-triquatrous and twisted (Jha et al. 2009).

Geographic distribution: Indian Ocean, Asia (China, Indonesia, the Philippines, Vietnam); Australia; Pacific Is (New Caledonia, Samoan Is).

Uses: Used as food in the Philippines and Vietnam, where the tender tops are used in salads; lining of pot when boiling fish (Michanek 1975, Agngarayngay et al. 2005, Leyte 2015).

This species is harvested from the wild and used for alginate production in India. Another note worthy product from *S. cinctum* in India is liquid seaweed fertilizer (Jha et al. 2009).

Sargassum confusum C. Agardh

Common names: Filipino: Aragan (Agngarayngay et al. 2005); Japanese: Fushisuji-moku (Tokida 1954).

Description: Thallus to 10–20 cm or more in length, with one to a few simple, terete to compressed, stipes 1–20 cm long arising from a discoid-conical holdfast. Stipes bearing radially or distichously borne, long primary branches, produced seasonally from the stipe apices and subsequently deciduous, leaving scars or other residues on the stipe. Primary branches 10 cm to 200 cm or more long, distichously, tristichously or radially branched with terete, angular, compressed or three-side axes; basal laterals simple or branched, compressed and relatively narrow to leaf-like; presence of bear spherical, non-apiculate vesicles are all referred to the present species. Their basal leaves, when present, are often more or less slightly serrated in part but rather rarely entire without an exception (Tokida 1954).

Geographic distribution: Asia (China, Japan, Korea, the Philippines, Vietnam).

Uses: Used as food in Japan and the Philippines, where the tender tops are used in salads (Tokida 1954, Arasaki and Arasaki 1983); lining of pot when boiling fish (Agngarayngay et al. 2005); is also especially used for the manufacture of potash salts (Tokida 1954).

Sargassum corderoi R.B. Modelo Jr, I. Umezaki & L.M. Liao

Description: Long, linear leaves which have the entire margin throughout, and having elliptic, slender vesicles (Modelo et al. 1998).

Geographic distribution: SE Asia (the Philippines); Australia.

Uses: Used as food in the preparation of main dishes (Lee 2008).

Sargassum gracillimum Reinbold

Common name: Aragan (Agngarayngay et al. 2005).

Description: Small discoid holdfast; stem very short, mostly less than 1 cm long, terete. Primary branches crowded at stem apex, appearing almost sessile on holdfast; filiform, terete. Secondary and terminal branches terete, slightly to irregularly lumpy near their bases because of leaf scars, which are arranged alternately along primary branches at irregular intervals, farther apart near base and closer and crowded toward distal ends. Leaves on primary and secondary branches on younger thalli relatively longer than those at distal portions, generally obovate-oblanceolate, up to 12 mm long, 4 mm wide; base asymmetrical, acuminate to long and narrow, segment gradually grading, giving the leaf a long-stalked appearance; margin of basal half of leaf generally entire, distal half coarsely and irregularly serrate; tip acute, obtuse to rounded; midrib apparent but disappearing below tip; cryptostomata distinctly elevated, scattered. Leaves of terminal branches and those associated with receptacles generally linear to linear-oblanceolate, appearing lumpy because of elevated cryptostomata; margin entire or with few teeth (Trono 2001).

Geographic distribution: Asia (China, Japan, Indonesia, the Philippines, Vietnam).

Uses: In the Philippines, the tender tops are used in salads; coat the pot when boiling fish (Agngarayngay et al. 2005).

Sargassum granuliferum C. Agardh

Common name: Indonesian: Arien-wari (Zaneveld 1955, Aaronson 2000).

Description: Stem winding, branched, terete; the young leaves of are linear-oblong, serrate, planes; presence of spherical vesicles, that are naked; receptacles linear numerous unarmed.

Geographic distribution: Indian Ocean; Asia (China); SE Asia (Indonesia, Malaysia, the Philippines, Singapore); Australia; Pacific Is (Micronesia).

Uses: Is consumed in Moluccas Is (Indonesia) as a vegetable (Zaneveld 1955, Johnston 1966, Aaronson 2000).

Sargassum ilicifolium (Turner) C. Agardh

Synonyms: *Sargassum cristaefolium, S. duplicatum*

Common name: Filipino: Aragan (Agngarayngay et al. 2005); Tamil: Kattaikkorai (Kaliaperumal et al. 1995).

Description: Fronds 40–65 cm high, brown; holdfast a plate-like disc, 1–1.4 cm wide; main axis terete, 2–4 cm long, 1.5 mm in diam., bearing several primary branches. Primary branches sub-cylindrical, slightly compressed, with smooth surface, 1 mm in diam.; secondary branches similar to primary branches, racemose arranged, generally 5–15 cm long; branchlet giving rise from secondary branches, shorter and more slender than secondary branches, 1.5–2.5 cm long. Leaves papyraceous, lower leaves obovate to long-elliptical, 2–2.5 cm long and 8–10 mm wide, with round apex and asymmetrical base, midrib evanescent in the middle part, margin with small, acute teeth, some leaves duplicated, forming a slightly concave, circular lobe perpendicular to the blade, cryptostomata irregularly scattered. Vesicles spherical or ovate, 4–6 mm long and 3–5.5 mm in diam., often provided with wings, stalk short and compressed. Androgynous receptacles giving rise from axils, mostly compressed to flattened, with spiny forks, with short stalks, racemes arranged, 3–7 mm long (Yang 2014b).

Geographic distribution: Indian Ocean; Asia (China, Japan, Taiwan, Indonesia, Malaysia, the Philippines, Singapore, Vietnam); Australia and New Zealand; Pacific Is (Micronesia, Fiji, Mariana Is, Solomon Is).

Uses: In the Philippines, the tender tops are used in salads; coat the pot when boiling fish (Agngarayngay et al. 2005).

Sargassum kushimotense Yendo

Common name: Aragan (Agngarayngay et al. 2005).

Description: Leaves on the lower portion of *S. kushimotense* have a simple, linear-lanceolate shape with dentate margins; the leaves of the upper portion are narrower and more branched (Shimabukuro et al. 2008).

Geographic distribution: Asia (China, Japan, Taiwan, the Philippines).

Uses: In the Philippines, the tender tops are used in salads; coat the pot when boiling fish (Agngarayngay et al. 2005).

Sargassum marginatum (C. Agardh) J. Agardh

Description: Thallus to 10–20 cm or more in length, with one to a few simple, terete to compressed, stipes 1–20 cm long arising from a discoid-conical holdfast. Stipes bearing radially or distichously borne, long primary branches, produced seasonally from the stipe apices and subsequently deciduous, leaving scars or other residues on the stipe. Primary branches 10 to 20 cm or more long.

Geographic distribution: SW Asia (India, Sri Lanka); SE Asia (Indonesia).

Uses: Used as food in India for the preparation of "Pasta" (Prabhasankar et al. 2009).

Sargassum mcclurei Setchell

Description: Fronds 60 cm or higher; holdfast discoid, 8 mm wide. Main axis terete, smooth, 3–5 mm long, 1–2 mm in diam., arising several primary branches; primary branches smooth, bases cylindrical, slightly compressed upward, 1.5–2 mm wide. Secondary branches giving rise from primary branches spirally and at interverals 1.5–3 cm, 2–8.5 cm long. Leaves coarse, lower leaves elongated-elliptical or lanceolate, 1.5–2 cm long, 5–8 mm wide, with entire or dentate margin, asymmetrical, cuneate below, percurrent, cryptostomata irregularly scattered on both sides of midrib, middle and upper leaves shorter than lower leaves, elliptical or ovate, smaller upward, sometimes lanceolate, with obtuse apices and dentate margins, 5–15 mm long, 1–10 mm wide, percurrent, part of upper leaves longitudinally inflated to vesicles, with the similar margin like other leaves, cryptostomata irregularly scattered (Yang 2014c).

Geographic distribution: Indian Ocean; Asia (China, Taiwan, Vietnam).

Uses: Used as food in Vietnam (Zemke-White and Ohno 1999, Hong et al. 2007).

Sargassum oligocystum Montagne

Description: Fronds 10–60 cm. Holdfast discoid, 5–10 mm in diam.; main axis terete, with warty processes, 0.5–1 cm long, 1.5–2 mm in diam., bearing 1 to 6 primary branches. Primary branches flattened, 10–60 cm long, opposite or alternate, 1.5–3.5 mm wide, giving rise to secondary branches from laterals. Secondary branches alternate, 5–20 cm long, shorter both lower and upper parts, longer on middle parts, 1.5–3 mm wide. Branchlets giving rise from the axils of secondary branches, alternately arranged, similar to that of primary branches, but more slender, less than 1.5 mm in diam., 2–4 cm long; leaves coarse, lanceolate or elongated-elliptical, 3–6 cm long, 6–10 mm wide, margins dentate, apices obtuse, bases cuneate, sessile, midrib vanishing below the apex (Yang 2014d).

Geographic distribution: Indian Ocean; Asia (China, Japan, Taiwan, Indonesia, Malaysia, the Philippines, Singapore, Thailand, Vietnam); Australia; Pacific Is (New Caledonia, Palau, Samoa, Solomon Is).

Uses: Used as food in Thailand (Zemke-White and Ohno 1999, Harrison 2013).

Sargassum polycystum C. Agardh

Synonym: *S. myriocystum*

Common names: Chinese: Agar-agar koepan (Kirby 1953); Indonesion: Agar-agar koepan (Kirby 1953), Arien harulu (Zaneveld 1955); Tamil: Kattaikkorai (Kaliaperumal et al. 1995).

Description: Thallus 35 cm tall, with yellowish brown color, attached with discoid holdfast; main axis cylindrical and rough due to the presence of large outgrowth, supporting alternately arranged branches bearing leaves and vesicles; in young thalli leaves are longer and broader measuring 13–42 mm long including the stalk and 2.5–11.5 mm wide; leaves are generally oblong slightly tapered, retuse (slightly rounded) or emarginate at the tip finally serrated throughout the margin; mature thalli fewer leaves smaller, 7–15 mm long including stalk and 17–40 mm wide, oblanceolate, oblong with tapered bases, the apices are rounded, obtuse to acute outer margin is coarsely serrate; prominent midrib at a short distance from apex of the leaves; cryptostomates are scattered on the surface of the blade. Pedunculate vesicles are ovate or spherical with a diam. of 1.5–3 mm; these are tipped with spinose or thin leaf-like extensions and with few cryptostomates; vesicles may be solitary or may form clusters attached to the primary or secondary branches and are more numerous but smaller in mature thalli (Guiry and Guiry 2014i, Paula 2014e).

Geographic distribution: Caribbean Sea; Indian Ocean; Asia (China, Taiwan, Xisha Is, Indonesia, Malaysia, Myanmar, Moluccas, the Philippines, Singapore, Thailand, Vietnam); Australia; Pacific Is (Polynesia, Micronesia, Fiji, Mariana Is, New Caledonia, Solomon Is, Tonga).

Uses: Used as food in India, China, Malaysia, Thailand, Moluccas, and Indonesia, eaten raw with lemon juice after being washed. Sometimes the seaweeds are first cooked with coconut milk. Used in vegetable soup and to make sweetened jellies (Zaneveld 1955, Rao 1970, Kaliaperumal et al. 1995, Istini et al. 1998, Zemke-White and Ohno 1999, Matanjun et al. 2009, Ahmad et al. 2012, Harrison 2013). The indigenous of Amboina (Indonesia) preserve the seaweeds smoke-dried (Zaneveld 1959, Lewmanomont 1978, Bangmei and Abbott 1987, Istini et al. 1998, Gray 2015, Irianto and Syamdidi 2015). Also used in Caribbean Is as food (Radulovich et al. 2013).

Used for alginate production, animal feed, fertilizer and for medical purposes (Istini et al. 1998, Erulan et al. 2009, Paula 2014). Extracts of this species have antioxidant (Anggadiredja 1997), hypoglycemic (Motshakeri et al. 2013), and antibacterial (Chiao-Wei et al. 2011) activity.

Sargassum serratifolium (C. Agardh) C. Agardh

Common name: Japanese: Nokogiri-moku (Tokida 1954).

Description: Thallus to 10–20 cm or more in length, with one to a few simple, terete to compressed, stipes 1–20 cm long arising from a discoid-conical holdfast. Stipes bearing radially or distichously borne, long primary branches, produced seasonally from the stipe apices and subsequently deciduous, leaving scars or other residues on the stipe. Primary branches 10 cm to 20 cm or more long; presence of duplicate-serrated leaves (Tokida 1954).

Geographic distribution: Asia (China, Hong Kong, Japan, Korea, Taiwan, the Philippines); New Zealand.

Uses: Used as food in Japan, China and Korea (Tokida 1954, Bonotto 1976, Chapman and Chapman 1980).

Sargassum siliquosum J. Agardh

Common name: Aragan (the Philippines); Yore-moku (Japan) (Zaneveld 1959).

Description: Fronds dark brown, attaining a height of 1 m, with discoid holdfasts, 1–1.5 cm broad. Main axes short, conical, 3–5 mm in height, 2–3 mm in diam., nearly smooth on the surface, bearing several primary branches. Primary branches terete in lower parts, gradually compressed upward, smooth, 1.5 mm in diam.; secondary branches alternate at intervals of 2–3 cm, shorter upward, somewhat glandulate in upper branches. Leaves papery, and which on lower secondary branches elongated-elliptical, lanceolate or spatulate, apices obtuse, 2–4 cm long, 5–10 mm wide, margins entire, undulate or shallowly irregularly dentate, midribs slender, vanishing below the middle parts, with an oblique asymmetrical cuneate base, sessile or stalked, irregularly scattered cryptostomata; leaves on the upper secondary branches spatulate, bat-shaped or linear-lanceolate, 7–20 mm long, 1–5 mm wide, margins with irregularly serrate at distal end and entire at proximal end, bases cuneate, sessile or stalked, irregularly scattered cryptostomata; branchlet giving rise from secondary branches, more slender than that of secondary branches, 1–4 cm long; vesicles spherical, obovoid to elliptical, apices obtuse, sometimes with sharply pointed or auricular wings, 3–7 mm long, 2–7 mm in diam., pedicels cylindrical or flattened, 3–8 mm long; plants dioecious; male receptacles terete, sharper upward, 5–8 mm long, 0.5–0.7 mm in diam., always mingled with vesicles, inflorescence cymosely or panicle; female receptacles flattened or triquetrous, upper portions with spiny or dentate processes, 2–3 mm long, about 0.5 mm wide, solitary or two-three forked, racemosely or paniculately arranged on the branchlets, usually with vesicles (Yang 2014e).

Geographic distribution: Indian Ocean; Asia (China, Japan, Taiwan); SE Asia (Indonesia, Malaysia, the Philippines, Singapore, Vietnam); Australia.

Uses: In the Philippines this seaweed is eaten either fresh as a salad or cooked as a vegetable (Zaneveld 1959, Harrison 2013). In Indonesia is used to make sweetened jellies and salad (Istini et al. 1998, Zemke-White and Ohno 1999, Irianto and Syamdidi 2015). It is also consumed in Malaysia (Zaneveld 1955).

Sargassum tenerrimum J. Agardh

Common name: Tamil: Kattaikkorai (Kaliaperumal et al. 1995).

Description: Plants pyramidal in form, delicate and with a disk shaped holdfast; yellowish-brown in color; reach a height of 30–40 cm; axis glabrous and rounded; ultimate branchlets modified into vesicles and receptacles; leaves thin, translucent, 2–6 cm long and 0.5–1.5 cm broad; alternately arranged, being larger and broader in the lower portion becoming smaller and narrower towards the apex; margins of the leaves somewhat dentate; mid rib more or less prominent; vesicles stalked and spherical; receptacles richly branched and spinose (Kaliaperumal et al. 1995).

Geographic distribution: Indian Ocean; Asia (China, Hong Kong, Taiwan); SE Asia (Malaysia, the Philippines, Vietnam); Australia; Pacific Is (Micronesia, Polynesia, Samoa).

Uses: Used for foodstuff in India and Pakistan; is also a source for production of sodium alginate (Hoppe 1979, Kaliaperumal et al. 1995, Rahman 2002).

Sirophysalis trinodis (Forsskål) Kützing

Synonym: *Cystoseira trinodis*

Description: The alga's main stripes arising from a holdfast are 1–4 cm long. The stripes bear a few to numerous primary branches (20–50 cm long) which are usually formed and lost seasonally. Branchlets are borne on the primary branches. These bear air bladders, egg-holding structures and male gametes. In summer the plant sends up fertile fronds, which float on top of the water at low tide and are easily visible. In late summer these disappear, leaving the basal holdfast (Sanderson 2000).

Geographic distribution: Arabian Gulf; Indian Ocean; SE Asia (Indonesia, the Philippines); Australia; Pacific Is (New Caledonia).

Uses: Considered an edible species in India (Sathya et al. 2013).

Sphacelaria indica Reinke

Description: Rather small (about 2–10 cm tall), brownish to olivaceous plants, forming complanate tufts or spreading mats.

Geographic distribution: SE Asia (Singapore).

Uses: Used as food (Gray 2015). Extracts have antiviral (Bandyopadhyay et al. 2011) activity.

Splachnidium rugosum (Linnaeus) Greville

Common names: English: Dead-man's finger (Braune 2008); Gummy weed (Wassilieff 2012).

Description: Sporophyte thalli medium brown near apices (living) to dark brown (dried), very mucoid, usually 4–20 cm high with several axes from a small, discoid, rhizoidal holdfast 3–10 mm across; epilithic or on barnacles. Axes with several to numerous, radially and irregularly placed laterals, 2 mm–3 cm apart, 4–8 mm in diam., basally constricted, becoming rugose (especially when dried), linear to clavate and usually increasing slightly in diam. to just below the rounded branch apex which has a slight depression. Clavate to pyriform "apical cells", each tapering to a rhizoid, are present in the cortex of apical depressions (and young conceptacles) but these "apical cells" rarely divide (Womersley 1987).

Geographic distribution: Atlantic Is (Tristan da Cunha); E Pacific (Chile); Indian Ocean; Australia and New Zealand.

Uses: Used as food in New Zealand (Wassilieff 2012).

Turbinaria conoides (J. Agardh) Kützing

Common names: Filipino: Samo (Leyte 2015); Indonesian: Labi-labi (Zaneveld 1951); Malayan: Agar-agar lesong (Kirby 1953, Santhanam 2015); Tamil: Pakkoda pasi (Kaliaperumal et al. 1995).

Description: The holdfast is branched in this species, and there are numerous secondary branches. The thin triangular leaves are concave at the center and have single margins with large sharp serrations. The plants grow on rocks in subtidal zones along relatively calm shorelines with water movement (JIRCAS 2012h, Santhanam 2015).

Geographic distribution: Indian Ocean; SW Asia (Arabian Gulf, India, Iran, Saudi Arabia, Sri Lanka); Asia (China, Japan, Taiwan); SE Asia (Indonesia, Malaysia, the Philippines, Singapore, Thailand, Vietnam); Australia; Pacific Is (Polynesia, Micronesia, Palau, Samoa, Solomon Is).

Uses: Occurs throughout Malaysia, this species is consumed raw, described as having a pleasant taste. They are made into a pickle in E Malaysia (Ahmad et al. 2012, Sidik et al. 2012). According to Rumphius (1750), the seaweeds belonging to this species are eaten raw by the native chiefs (Radja's) as a salad with a little vinegar (Zaneveld 1959). This species is also eaten in India (Rao 1970), the Philippines (Leyte 2015), Indonesia (Zaneveld 1951), and Thailand (Lewmanomont 1978, Irianto and Syamdidi 2015).

Turbinaria decurrens Bory de Saint-Vincent

Common name: Tamil: Pakkoda pasi (Kaliaperumal et al. 1995).

Description: Plants yellowish-brown or brown in color; 20–25 cm in height; possesses elongated cylindrical branches; leaves possessing angular margins (Kaliaperumal et al. 1995, JIRCAS 2012i).

Geographic distribution: Indian Ocean; Asia (China); SE Asia (Indonesia, the Philippines, Singapore, Thailand, Vietnam); Australia; Papua New Guinea; Pacific Is (Micronesia, Palau).

Uses: Used as food in Thailand (Lewmanomont 1978).

Turbinaria ornata (Turner) J. Agardh

Common names: English: Ornate seaweed, Spiny-leaf seaweed, Spiny tops; Filipino: Tubol (Leyte 2015); Japanese: Rappamoku (Oshie-Stark et al. 2003); Samoan: Limu lautalatala (Novaczek 2001, Ostraff 2003); Pacific name (Cook Is, New Zealand): Rimu taratara (Novaczek 2001, Ostraff 2003); Tamil: Pakkoda pasi (Kaliaperumal et al. 1995).

Description: Thallus light brown to yellow, stiff and erect with distinctive angular turban-like blades, up to 15 cm tall and 5.5 cm broad. Lateral branchlets generally concave on the top, 1–2 cm wide, with terete stalks for about half their length, terminally distended in a rounded to obpyramidal manner with obtuse ridges; a large vesicle usually occupying the central portion. Up to 13 marginal crown teeth of 3 mm high, on periphery of the top of the leaves, and often up to six paired erect teeth arranged at about 120° angle over the peripheral surface of the blades. Up to 18 leaves per plant; principal axes moderately branched, to 3 mm wide. Thallus attached to the substratum by stilt-like haptera up to 2 mm in diam. and 25 mm long (N'Yeurt 1999).

Geographic distribution: Indian Ocean; Asia (China, Japan, Korea, Taiwan, Indonesia, Malaysia, Myanmar, the Philippines, Singapore, Thailand, Vietnam); Australia; Pacific Is (Samoa, Polynesia, Cook Is, Micronesia, Fiji, Hawaiian Is, Mariana Is, Palau, Marshall Is, Solomon Is).

Uses: Used as food in India (Rao 1970), Indonesia, the Philippines, and Thailand (Lewmanomont 1978, Gray 2005, Irianto and Syamdidi 2015, Leyte 2015). *Turbinaria* is used in the Pacific Is cuisine as a pickle, fritter, soup or stew, omelets or stir fried. When chopped fine it can be added to salads. Dried plants can be crisped and pounded into a powder for use as seasoning (Novaczek 2001, Ostraff 2003). In the Moluccas the seaweeds are most frequently eaten raw with lemon juice but sometimes are cooked with coconut milk. *T. ornata* is made into a pickle in eastern Malaysia (Zaneveld 1959).

Turbinaria turbinata (Linnaeus) Kuntze

Common name: English: Turbinweed (De Kluijver et al. 2015).

Description: Erect central column with branches bearing clumps of triangular, cone-shaped blades with saucer-like tips; axial stem up to 30 cm. A blister-like swelling at the center of the leaf tips is the result of an embedded air bladder that holds the leaves and plant erect; brownish cream to tan to brown, often with dark brown speckles (De Kluijver et al. 2015).

Geographic distribution: Caribbean; Indian Ocean; W Pacific (the Philippines).

Uses: This species is eaten raw or marinated and may be used as fertilizer in coconut plantations (Braune and Gury 2011).

2.4.3 Rhodophyta

Agardhiella subulata (C. Agardh) Kraft & M.J. Wynne (formerly *Agardhiella tenera*)

Common name: Indonesian: Gulaman-Dagat (Hoppe and Levring 1982).

Geographic distribution: Tropical and subtropical W and E Atlantic; Asia (Indonesia, the Philippines).

Uses: The species is probably the most generally used one in Manila. The algae are brought into the market during the rainy season by fishermen. Gulaman is a very popular gelatin-like sweetmeat among the Filipinos. It is prepared by boiling the seaweeds with sugar and various spices (Zaneveld 1959, Johnston 1966).

Amphiroa fragilissima (Linnaeus) Lamouroux

Description: Thalli with dense cushion-like tufted growth, very brittle to strong calcification; branches cylindrical, thin, segmented, rather regularly forking, sometimes also trichotomous, the angles between two fork branches usually rather wide (broadly Y-shaped); the segments are slightly swollen at the endings; yellowish red to whitish pink. *A. fragilissima* is extremely fragile, as the name implies. The calcified branches will often crack and break upon collection and handling. This species could be confused with the similarly sized *Jania* spp. (Braune and Guiry 2011).

Geographic distribution: Widely distributed in tropical and subtropical seas.

Uses: Used as functional foods (Dhargalkar and Pereira 2005).

Extracts of this species have antiviral, antibacterial, cytotoxic and antioxidant, oxytocic and espamogenic activity (Pereira 2015a).

Antrocentrum nigrescens (Harvey) Kraft & Min-Thein (formerly *Solieria mollis*)

Description: Thallus medium to dark red-brown, 5–20 cm high, much branched irregularly with percurrent axes or main branches and terete, slender, laterals, 0.7–1.5 mm in diam., tapering to a point gradually or abruptly, often proliferous when damaged, occasionally with recurved branch ends. Holdfast densely fibrous, branched, 0.5–2 cm across, bearing one to several fronds; epilithic or epiphytic (Womersley 1994).

Geographic distribution: Asia (Japan, Vietnam); Australia.

Uses: Used as food in China (Bangmei and Abbott 1987).

Betaphycus philippinensis Doty

Common name: Filipino: Kanutkanot (Agngarayngay et al. 2005).

Description: Thalli cartilaginous, decumbent, or prostrate with generally irregular to pseudo-dichotomous branching, borne from a crustose base that immediately issues a short, erect, cylindrical, basal segment are noted. Fronds are dorsiventral, terete to compressed and measuring 0.2–1.0 cm broad and 1–2.5 mm thick. The ventral surface has a deep purple red color on dorsal surface; the color ranges from deep purple red to yellowish green or a mixture thereof. Margin may be smooth but usually dentate on both or one of the either segment margins. Apical tips may be simple and/or forked, acuminate to spinose. Numerous tubercles are consistently found on the ventral surface where protuberances are few or undeveloped. Simple, spinose, and lanceolate to spindle-shaped protuberances develop mainly on but not limited to the ventral and marginal surfaces (Dumilag et al. 2014).

Geographic distribution: Asia (the Philippines).

Uses: Used as food in salads, soups, and mixed with meat (Agngarayngay et al. 2005).

Betaphycus speciosus (Sonder) Doty ex P.C. Silva (Fig. 81)

Synonym: *Eucheuma speciousum*

Description: Fronds are apically flat, pliable and arising from marginal cylindrical teeth. Tolerates direct wave action.

Geographic distribution: Indian Ocean; Australia.

Uses: Used as food in Australia (Tasmania), to make jelly (Irving 1957, Michanek 1975).

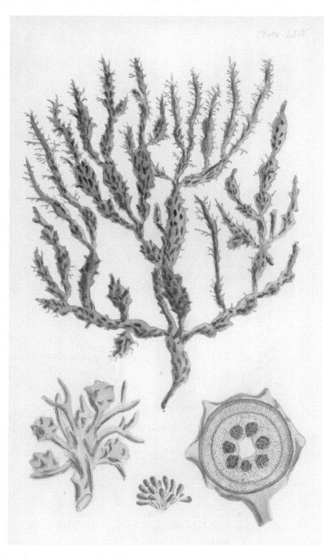

Figure 81. *Betaphycus speciosus* (Sonder) Doty ex P.C. Silva (Rhodophyta). Illustration from William H. Harvey (Plate LXIV, Phycologia Australica, 1858–1863).

Boergeseniella thuyoides (Harvey) Kylin (Fig. 82)

Common name: English: Tufted conifer weed (Bunker et al. 2010).

Description: Cylindrical, cartilaginous, tufted, deep brownish purple fronds, to 150 mm high, from creeping rhizoidal base; fronds distichously bi-tri-pinnate, patent, short, of nearly uniform length giving branches a linear appearance; ramuli short, spine-like; polysiphonous, central siphon with

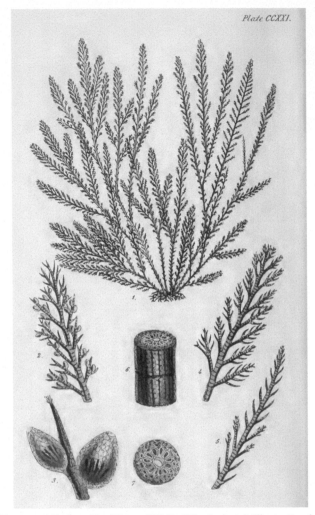

Figure 82. *Boergeseniella thuyoides* (Harvey) Kylin (Rhodophyta). Illustration from William H. Harvey (Plate CCXXI, Phycologia Britannica, 1846–1951).

eight–12 pericentral siphons and outer cortication of small, colored cells; articulations as broad as long, barely visible (Guiry and Guiry 2014b).

Geographic distribution: NE Atlantic.

Uses: Used as food (Gray 2015). Extracts has antiviral and antioxidant activity (Rhimou et al. 2011, Rhimou et al. 2013).

Bostrychia radicans (Montagne) Montagne

Description: The filiform, cartilaginous plants are pinnately branched. Towards the apex these branches are usually hook-pointed (Zaneveld 1959).

Geographic distribution: Tropical and subtropical W and E Atlantic; Indian Ocean; Asia (China, Japan, Korea, Indonesia, Malaysia, Myanmar, the Philippines, Singapore, Thailand, Vietnam); Australia; Pacific Is (Micronesia, Fiji).

Uses: This species is eaten in Indonesia, Thailand and Myanmar either raw or boiled, in salads and soups (Zaneveld 1959, Lewmanomont 1978, Irianto and Syamdidi 2015).

Caloglossa bengalensis (G. Martens) R.J. King & Puttock

Synonym: *C. adnata*

Description: The thallus of the plants is flat and dichotomously branched with a pronounced midrib. The secondary branches arise from this midrib (Zaneveld 1959).

Geographic distribution: Indian Ocean; Asia (Japan, Indonesia, Malaysia, Myanmar, the Philippines, Singapore, Vietnam); Australia; Pacific Is (Micronesia, Fiji).

Uses: It is eaten raw or boiled (Zaneveld 1959); is used in Indonesia for making salads and vegetable soup (Istini et al. 1998, Zemke-White and Ohno 1999, Roo et al. 2007, Harrison 2013); is also used in Myanmar (Zaneveld 1955).

Caloglossa leprieurii (Montagne) G. Martens

Common name: Chinese: Zhegucai (Globinmed 2011a).

Description: The blades are strap-shaped, reddish-violet, reddish-brown to pale-brown, thin, membranous, measure 1–3 mm across, occasionally arising in rosettes from a stipe, alternately branched and forked at the apices, constricted and often with rhizoids at the constrictions (nodes). The internodal segments are elongate-ovate to linear-lance-shaped, measure 1–4 mm long, composed of a midrib of large rectangular cells and with lateral blade.

The secondary branches or segments are leaf-like, proliferous from the forkings or midribs of blades and with small sub-hexagonal cells in oblique series from the midrib to the margin. Epiphytic on mangroves or epilithic, usually lower to mid eulittoral (Womersley 2003).

Geographic distribution: *C. leprieurii* is widely distributed in tropical and warm temperate waters throughout the Atlantic, Indian and Pacific Oceans.

In Southeast Asia, it has been recorded in Myanmar, Thailand, peninsular Malaysia, Singapore, Indonesia, Brunei, the Philippines and northern Papua New Guinea.

Uses: Used as food in India, Myanmar, Korea and Indonesia (Subba Rao 1965, Johnston 1966, Bonotto 1976, Irianto and Syamdidi 2015); this species is eaten either raw or boiled (Zaneveld 1959).

Extracts of *C. leprieurii* have vermifuge activity (Tseng 1983).

Catenella impudica (Montagne) J. Agardh

Common names: Burmese: Kyauk pewint (Kirby 1953); English: Burmese moss (Kirby 1953).

Description: Frond decumbent, slightly flabellately expanded, articulate with deep constriction between the segments, irregularly di- or trichotomously branched, branches sub-terminally formed, tips of ultimate segments being prolonged into terete acuminate apices, segments obovate, young segments scattered (Rath and Adhikary 2006).

Geographic distribution: Caribbean Sea; SE Atlantic; SW Asia (Bangladesh, India); Asia (China, Japan); SE Asia (Indonesia, Malaysia, the Philippines, Singapore, Vietnam).

Uses: Used as food in Asia (Zaneveld 1955, Rao et al. 2008, Sidik et al. 2012, Irianto and Syamdidi 2015); *C. impudica* is prepared for eating by pouring boiling water over it; then it added to salads (Zaneveld 1959).

Catenella nipae Zanardini

Common name: Visayan: Kolongkong (Baldock 2009b).

Description: Plants dark brown-red, fading to yellow, forming tangled tufts or turfs 20–100 mm tall segments thin, elongate cylindrical in upper parts of the plant, compressed below, branch irregularly small, hooked attachment branches (haptera) occur just below branch tips (Baldock 2009b).

Geographic distribution: Indian Ocean; Asia (Bangladesh, India, China, Japan, Indonesia, Malaysia, Myanmar, the Philippines, Singapore, Thailand, Vietnam); Australia and New Zealand.

Uses: In Myanmar, some like it raw, mixed with the oil of *Sesamum indicum*, salt, powdered fruit of sweet pepper (*Capsicum annuum*), dried rhizome of ginger (*Zingiber officinale*), onion (*Allium cepa*), and garlic (*Allium ursinum*); others boil it for an hour and then mix it with the above ingredients (Zaneveld 1959, Chapman and Chapman 1980, Arasaki and Arasaki 1983, Soe-Htum 1998, Zemke-White and Ohno 1999, Roo et al. 2007); also eaten in Indonesia and Thailand, in salads and soups (Lewmanomont 1978, Irianto and Syamdidi 2015).

Centroceras clavulatum (C. Agardh) Montagne

Common names: Hawaiian: Limu huluilio, limu hulu, Limu hulu wawae-iole (Reed 1906).

Description: Thalli are filamentous turfs, usually in entangled mats, 3–5 cm in height, and bright red to dark brown in color. Erect branches are un-branched or sub-dichotomously branched with straight, incurved apices. Axial cells are completely corticated with covering axial cells, 15–18 µ in diam. arranged in one layer rectangular in shape, between successive segments. The covering cortical cells at nodes are usually in two layers and with spines. Tetrasporangia are borne on outer nodes, 45–55 µ in diam. in whorls (Lin 2015a).

Geographic distribution: Tropical and subtropical Atlantic; Caribbean; Adriatic Sea; Mediterranean; Atlantic Is (Ascension, Azores, Canary Is, Cape Verde Is, Madeira, Selvage Is, Tristan da Cunha); Indian Ocean; Asia (China, Japan, Korea, Taiwan); SE Asia (Indonesia, Malaysia, the Philippines, Singapore, Thailand, Vietnam); Australia and New Zealand; Pacific Is (Polynesia, Easter Is, Micronesia, Fiji, Hawaiian Is, Mariana Is, Marshall Is, Samoa, Solomon Is).

Uses: Used as food in Hawaii (Reed 1906), India (Rao 1970, Kalesh 2003), and Nigeria (Fakoya et al. 2011).

Ceramium boydenii E.S. Gepp

Common names: Japanese: Amikusa (Arasaki and Arasaki 1983); English: Fairy herb (Hu 2005).

Description: Crimson-red to dark red seaweed, forming a mass of interwoven filaments 10 cm in diam.; epiphytic on other algae in intertidal zones, young plants dichotomously branched, branchlets three–seven, in whorls (Hu 2005).

Geographic distribution: Asia (China, Japan, Korea).

Uses: Used as food in China and Japan (Hu 2005).

Ceramium deslongchampsii Chauvin ex Duby

Common name: English: Estuary banded picer weed (Bunker et al. 2010).

Description: Thallus bushy, rose-red or dark brown, to 8 cm high; erect axes tufted, 0.1–0.4 mm diam., narrowing four-five times in the upper part; nodes and internodes clearly marked; heavily branched, branching dichotomous or pseudo-dichotomous; branches relatively long, simple, lower part sparsely at arranged, divaricated (45–60°); apices forked, with straight or slightly in rolled apices; main branches with many branchlets, usually with colorless hair on young branches; segments in the upper and lower parts part of thallus four–six times and 1.5–3 times longer than wide, respectively; corticated nodes slightly bulged, of smooth-edged bands, nearly as high as wide, separated by narrow internodes (Milchakova 2011).

Geographic distribution: NE Atlantic; Atlantic Is (Azores, Canary Is, Iceland, Madeira); Tropical W Atlantic; Adriatic Sea; Baltic Sea; Black Sea; Mediterranean; Caribbean Sea; Indian Ocean; Arabian Gulf; Asia (Russia, Taiwan); SE Asia (Indonesia, Vietnam); NE Pacific (British Columbia, Texas).

Uses: *C. deslongchampsii* is the source of polysaccharides, agarose, agaroids, carrageenan, proteins, BASs with antimicrobial, antiviral, fungicidal and antitumor activity; food in eastern and SE Asia (Milchakova 2011).

Callophyllis adnata Okamura

Common name: Lablabig (Agngarayngay et al. 2005).

Description: Erect, flat, soft-membranous to cartilaginous, usually flabellately, laciniately or more or less pinnately branched in one plane; stipe short; holdfast discoid. Margin smooth, ruffled, or with protuberances (Parker and Guiry 2014b).

Geographic distribution: Asia (China, Japan, Korea, the Philippines).

Uses: Used as food in salads and mixed with meat (Agngarayngay et al. 2005).

Chondrophycus cartilagineus (Yamada) Garbary & J.T. Harper (formerly *Laurencia cartilaginea*)

Description: This cartilaginous species is erect, grows to about 10–15 cm tall and is reddish to purple in color. The holdfast is small and discoid. The branches are cylindrical, and branchlets are arranged alternately. The apices of ultimate branchlets are truncated. This plant is found on rocks in subtidal zones along strongly wave exposed shorelines (JIRCAS 2012).

Geographic distribution: Asia (Xisha Is, Japan, Indonesia, the Philippines, Thailand, Vietnam); Australia; Pacific Is (Micronesia, Fiji, Hawaiian Is, Mariana Is, Solomon Is).

Uses: Commonly used in vegetable salads in the Philippines (Trono 1998).

Chondrus elatus Holmes

Common names: Japanese: Kotoji-tsunomata (Arasaki and Arasaki 1983); English: Dee-horn vegetable (Hu 2005).

Description: Purplish-red caespitose seaweed with sub-cylindrical dichotomously branched thalli, 10–30 cm long, cartilaginous; growing on intertidal rocks by small disk-like holdfast (Hu 2005).

Geographic distribution: Asia (China, Japan, Korea).

Uses: Occurring in Taiwan and Fujian, pickled material available in American Chinese stores (Hu 2005); is also eaten in China (Johnston 1966).

Chondrophycus undulatus (Yamada) Garbary & Harper
(formerly *Laurencia undulata*)

Common names: Filipino: It-ittip (Agngarayngay et al. 2005); Japanese: Kohu-sozo (Saito 1967).

Description: The fronds are caespitose, erect, cartilaginous and rigid in texture and generally dark purple or purplish brown in color. They stand on a discoid base and never adhere to paper when dried. Stoloniferous basal branches are present but not abundant. The erect main branches are cylindrical near the base but becoming complanate upward, 3–6.5 cm high. They are pinnately, sometimes sub-dichotomously, branched. The secondary branches generally issue many short and knobby ultimate branchlets. The ultimate stichidial branchlets are not compressed but truncate or wart-like (Saito 1967).

Geographic distribution: W Mediterranean; Indian Ocean; Asia (Japan, Korea, Taiwan, the Philippines, Vietnam); Pacific Is (Hawaiian Is).

Uses: Used as food in salads, and mixed with meat in the Philippines and other countries of Asia (Agngarayngay et al. 2005, Yong-Xin et al. 2009).

Chnoospora implexa J. Agardh

Description: Thalli consisting of densely aggregated interconnected dichotomous axes anchored to rock and coral rubble by numerous fibrous holdfasts composed of dense aggregates of free undivided filaments derived from cortical cells mostly at the apices of dichotomous axes; under moderate water-movement conditions large numbers of erect axes forming a cushion-like configuration, these producing widely spaced (to 12 mm apart) dichotomies and sub-dichotomies below and more densely crowded forks above. Fronds very pale to very dark brown, the former usually displaying a characteristic farinaceous texture, 3–9 cm long, growing from broadly rounded tips 200–400 µ wide, with tufts of anchoring fibers commonly arising from the tips or laterally (Kraft 2009).

Geographic distribution: Tropical and subtropical W Atlantic; Indian Ocean; Asia (China, Japan, Taiwan, Indonesia, the Philippines, Singapore, Vietnam);

Pacific Is (Polynesia, Micronesia, Hawaiian Is, Mariana Is, Marshall Is, New Caledonia, Samoan Is); NE Pacific (Galapagos Is, Mexico).

Uses: Eaten in Hawaii (Levring et al. 1969), and the Philippines (Trono 1998).

Chrysymenia enteromorpha Harvey

Description: The length of this plant usually varies between 10 cm and 30 cm but it can sometimes attain the height of 50 cm. In such a large plant the frond becomes nearly 7 mm thick in the lower part. Throughout the whole length the frond is cylindrical, the inside always making a central cavity and there are no diaphragms at all. The branching is rather irregular, branches being issued from every side and alternately and this mode of ramification is repeated, so that the whole plant is of a paniculate appearance. Branches as well branchlets are constricted at the base and attenuate toward the end (Yamada 1932).

Geographic distribution: Atlantic Is (Bermuda, Canary Is); Tropical and subtropical W Atlantic; Caribbean; Indian Ocean.

Uses: Used as food in Bangladesh (Aziz 2009).

Dermocorynus dichotomus (J. Agardh) Gargiulo, M. Morabito & Manghisi

Synonym: *Grateloupia dichotoma*

Description: NE Atlantic; Atlantic Is (Azores, Canary Is); Black Sea; Mediterranean; Tropical and subtropical W Atlantic; Caribbean Sea; Asia (China); SE Asia (the Philippines, Vietnam).

Geographic distribution: Thallus flat, dark-brown, basal part cuneiform, with short cylindrical stem, to 7 cm high, attachment to substratum with holdfast; branching regularly or irregularly dichotomous; branches linear, linear-cuneated or lance-oblong, 0.5–2 mm wide, with blunt or pointed apices often arranged at same level; laterals may develop on both sides of branches (Milchakova 2011).

Uses: Used as foodstuff in some Asian countries (Milchakova 2011).

This species is a source of agar, vitamins B and E (thiamine, riboflavin, α-tocopherol), pantothenic and nicotine acids, mineral salts, antihelmintic and antitumor BASs (Milchakova 2011).

Dermonema pulvinatum (Grunow) Fan
(formerly *Nemalion pulvinatum*)

Description: Thallus cylindrical, dark pink red in color, up to 17 cm tall, opaque, dichotomously branched.

Geographic distribution: Asia (China, Japan, Korea, Taiwan, Vietnam); Pacific Is (Hawaiian Is).

Uses: In China, this species is raked from the rocks and sold fresh, or later dried or cured in salt (Tseng 1983, Bangmei and Abbott 1987). It is pressed into cakes, dried and eaten with wine, used in soup, or prepared by other methods. This species is also eaten by the Yemei tribe in Taiwan (Bangmei and Abbott 1987).

Dermonema virens (J. Agardh) Pedroche & Ávila Ortíz

Synonyms: *D. dichotomum, D. frappieri, D. gracile*

Description: Erect, red-brown, robust thalli, lubricated, up to 8 cm, cylindrical, with profusely dichotomous branches from this base, tapering towards the apex (Pedroche and Ortiz 1996).

Geographic distribution: Caribbean; Indian Ocean; Asia (China, Japan, Taiwan, Indonesia, the Philippines, Vietnam); Pacific Is (Mariana Is, Micronesia, Polynesia, Solomon Is).

Uses: Used in Vietnam for making jellies (Nang and Dinh 1998); and eaten in Taiwan and Caribbean—Tobago Is (Hansen et al. 1981, Bangmei and Abbott 1987), and China (Lee 1965).

Eucheuma denticulatum (N.L. Burman) F.S. Collins & Hervey

Synonyms: *E. spinosum, E. muricatum*

Common names: English: Macassar; East Indian carrageen (Guiry and Guiry 2014), Spinosum (Santhanam 2015), Unicorn seaweed (Hu 2005); Filipino: Kanutkanot (Agngarayngay et al. 2005), Tambalang (Fortner 1978), Canot-canot (Velasquez 1972); Indonesian: Agar gésér (Kirby 1953), Agae-poeloe (Ohmi 1968), Agar séran laoet (Kirby 1953); Japanese: Kirinsai, Ryukyu-tsunomata (Arasaki and Arasaki 1983); Malayan: Agar-agar kasar, Agar-agar haloes (Kirby 1953).

Description: Perennial red seaweed composed of rigid clumps of low-growing, cartilaginous thalli. These clumps can range in size depending on growing conditions, but are able to reach considerable size, up to 50 cm in length and weighing over 1 Kg. The primary axis and regularly spaced branches are cylindrical, each bearing whorled spinose (spine-like) branchlets which sometimes develop into secondary lateral branches. This one species can be found in a variety of shades from brown to green to red.

Geographic distribution: Native to the Indian Ocean, but since being discovered as a source of iota carrageenan (Pereira and Ribeiro-Claro 2014), it has been spread elsewhere through cultivation. It can be found in the following countries: Malaysia, the Philippines, Indonesia, Vietnam, Cambodia, India, Madagascar, Tanzania, Fiji, Kiribati, Tonga, and Vanuatu.

Uses: Tambalang, as it is known in the Philippines, is prized as a salad vegetable and can be picked whole or grated as a relish. It is quite succulent and crunchy and adapts well to "Kim chee" and "Namasu" recipes (Velasquez 1972, Fortner 1978, Trono 1998, Harrison 2013); used as food in China, Japan and Korea in salads, and mixed with meat (Arasaki and Arasaki 1983, Bangmei and Abbott 1987, Trono 1998, Lee 2008, Agngarayngay et al. 2005, Hu 2005, Sidik et al. 2012). This species is eaten as sweetened jellies in Indonesia (Subba Rao 1965, Zemke-White and Ohno 1999, Istini et al. 2006). In W Australia *E. denticulatum* is used for the same purpose by housewives (Zaneveld 1959). In Indonesia is used for making a salad, soup, pickle, and as raw material for carrageenan extraction (Irianto and Syamdidi 2015). Used also as food and for carrageenan extraction in Malaysia (Zaneveland 1955, Ahmad et al. 2012).

This species is used for goiter treatment, cough, asthma and bronchitis (Istini et al. 1998, Roo et al. 2007).

Eucheuma edule (Kützing) Weber-van Bosse

Common names: Chinese: Hai jen ts'ai, Tser koo ts'ai (Arasaki and Arasaki 1983), Hai ts'ai mu (Tseng 1983); Indonesian: agar-agar-besar (Zaneveland 1955); Japanese: Makuri, Kaijinso (Arasaki and Arasaki 1983).

Description: Plants of the largest species are bushy, reach over 50 cm in length, and can weigh over a Kg. Thalli are often very fleshy and rigidly cartilaginous when fresh, and are erect from an encrusting base or prostrate to entangled and anchored at many points by haptera. Most bear whorled to scattered, simple to compound spines.

Geographic distribution: Indian Ocean; Asia (China, Indonesia, the Philippines, Singapore); Pacific Is (Hawaiian Is, New Caledonia).

Uses: Used as food in China (Arasaki and Arasaki 1983, Tseng 1983, Simoons 1990), and Indonesia (Zaneveld 1955).

Eucheuma edule f. *majus* Weber-van Bosse

Common name: Indonesian: Agar-agar besar (Zaneveld 1959).

Description: Cylindrical to flattened main axis and branches of a pale reddish color, drying to yellowish brown; branching irregularly alternate to somewhat di- or trichotomous in appearance. Main axis and branches roughened by blunt nodules or spines which harbor the gametangia (Zaneveld 1959).

Geographic distribution: Asia (Indonesia); Pacific Is (New Caledonia).

Uses: The species is used by the inhabitants of Java in the preparation of jellies (Zaneveld 1959).

Eucheuma horridum J. Agardh

Description: Plants of the largest species are bushy, reach over 50 cm in length, and can weigh over a Kg. Thalli are often very fleshy and rigidly cartilaginous when fresh, and are erect from an encrusting base or prostrate to entangled and anchored at many points by haptera. Most bear whorled to scattered, simple to compound spines (Mshigeni and Jahn 2014).

Geographic distribution: Indian Ocean; SE Asia (Indonesia, Malaysia, the Philippines, Singapore).

Uses: Used as food in Indonesia, Malaysia and the Philippines (Zaneveld 1955, 1959, Johnston 1966, Sidik et al. 2012, Irianto and Syamdidi 2015).

Eucheuma serra J. Agardh

Common names: Bulung, Bulung lipan, Bulung djukut lelipan (Bali) (Zaneveld 1955, 1959).

Description: Thallus flat, serrated edges, smooth surface, cartilaginous, red or pale red color. Special features are morphologically similar form that resembles the shape of a centipede, with intermittent and irregular branching.

Geographic distribution: Indian Ocean; Asia (China, Japan, Taiwan); SE Asia (Indonesia, Malaysia, the Philippines); Australia.

Uses: Considered an edible species in Japan (Lideman et al. 2011), and used as food in Indonesia, Malaysia, Borneo, and the Philippines (Zaneveld 1955, Johnston 1966, Sidik et al. 2012).

Ganonema farinosum (J.V. Lamouroux) K.C. Fan & Yung C. Wang

Synonym: *Liagora farinosa*

Common name: Filipino: Baris-baris (Zaneveld 1955, 1959).

Description: Thalli brownish red, moderately calcified, 5–15 cm high, terete, sub-dichotomously branched every 0.5–2 cm with occasional lateral branches; lower thallus 1–2 mm in diam., tapering slightly to apices; holdfast discoid, 1–5 mm across; epilithic or epiphytic on larger brown algae. Structure of a medulla of large, elongate cells 50–150 µ in diam., mixed with slender rhizoids in older parts, and a cortex 350–450 µ broad, of branch tufts of filaments consisting of a basal cell, cut off from a medullary cell, which produces several straight to slightly curved cortical filaments, branched near the base 1–4 times, 15–20 µ in diam., cells with rhodoplasts irregularly stellate, each with a central pyrenoid; basal cells of cortical branch tufts producing slender rhizoids which grow over and between the medullary cells. Tetrasporophyte minute, filamentous, branched (Womersley 1994).

Geographic distribution: Found in tropical to warm-temperate seas worldwide.

Uses: Used as food in Asia, particularly in the Philippines (Zaneveld 1955, 1959, Johnston 1966).

Gelidiella acerosa (Forsskål) Feldmann & G. Hamel

Synonyms: *Gelidium regidum, Gelidiopsis rigida*

Common names: English: Little wire weed (Novaczek 2001); Filipino: Lali-lali (Domingo and Corrales 2001); Linggau: Sangan (Chapman and Chapman 1980), Samoan: Limu uaea (wire-like seaweed) (Skelton 2003).

Description: Thallus greenish-yellow, forming tough, wiry decumbent clumps to 10 cm across, composed of oppositely-branched axes 0.5–1 mm in diam.; internal structure cellular (N'Yeurt 1999). Thalli erect to decumbent, wiry in texture, light to dark brown in color, regular to irregular branching evident (Domingo and Corrales 2001).

Geographic distribution: NE Atlantic; Tropical and subtropical W Atlantic; Indian Ocean; SW Asia (Arabian Gulf, Bahrain, India, Iran, Oman, Sri Lanka, Yemen); Asia (China, Japan, Taiwan); SE Asia (Indonesia, the Philippines, Singapore, Thailand, Vietnam); Australia; Pacific Is (Cook Is; Fiji, New Caledonia, Samoa, Polynesia).

Uses: Add to soups and stews or blanch and then add to pickles. Make into a salad with tomatoes, onion, vinegar and salt. Boil with fruit juice or tomato juice to make jelly, strain and chill in a pan or mold (Novaczek 2001). Used also for dessert preparations (Chapman and Chapman 1980, Lee 2008), and as food in India, China, Indonesia, Malaysia, Vietnam and the Philippines (Zaneveld 1959, Subba Rao 1965, Arasaki and Arasaki 1983, Bangmei and Abbott 1987, Trono 1998, Zemke-White and Ohno 1999, Hong et al. 2007, Roo et al. 2007, Sidik et al. 2012, Harrison 2013, Mole and Sabale 2013, Leyte 2015).

Gelidiella indica Sreenivasa Rao

Description: Plants purple in color, 3 to 4.5 cm high in entangled tufts, giving rise to stolon-like creeping branches from the base, these attached by discoid rhizoids; creeping stolons give rise to erect axes above; erect axes cylindrical or slightly complanate below, gradually expanding to a flattened frond with obtuse apex, entire or branched above (Kaliaperumal et al. 1995).

Geographic distribution: Indian Ocean.

Uses: Used for foodstuff in India, as potential raw material for agar production (Kaliaperumal et al. 1995).

Gelidium amansii (J.V. Lamouroux) J.V. Lamouroux

Common names: Chinese: Niu mau tsai (Madlener 1977); English: Ceylon moss (Kim et al. 2013); Japanese: Tengusa, Makusa, Genso (Tokida 1954, Madlener 1977), Kanten (Rhoads and Zunic 1978), Oyakusa (Chapman and Chapman 1980); Korean: Umutkkasari (Sohn 1998), Umutgasari (Song et al. 2013).

Description: Thallus more than 10 cm tall, flattened to compressed, at least in the lower portion, branches acute at apex. Plants irregularly branched, branches very regularly and elegantly pinnated with equally long branchlets (Santelices and Stewart 1985).

Geographic distribution: *G. amansii* is an economically important species of red algae commonly found in the shallow coast of many E and SE Asian countries.

Uses: This alga is used to make agar, and is sometimes served as part of a salad in producing regions, and to make sweetened jellies in China, Japan, Korea and East Asia (Tokida 1954, Zaneveld 1959, Bangmei and Abbott 1987, Istini et al. 1998, Zemke-White and Ohno 1999, Roo et al. 2007, Sidik et al. 2012, Harrison 2013, Song et al. 2013).

   Araki (1965) reported that agar of *G. amansii* is a mixture of two different polysaccharides, one a neutral agarose which consists of alternating 1,3-linked 3-D-galactopyranose and a 1,4-linked 3,6-anhydro-α-L-galactopyranose and the other a charged agaropectin. Agaropectin contains galactopyranose residues with sulfate and other charged groups present in varying degrees in the molecule. Hirase (1957) reported the presence of pyruvic acid in the agar of *G. amansii* as a ketal attached to the C4 and C6 of the 1,3-linked-3-D-galactopyranose residues. Araki (1965) also proved that the D-galactose residues are 6-0-methylated to certain degrees (Santos 1990).

Gelidium pusillum (Stackhouse) Le Jolis

Common name: Japanese: Hai-tengusa (Chapman and Chapman 1980).

Description: Turf formation in the upper intertidal, blackish-red color when slightly dry, and extensive rhizoids; cartilaginous, purplish or blackish red, turf-forming, 2–10 mm high, arising from extensive creeping base and incorporating shell debris and small mollusks. Axes flat, compressed, up to 3 cm high and 2 mm broad; simple or irregularly branched along the margins or with lanceolate proliferations from the truncate upper end of the erect axes; branches lanceolate or spatulate, contracted at their bases; tetrasporangia in rounded sori on ovoid or rounded fertile branches (Santelices 1977, Guiry and Guiry 2014f).

Geographic distribution: NE Atlantic; Adriatic Sea; Atlantic Is (Ascension, Azores, Bermuda, Canary Is, Cape Verde Is, Madeira, Selvage Is); Tropical and subtropical W and E Atlantic; Mediterranean Sea; Indian Ocean; Asia (China, Japan, Korea, Taiwan, Indonesia, Malaysia, Myanmar, the Philippines, Singapore, Vietnam); Australia and New Zealand; Pacific Is (Polynesia, Easter Is, Micronesia, Hawaiian Is, Mariana Is, Marshall Is, Solomon Is).

Uses: In Bangladesh this species is used as food (Islam 1998, Zemke-White and Ohno 1999, Roo et al. 2007, Harrison 2013, Siddique et al. 2013), and for foodstuff in India, as potential raw material for agar production (Kaliaperumal et al. 1995); is also eaten in the Hawaiian Is (Reed 1906).

Gelidium spinosum (S.G. Gmelin) P.C. Silva

Synonym: *Gelidium latifolium*

Common names: English: Spiny straggle weed (Bunker et al. 2010); Hawaiian: Limu loloa (Reed 1906).

Description: Small alga, cartilaginous, crimson to purplish red, 2–6 cm long; main axes distinctly flattened, often narrower at base, ultimate branches short, often opposite, spine-like or spatulate (Neto et al. 2005, Pereira 2014).

Geographic distribution: NE Atlantic; Mediterranean; Asia (Japan, Korea); SE Asia (Indonesia, Malaysia, the Philippines); Australia; Pacific Is (Hawaiian Is).

Uses: Used as food in Indonesia, Borneo and the Philippines (Johnston 1966, Istini et al. 1998, Zemke-White and Ohno 1999, Roo et al. 2007, Milchakova 2011, Sidik et al. 2012, Harrison 2013); is also eaten in Hawaii (Reed 1906).

G. spinosum is a valuable commercial species, the source of high-quality agar, agarose, agaropectin, polysaccharides and sulfates, vitamins B, PP and E; BASs with antitumor, antiviral, antimicrobial and immunoprotective activity (Milchakova 2011).

Gracilaria Greville

Description: Thallus erect, pink-reed, branching tufted; branching radial, alternate, distant or slanting towards the front, to the second to fourth degree; all axes cylindrical, lateral axes tapering, pointed at the tip and constricted at the base; texture meaty to cartilaginous (Braune and Guiry 2011).

Geographic distribution: Caribbean; Indian Ocean; Asia (Japan, China); SE Asia (Indonesia, the Philippines); Australia.

Uses: *Gracilaria* is eaten in salads, cooked as a vegetable, made into candy and pickles, fried in batter or boiled together with lemon juice to provide a soft jelly. Because the fresh plant has a rubbery texture it is often blanched before being eaten. Use it to thicken soups and stews. Different types of *Gracilaria* may have more or less jelly in them, and the amount of jelly will also vary with the season. When making puddings you may have to adjust the amounts suggested in recipes to suit your local resource (Novaczek 2001).

Gracilaria arcuata Zanardini

Common name: Filipino: Kanutkanot (Agngarayngay et al. 2005).

Description: The thallus is erect, rigid and cartilaginous having a dull green or purple color when fresh and attached to the solid substratum by a discoid holdfast; main axis are somewhat irregularly branched; cylindrical an 2.5–5 mm in diam.; primary and secondary branches are generally second, slightly constricted at the base, slightly enlarged above the constrictions and attenuated towards the mucronate tips; ultimate branchlets are 2–7 mm long, arcuate, appearing like long simple coarse spines or may be forked into two short stubby spinose branchlets at the terminal portion of the thallus (Paula 2014a).

Geographic distribution: Indian Ocean; SW Asia (India, Jordan, Oman, Pakistan, Sri Lanka); Asia (China, Japan, Korea, Taiwan, Xisha Is); SE Asia (Indonesia, the Philippines, Singapore, Vietnam); Australia (Papua New Guinea); Pacific Is (Micronesia, Fiji, Guam, Mariana Is).

Uses: Used as food in Indonesia, in soups (Lee 2008), salads, mixed with meat, and for desserts (Agngarayngay et al. 2005, Irianto and Syamdidi 2015).

Gracilaria articulata C.F. Chang & B.M. Xia

Description: Thallus erect, solitary or caespitose, arising from small disks, cylindrical throughout, usually 10–18 cm in length, alternately, secondly or irregularly dichotomously branched; common mango in color, succulent, brittle, easily broken, adhering rather well to paper on drying; branches and branchlets conspicuously articulate-constricted to a sharp spine; articulations elongated club-shaped, arcuated, upwards, 10–20 times as long as broad, ending in blunt, almost equal in breadth, abruptly constricted at the base (FAO 1996).

Geographic distribution: Asia (China, Japan, Vietnam).

Uses: Used as food in China (Bangmei and Abbott 1987).

Gracilaria blodgettii Harvey

Common name: Guraman (Agngarayngay et al. 2005).

Description: This species is soft in texture, slender, terete, irregularly branched, and acuminate at the apical end. There is distinct constriction at the basal portion of the branchlets. The plants are found in muddy areas attached to gravel or mangrove roots in mid to lower intertidal zones along calm shorelines (JIRCAS 2012c).

Geographic distribution: Tropical and subtropical W Atlantic; Indian Ocean; SW Asia (India, Sri Lanka); Asia (China, Japan, Taiwan); SE Asia (Indonesia, Malaysia, the Philippines, Singapore, Thailand, Vietnam); Australia.

Uses: Used as food in Indonesia and tropical zones of W Atlantic, in salads and desserts; used also for agar extraction (Soegiarto and Sulustijo 1990, Agngarayngay et al. 2005, Radulovich et al. 2013, Deane 2014b, Irianto and Syamdidi 2015).

Gracilaria canaliculata Sonder (formerly *Gracilaria crassa*)

Common names: Filipino: Susueldot-baybay (Zaneveld 1955); Japanese: Taiwan-ogonori (Arasaki and Arasaki 1983).

Description: Thalli are terete, 5–13 cm in height, attached to the substratum by a large discoid holdfast and by secondary holdfasts formed on lower portion of branches. Branches are sub-dichotomously, unconstricted with obtuse apices. Fresh plants firm in texture and are light to bright red in color. Surface of the branches are usually covered with many translucent white spots throughout. Hair cells are present, borne in clusters visible as white spots on the frond surface (Lin 2014a).

Geographic distribution: Asia (China, Japan, Taiwan); SE Asia (the Philippines, Singapore, Vietnam); SW Asia (India, Sri Lanka); Australia; Pacific Is (Micronesia, Guam).

Uses: Used as food in soups, and for agar extraction, in India, Hawaii and SE Asia (Arasaki and Arasaki 1983, Kaliaperumal et al. 1995, Lee 2008, Sidik et al. 2012). The species is collected for food in Vietnam and is eaten in the Philippines (Zaneveld 1955, 1959).

Gracilaria changii (B.M. Xia & I.A. Abbott) I.A. Abbott, J. Zhang & B.M. Xia

Description: Plants bushy, 5–7 cm high, branches cylindrical, main branches 1.5–2.5 mm in diam.; branching alternate or irregular to four orders, most of the branches are abruptly constricted at the base and taper towards apices. Plants erect to spreading, attached to substratum by disk-like holdfast;

branching irregular; main axis percurrent, branches cylindrical, lateral branches distinctly constricted and somewhat enlarged near the base, and tapering toward the apices; cortex with two to three layers of cells; medulla with large, thin-walled parenchymatous cells; cell size transition from cortex to medulla very abrupt; cystocarps hemispherical, some slightly rostrate, base not constricted (FAO 1996, Guiry and Guiry 2014g).

Geographic distribution: Asia (China); SE Asia (Malaysia, Myanmar, the Philippines, Singapore, Thailand, Vietnam).

Uses: Used as food and for agar extraction in Malaysia (FAO 1996, Lewmanomont 1998, Norziah and Ching 2000, Roo et al. 2007, Harrison 2013), and in Thailand (Zemke-White and Ohno 1999).

Extracts of this species have antifungal (Sasidharan et al. 2011), and antibacterial (AL-Haj et al. 2009) activity.

Gracilaria corticata (J. Agardh) J. Agardh

Description: Thallus erect, up to 14 cm in length, arising singly from a discoid holdfast; stipe very short, terete, up to 5 mm long, often inconspicuous. Branching frequently, becoming more dense in upper parts of the plant; mostly dichotomous, up to many orders; rarely, fine laterals extend from the blades producing a bushy appearance; multipartite at points of damage. Axes compressed, almost cartilaginous; constricted at the base in basal branches. Blades linear, up to 15 cm long, up to 4 mm wide; apices generally obtuse, acute in finer branches; blade surface and margins smooth; fresh specimens purple to green and firm but pliable; dried specimens black to green; large galls/tumors often evident on the surface of mature blades (Iyer et al. 2004).

Geographic distribution: Persian Gulf; Indian Ocean; Asia (Korea, Singapore); SE Pacific (Peru).

Uses: Considered an edible species (Rahman 2002, Kumar and Kaladharan 2007, Mohammadi et al. 2013, Mole and Sabale 2013).

Gracilaria edulis (S.G. Gmelin) P.C. Silva
(formerly *Gracilaria taenioides*)

Common names: English: Ceylon moss; Indonesian: Doejoeng, Djanggoet (Chapman and Chapman 1980); Filipino: Goso-goso (Leyte 2015).

Description: *G. edulis* can grow up to 27 cm tall. It is brownish-red and arises from a discoid holdfast. The branching is dense and fastigiated, extremely divergent, dichotomous to trichotomous, in up to seven orders and with branches at long intervals. The branches are 1–1.5 mm in diam., cartilaginous, flexuous, with or without a constriction at their bases or with only a slight constriction, which is cylindrical and ends in pointed apices (Globinmed 2011b).

Geographic distribution: Indian Ocean; SW Asia (India, Sri Lanka); Asia (China, Japan, Korea, Taiwan); SE Asia (Indonesia, Malaysia, Myanmar, the Philippines, Singapore, Thailand, Vietnam); Australia; Pacific Is (Hawaiian Is).

Uses: Used as food in India, Indonesia, the Philippines, Japan, Malaysia, and Hawaii (Subba Rao 1965, Johnston 1966, Chennubhotla et al. 1987, Sidik et al. 2012, Irianto and Syamdidi 2015, Leyte 2015).

Gracilaria gigas Harvey

Description: Thalli are 10–20 cm in height, erect, fragile to cartilaginous, greenish, purple to red in color, arising from a discoid holdfast. Main and lateral branches are mostly arcuate, terete throughout, sometimes slightly constricted at the base (Lin 2015b).

Geographic distribution: SW Asia (Sri Lanka); Asia (China, Japan, Korea, Taiwan); SE Asia (Indonesia, the Philippines, Vietnam).

Uses: Used as food in Indonesia for making pickles and salads (Irianto and Syamdidi 2015).

Gracilaria heteroclada (Montagne) J. Feldmann & G. Feldmann

Description: Plants erect, up to more than 60 cm tall, dark brown to purple, sometimes olive-green in color, brittle when fresh; thallus terete or cylindrical throughout; branching irregularly alternate; main branches percurrent with many spinose determinate branchlets; branches distinctly larger than the branchlets (FAO 1996).

Geographic distribution: Mediterranean; Asia (the Philippines, Vietnam).

Uses: Used in the Philippines and Vietnam to make jellies, sweet soups, sour vegetables, eaten raw, and also used for agar extraction (Nang and Dinh 1998, Trono 1998, Zemke-White and Ohno 1999, Hong et al. 2007, Harrison 2013).

Gracilaria firma Chang & Xia

Common name: Kawkawayan (Agngarayngay et al. 2005).

Description: Thalli are up to 30 cm in height, erect arising from a conspicuous, discoid holdfast attached to oyster shells or oyster bamboo shelves in the water. Thalli are terete, bearing slender lateral branches with tapering or hooked apices (Showe-Mei 2014).

Geographic distribution: Asia (China, Japan, Malaysia, the Philippines, Thailand, Vietnam).

Uses: Used in Vietnam as food in salads, mixed with meat, and in desserts (Zemke-White and Ohno 1999, Agngarayngay et al. 2005, Nang 2006, Roo et al. 2007, Harisson 2013); Used also for Agar extraction (Trono 1998).

Gracilaria hainanensis C.F. Chang & B.M. Xia

Description: Thallus solitary or caespitose, arising from a small disk, cylindrical throughout, usually 15–35 cm, up to 45 cm in length, one to two orders of branches; chestnut purple or kelp green in color, succulent in substance, becoming soft and adhering imperfectly to paper upon drying; branches generally supple, elongated, up to 30–40 cm, 2–3 mm in diam., attenuated to a fine apex, flagelliform, irregularly alternately or secondly and sparingly branched, abruptly constricted at the base (FAO 1996).

Geographic distribution: Asia (China, Vietnam).

Uses: Used as food in China (Bangmei and Abbott 1987).

Gracilaria incurvata Okamura

Common name: Marakawayan (Agngarayngay et al. 2005).

Description: Plants are solitary or caespitose, generally up to 10 cm long, reddish brown in color, not adhering to paper in drying; foliose arises from a holdfast (basal disk) with a short stalk, more or less curved and twisted, branching dichotomously or trichotomously up to six times, sometimes flabellately; the apices and margins of the entire blade are smooth (Terada and Ohno 2000).

Geographic distribution: Asia (Japan, Korea, Taiwan, Indonesia, the Philippines).

Uses: Used as food in salads, mixed with meat, and in desserts (Agngarayngay et al. 2005).

Gracilaria minor (Sonder) Durairatnam
(formerly *Corallopsis salicornia* var. *minor*)

Common names: Indonesian: Boeloeng (Chapman and Chapman 1980); Bulung buku (Bali) (Zaneveld 1959).

Description: Cylindrical thallus of a cartilaginous substance which is however, somewhat folded and constricted at several places. The thallus is irregular, dichotomously branched, the branches taking rise in the constricted places. However, some of the branchlets are only developed as a kind of spine. The plants are light brown (Zaneveld 1959).

Geographic distribution: Asia (Indonesia, Taiwan, China, the Philippines); Pacific Is (Micronesia).

Uses: The plants are eaten as a vegetable in Bali, Indonesia (Zaneveld 1959, Arasaki and Arasaki 1983). This species can be used for agar extraction (Zaneveld 1959).

Gracilaria pudumadamensis V. Krishnamurthy & N.R. Rajendran

Description: Thallus shape with cylindrical-round, compressed, pink-red, texture cartilaginous.

Geographic distribution: Indian Ocean (India, Pakistan).

Uses: Considered and edible species (Gomez-Gutierrez et al. 2011).

Gracilaria tenuistipitata C.F. Chang & B.M. Xia

Description: Characterized by its extreme slenderness near the base. The thallus arises from a small disk-like holdfast, and is simply, moderately, or alternately branched near the base. The branches tend to become like the main axis (FAO 1990; Guiry and Guiry 2014h).

Geographic distribution: Asia (China, Japan); SE Asia (Indonesia, Malaysia, Singapore, Thailand, Vietnam).

Uses: Used in SE Asia as food and for agar extraction (Bangmei and Abbott 1987, FAO 1990, Lewmanomont 1998, Nang and Dinh 1998, Hong et al. 2007, Benjama and Masniyom 2012).

Extracts of this species have antiviral (Chen et al. 2013), and antioxidant (Yang et al. 2012) activity.

Gracilaria tenuistipitata var. *liui* Zhang & Xia

Common name: Lumot (Agngarayngay et al. 2005).

Description: Variety *liui* differs from the var. *tenuistipitata* by the slender thalli bearing numerous, delicate, short to long lateral branchlets, branching mostly from the percurrent axes. Its cystocarpic structure and shallow spermatangia are similar to those of var. *tenuistipitata* (FAO 1990).

Geographic distribution: Asia (China, Taiwan, the Philippines, Thailand, Vietnam).

Uses: Used as food in salads, and mixed with meat, in Thailand and Vietnam (Zemke-White and Ohno 1999, Agngarayngay et al. 2005, Roo et al. 2007, Harrison 2013).

Gracilaria textorii (Suringar) De Toni

Common name: Filipino: Lablabig (Agngarayngay et al. 2005); Japanese: Kabonori (Arasaki and Arasaki 1983).

Description: The plants range from brownish-red to yellowish red in color. Thalli are coriaceous to membranous; the fronds are flattened with cylindrical stipes, and attach to the substratum by small discoid holdfasts. The fronds are irregularly dichotomous, with margins entire or with proliferations; apices blunt, bifurcate, or ligulate; branching in one plane, profuse, alternate or second (Lim and Phang 2004).

Geographic distribution: Caribbean Is (Trinidad and Tobago); Indian Ocean; NE and E Pacific; SW Asia (India, Sri Lanka, Yemen); Asia (China, Japan, Korea, Russia, the Philippines, Taiwan, Vietnam); Australia.

Uses: In Korea is used for foodstuff and for agar production (Kang 1968, Bonotto 1976, Tseng 1983, Bangmei and Abbott 1987, Hotta et al. 1989, Trono 1998, Agngarayngay et al. 2005, Hu 2005).

Gracilaria vermiculophylla (Ohmi) Papenfuss

Synonym: *G. asiatica*

Common name: Japanese: Ogonori (Yoshie-Stark et al. 2003).

Description: This alga grows to a length of 15–100 cm, with branches around 2–5 mm in diam.; it is irregularly but quite richly branched (resembles a wig), the branching either profuse or sparse depending on how the plant grows and whether it is unattached. Large specimens may be hollow lower down the plant. The color varies, from brownish to a grayish wine-red, depending on the availability of sunlight. Loose-lying specimens often lack epiphytic growth (Naylor 2006).

Geographic distribution: N Atlantic; Tropical and subtropical W Atlantic; Asia (China, Japan, Korea, Russia, Vietnam).

Uses: Used as food in China, Japan and Vietnam (Bangmei and Abbott 1987, Nang and Dinh 1998, Zemke-White and Ohno 1999, Hong and Hien 2004, Hong et al. 2007, Terasaki et al. 2012, Harrison 2013). In Asia *G. vermiculophylla* is cultivated as a raw material for the production of agar, a jelly-like thickening agent. Among other things, agar is used in *Petri* dishes and test tubes as a culture medium for bacteria (Naylor 2006).

Gracilariopsis lemaneiformis (Bory de Saint-Vincent) E.Y. Dawson, Acleto & Foldvik (formerly *Gracilaria lemaneiformis*)

Common name: English: Red wiry weed (Baldock 2006).

Description: Plants red to red-brown, 200 mm–1.2 m tall, gristly when dry, main branches cylindrical or slightly flattened, about 1 mm across with fine, hair-like side branches, small, round, dark, dot-sized swellings (cystocarps) on all sides of the branches in mature female plants (Baldock 2006).

Geographic distribution: Tropical and subtropical W Atlantic; E Pacific; Asia (China, Japan); SE Asia (the Philippines); Pacific Is (Hawaiian Is).

Uses: Used as food in China and Japan (Bangmei and Abbott 1987, Zemke-White and Ohno 1999, Zhou et al. 2006, Ohno and Largo 2006, Roo et al. 2007, Harrison 2013), and for agar production (Zhou et al. 2006).

Grateloupia doryphora (Montagne) M.A. Howe

Common name: Spanish: Cochayuyo (Polo 1977).

Description: This species grows in both sheltered and exposed places, and attaches to various hard surfaces. Its thallus is red purple to brown, presents a gelatinous texture and is usually submerged by 15–45 cm seawater. However, thalli measuring 30 cm in length have been found on intertidal rocks. *G. doryphora* has broad salinity and temperature tolerances, and adapts to waters disturbed by eutrophication. It can develop to a length of 3 m and has been considered the biggest red alga in the world (Bouyssou 2012).

Geographic distribution: Adriatic Sea; NE Atlantic (Britain; France, Spain); S Atlantic (Brazil, Venezuela, Africa); Atlantic Is (Canary Is); Caribbean Is; Indian Ocean; Asia (China); SE Asia (the Philippines); New Zealand.

Uses: Used as food in SE Asia (Arasaki and Arasaki 1983), and in Peru (Polo 1977).

Grateloupia indica Børgesen

Description: Plant attached to substratum by a very small basal disk. Stipe short, compressed, cuneate, expanding in broad thallus; thallus flat, oblong to linear, extensively divided, tough, slippery, large, reaching considerable dimension to more than 1 m, 7–30 cm or more broad at broadest part, 300–350 µ thick; lobes of thallus many, margin irregularly sinuate, tips gradually tapering. Rhizoids 3–4 µ thick; outer medulla more compact than inner; cortex many layered, cystocarp dispersed throughout the thallus, sub-spherical to spherical, occasionally flattened, 250 µ in diam. (Paula 2014b).

Geographic distribution: Indian Ocean (India).

Uses: Used as food (Gray 2015). Extracts have anti-coagulant (Kumar et al. 1994), and antiviral (Chattopadhyay et al. 2007) activity.

Grateloupia turuturu Yamada

Common names: English: Devil's tongue weed (Bunker et al. 2010), Red lettuce (Kai Ho 2015b).

Description: Thallus flat, membranous, with short stipe, the single fronds linear to broad-lanceolate, undivided or irregularly dividing from the base, narrowing towards the base as well as the stipe; sometimes proliferating on the margins and the surface; consistency gelatinous-slippery but firm; discoid holdfast; violet- to crimson-red, often greenish at the top thallus (Braune and Guiry 2011, Kai Ho 2015b).

Geographic distribution: SE Pacific (Chile), SW Atlantic (Peru), *G. turuturu* is considered native to Japan, China and Korea, but has spread to the NE Atlantic, the Mediterranean, S America, Australia and New Zealand.

Uses: *G. turuturu* is an introduced seaweed to the Tasmanian coast. This seaweed is nutritious and a colorful addition to all meals. Serving suggestions include chopped and shredded and added to: omelets, soups, savory biscuits, bread, fried rice, salads and can also been used as a garnish (Kai Ho 2015b).

Grateloupia livida (Harvey) Yamada

Synonym: *G. ligulata*

Common names: Chinese: Hai-ts'ai (Tseng 1935), Tongue centipede algae (Huang 2014).

Description: Thalli light pink to purplish red, erect, firmly cartilaginous, 10–30 cm tall, 1 cm broad, simple blade, usually not branched, compressed and attenuate at the base, the margins with pinnate branches or bearing proliferation on the surface of the blade, with cylindrical stalk and small discoid holdfast. In cross section, cortical region consists of seven to 10 layers of cells, the innermost are rectangular to stellate on bordering medulla. Medulla is made of periclinally directed colorless filaments. Sporangia spherical, cruciately divided, scattered and embedded on surface of the frond. Cystocarps also immersed and in groups in the branches (Huang 2014c).

Geographic distribution: Asia (China, Japan, Korea, Taiwan, Vietnam).

Uses: In China (Huilai City, Taishan, and Hainan), is boiled for a long time. The liquid that is formed is gelatinous on cooling and can be flavored, either sweet or salty. In Rongcheng, Shandong, these algae are boiled for a short time in a soup (Tseng 1935, Bangmei and Abbott 1987, Simoons 1990); used as food also in Japan (Terasaki et al. 2012).

Griffithsia corallinoides (Linnaeus) Trevisan (formerly *Griffithsia corallina*) (Fig. 83)

Common name: English: Mrs Griffiths's coral weed (Bunker et al. 2010).

Description: Gelatinous, tufted, uniseriate, ecorticate, crimson filaments, to 200 mm long, repeatedly dichotomous, axils wide. Articulations club-shaped, four-five times as long as broad in lower parts, short and beadlike distally; known to have a strong smell (Guiry and Guiry 2015b).

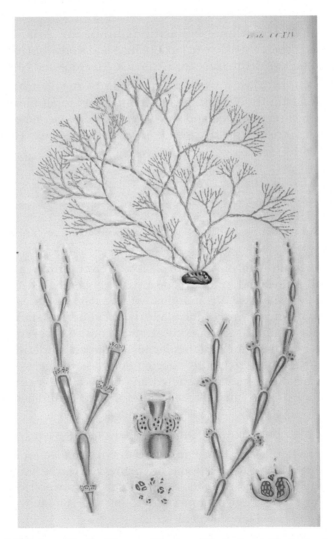

Figure 83. *Griffithsia corallinoides* (Linnaeus) Trevisan (Rhodophyta). Illustration from William H. Harvey (Plate CCXV, Phycologia Britannica, 1846–1951).

Geographic distribution: NE Atlantic, Atlantic Is (Azores, Canary Is); Mediterranean; Indian Ocean; Asia (Japan, Korea, Vietnam).

Uses: Used as food in Vietnam (Davidson 2004).

Halymenia dilatata Zanardini

Common name: Gayunggayong (Agngarayngay et al. 2005).

Description: Thalli purplish red in color, blade-like, up to 20 cm long and 5–12 cm wide, mucilaginous and gelatinous. The blade is generally sub-orbicular or transversely expanded, simple or lobed, undulate curled, margin entire or dentate, base mostly reniform, sessile or with short stipe, attached to substratum by a small disk holdfast (Huang 2014d).

Geographic distribution: Indian Ocean; Asia (China, Japan, Korea, Taiwan, Indonesia, Malaysia, the Philippines, Singapore, Thailand, Vietnam); Australia; Pacific Is (Micronesia, Fiji, Mariana Is, Palau, Solomon Is).

Uses: Used as food in salads and mixed with meat in the Philippines (Domingo and Corrales 2001, Agngarayngay et al. 2005).

Halymenia durvillei Bory de Saint-Vincent

Common name: Filipino: Aragan-ilek, Gayunggayong (Agngarayngay et al. 2005, Domingo and Corrales 2001), Gayong-gayong (Zaneveld 1955); Hawaiian: Limu lepeahina (Zaneveld 1955); Samoan: Limu mumu, Limu a'au (Skelton 2003).

Description: The plants have flattened soft, cartilaginous branches which are supported by a short cylindrical stipe. The branches are pinnate or alternate four to five times. The terminal branchlets are slender and linear with acuminate tips. Numerous spine-like projections are found on the surface of the main branches. This seaweed is orange to red in color, and forms bushy large clumps on rocks in subtidal zones along moderately wave exposed shorelines (JIRCAS 2012e).

Geographic distribution: Indian Ocean; SE Asia (Indonesia, Malaysia, the Philippines, Singapore, Thailand); Australia; Pacific Is (Samoa, Polynesia, Micronesia, Fiji, Guam, Mariana Is, Palau, Solomon Is).

Uses: Used as food in salads and mixed with meat in the Philippines (Zaneveld 1955, Zemke-White and Ohno 1999), Thailand and Indonesia (Johnston 1966, Lewmanomont 1978, Trono 1998, Agngarayngay et al. 2005, Domingo and Corrales 2001, Roo et al. 2007, Irianto and Syamdidi 2015); this species is also eaten in Samoa (Ostraff 2003, Skelton 2003) and Hawaii (Zaneveld 1955).

Halymenia floresii (Clemente) C. Agardh (Fig. 84)

Common names: English: Red sea lettuce, Dragons tongue (GCE 2010b).

Description: Striking flame shaped blades and pinkish/red coloration. Like all species of *Halymenia*, it is gelatinous and smooth in texture and very

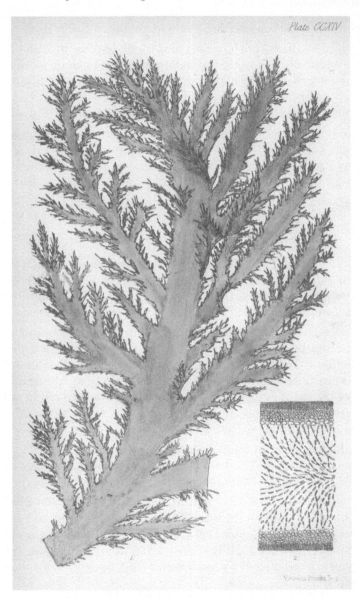

Figure 84. *Halymenia floresii* (Clemente) C. Agardh (Rhodophyta). Illustration from William H. Harvey (Plate CCXIV, Phycologia Australica, 1858–1863).

delicate. It is most often collected as free floating specimens or attached to harvested live rocks as it grows quite deep. Single plants can grow very large and reach heights of almost 50 cm or more. *H. floresii* requires good water quality and moderate to high flow to maintain its slippery fleshy blades. In an aquarium they will seldom attach so they must be anchored or allowed to tumble as free floating specimens. *H. floresii* is very palatable and is readily eaten by both Tangs and Angel fish (GCE 2010b).

Geographic distribution: Warm and temperate Atlantic; Indian Ocean; Asia (China, Japan, Taiwan); SE Asia (Indonesia, Malaysia, the Philippines, Singapore); Australia; Pacific Is (Micronesia, Fiji, Mariana Is).

Uses: Used as food (N'Yeurt 1999, Gray 2015); in Samoa, *Halymenia* is included with fish in the ground oven, or cooked as a vegetable with coconut milk. In Vanuatu it is eaten fresh or sundried, and used as a thickener and flavoring in seafood soup or stew. The fronds may also be cooked in coconut milk with local herbs and eaten as a vegetable. Dry fronds can be eaten as a snack. Fresh *Halymenia* is also sliced thin and eaten as a salad vegetable (N'Yeurt 1999, Novaczek 2001); also eaten in India (Rao 1970).

Extracts of this species have anticoagulant (Amorim et al. 2011), and anti-cancer (Novaczek 2001, Freile-Pelegrín et al. 2011) activity.

Halymenia floridana J. Agardh

Description: Pinkish, orange-reddish to red blades; a little thicker than membranous blades; very slick or slimy to the touch.

Geographic distribution: Atlantic Is (Bermuda, Selvage Is); Tropical and subtropical W Atlantic; Caribbean; Indian Ocean.

Uses: Used as food in Bangladesh (Aziz 2009).

Halymenia harveyana J. Agardh (formerly *Halymenia floresii* subsp. *harveyana*)

Common name: Gayunggayong (Agngarayngay et al. 2005).

Description: Thallus light to medium red or red-brown, 10–45 cm high, soft and mucilaginous (adhering strongly to paper), complanately and profusely branched with four-five orders of tapering laterals, occasionally with small surface leaflets; axes 1–3 cm broad, main laterals 1–2 cm broad and 300–400 µ thick, lesser laterals 3–10 mm broad, ultimate ones 1–2 mm broad and tapering to an acute apex; holdfast discoid (Womersley and Lewis 1994).

Geographic distribution: S Africa; SE Asia (Indonesia, the Philippines); Australia.

Uses: Used as food in salads and mixed with meat (Agngarayngay et al. 2005).

Halymenia maculata J. Agardh

Common name: Gayunggayong (Agngarayngay et al. 2005).

Description: Blade with about 8–10 cm long, not translucent, leathery (not slippery) and somewhat stiff. The portion near the base is flat but towards the edges it forms short flat branches with very frilly edges. The surface has little, regular bumps all over it. Colors from dull grayish brown to chocolate brown and purplish; also bright yellowish or orangey brown (Tan 2008c).

Geographic distribution: Asia (China, Indonesia, Malaysia, the Philippines, Singapore, Vietnam, Thailand); Australia (Papua New Guinea); Pacific Is (Fiji, Palau).

Uses: Used as food in salads and mixed with meat (Agngarayngay et al. 2005).

Halymenia microcarpa (Montagne) P.C. Silva

Description: Thalli purplish red, erect, bladelike, large and bushy, 10–35 cm tall, soft gelatinous and slimy, attached to rocky substrates by a discoid holdfast. The fronds are flattened, with a short stalk supporting 2–4 main axes which branching pinnately-alternately four-five times; the widest portion of main rachis are 5–15 mm wide; the diam. of branches decreases with increasing degree of branching. The ultimate branchlets are slender and linear with acuminate tips. The margins of the fronds are serrate. Surfaces of the axis are beset with few spine-like projections (Huang 2014).

Geographic distribution: Asia (China).

Uses: Used as food in China (Bangmei and Abbott 1987).

Hydropuntia edulis (S.G. Gmelin) Gurgel & Fredericq

Synonym: *Gracilaria lichenoides*

Common names: English: Ceylon moss (Chapman and Chapman 1980); Indonesian: Jafna moss, Agar-agar, Atjar (Zaneveld 1955), Ceylon (Kirby 1953), Doejoeng (Chapman and Chapman 1980).

Description: Thallus solitary, erect, cylindrical throughout, arising from a small discoid holdfast; main branches dichotomous, up to 14 cm high and less than 1.5 mm diam., with up to five orders of short branches less than 10 mm long toward the apices; branches irregularly arranged, acuminate; lower portions brown to purple; cartilaginous (Tsuda 1985).

Geographic distribution: Indian Ocean; SE Asia (Indonesia, Malaysia, the Philippines); Australia; Pacific Is (Hawaiian Is).

Uses: In Sri Lanka this species is eaten raw or often cooked with salt and tomatoes. This is done with fresh seaweed or after dipping it in boiling water. It is also served with onions and vinegar. For later use is dried without washing. When wanted it is soaked for a day, cleaned, chopped into small pieces and eaten. Also serves as food in the Philippines, Indonesia, Borneo, and the Hawaiian Is (Zaneveld 1959, Johnston 1966, Arasaki and Arasaki 1983, Sidik et al. 2012, Irianto and Syamdidi 2015).

See recipies to made agar and other dishes with *Gracilaria* (Annex I).

Hydropuntia eucheumatoides (Harvey) Gurgel & Fredericq (formerly *Gracilaria eucheumatoides*)

Common names: Filipino: Anggapang (Agngarayngay et al. 2005), Ambaang (Leyte 2015), Cauot-cauot, Kauatkauat (Zaneveld 1959), Cavot-cavot (Zaneveld 1955).

Description: Thalli are prostrate, cartilaginous, forming loose or thick clumps attached to coral reefs by means of discoid haptera, up to 15 cm in length. Thalli are coarse, consisting of irregular flattened branches, up to 1˙cm in width. The margins of flattened branches are mostly with coarse teeth of short spines. Reproductive structures were not found, but the spermatangial sori have been reported to be polycavernosa-type (Lin 2014b).

Geographic distribution: Indian Ocean; Asia (China, Japan, Taiwan, Indonesia, the Philippines, Thailand, Vietnam); Australia; Pacific Is (Micronesia, Fiji, Guam, Polynesia, Palau, Solomon Is).

Uses: In Vietnam this species is considered edible, but it does not occur abundantly enough to be especially collected. In the Philippines, Indonesia and Vietnam this species is eaten raw as a salad or cooked as a vegetable (Zaneveld 1955, 1959, Nang and Dinh 1998, Zemke-White and Ohno 1999, Hong and Hien 2004, Agngarayngay et al. 2005, Hong et al. 2007, Roo et al. 2007, Harrison 2013, Leyte 2015).

Hypnea aspera Kützing (formerly *H. boergesenii*)

Description: Thalli form dense tufts, consisting of more or less upright branches, brownish green always associated with other species of algae. Main cylindrical shafts, tapered in the summits, proliferated on top. Branching densely coated with very small lateral branches that are farther apart at the back than at its base; lateral branches can be simple or compound and pointed, forked or surrounded with short spiny projections; thallus to about 13 cm tall.

Geographic distribution: Indian Ocean; Asia (China, Korea, Taiwan); SE Asia (Indonsia, the Philippines, Vietnam); Australia.

Uses: Used as food in China and the Philippines (Bangmei and Abbott 1987, Leyte 2015).

Hypnea charoides J.V. Lamouroux

Common names: English: Spiny red weed (Santhanam 2015); Filipino: Kulot ti pusa (Agngarayngay et al. 2005); Kabutsu (Braune and Guiry 2011); Pacific name (Tonga Is): Limu vai (Ostraff 2003).

Description: Thallus cylindrical or filamentous, deeply dark red, also greenish-red, irregularly branching, loosely intertwined at the base. Branches pointed at the tip, covered on all sides with short, spike-like lateral branchlets (Braune and Guiry 2011).

Geographic distribution: W Indian Ocean (E Africa, Madagascar, Mauritius); NW Pacific (China, Japan) and W Pacific (Indonesia, Thailand, the Philippines); Pacific Is and Australia.

Uses: Used as food in preparation of salads and mixed with meat in China, Korea and Thailand (Kang 1968, Lewmanomont 1978, Tseng 1983, Bangmei and Abbott 1987, Hotta et al. 1989). This species is used also for agar production (Kang 1968, Okazaki 1971, Tseng 1983, Hotta et al. 1989, Agngarayngay et al. 2005).

Hypnea chordacea Kützing

Common name: Hawaiian: Limu huna (Fortner 1978).

Description: *H. chordacea* grows exposed on the low intertidal sections of lava or limestone flats. It is small, less than 15 cm in height, and branches tightly around a main stem. It looks like a forest of tall and narrow pine trees that have lost their needles (Fortner 1978).

Geographic distribution: Asia (China, Japan, Taiwan, Indonesia); Pacific Is (Hawaiian Is).

Uses: In Hawaii this species can be prepared fresh or cooked by methods closely following those for "ogo". Both can be used to make tempura or to thicken a soup or stew; in addition, can be used fresh in a "limu" salad (Fortner 1978).

Hypnea divaricata (C. Agardh) Greville

Common name: Arien (Amboina) (Zaneveld 1959).

Description: Thalli is distinguished by brownish-red or greenish-yellow branched thalli with short lateral branchlets (thorn-like) (Braune and Guiry 2011, Miguel et al. 2014).

Geographic distribution: Atlantic Is (Cape Verde Is); Indian Ocean; SW Asia (India); SE Asia (Indonesia, Myanmar, the Philippines); Australia; Pacific Is (Polynesia, Fiji, Samoan Is).

Uses: Used as food in Borneo, Indonesia, the Philippines, and Hawaii (Johnston 1966, Irianto and Syamdidi 2015). The algae are also eaten in Amboina (Moluccas Archipelago); this species is used in Timor for the preparation of domestic agar (Zaneveld 1959). In Myanmar a company produces instant noodle-like sheets using *Hypnea* for sale in the domestic market. These sheets are used as a dietary supplement to combat goiter, inhibit the progress of hypertension (Soe-Htum 1998).

Hypnea esperi Bory de Saint-Vincent (formerly *Gracilaria musciformis* var. *muscoides*)

Common name: Filipino: Ragutirit (Domingo and Corrales 2001).

Description: Thalli soft, forming dense tufts, attached to the substratum by small haptera. Main branches very fine, bearing short, spinose, determinate branchlets; *H. esperi* is the finest of the three *Hypnea* species collected (Domingo and Corrales 2001).

Geographic distribution: Indian Ocean; Asia (Japan, Taiwan, the Philippines, Singapore, Vietnam); Australia (Papua New Guinea); Pacific Is (Polynesia, Easter Is, Micronesia, Fiji, Polynesia, Marshall Is, Hawaiian Is); Antarctic Is (Fuegia); SE Pacific (Chile).

Uses: Used for making jellies in Vietnam (Nang and Dinh 1998, Zemke-White and Ohno 1999, Roo et al. 2007, Harrison 2013).

Hypnea hamulosa (Esper) J.V. Lamouroux

Description: This species has percurrent main branches and has relatively few spiny branchlets that are directed upward rather than at right angles.

Geographic distribution: SW Atlantic (Brazil); Indian Ocean; Asia (Taiwan, Indonesia, Thailand); Australia; Pacific Is (Polynesia).

Uses: Used as food in Thailand (Lewmanomont 1978).

Hypnea japonica Tanaka

Common name: English: Japanese red algae (Santhanam 2015).

Description: Thalli are purple-red in color, fleshy, terete, erect, bushy and entangled, irregularly branching with wide angles, branches in 0.1–0.2 cm in diam.; tips of the branches often elongate, swollen and hooked. Construction is uni-axial, and cross sections are pseudo-parenchymatous throughout, the medulla surrounding the distinct central axial filament and containing numbers of cells with lenticular secondary wall thickenings. Inner cells are multinucleate and linked by secondary pit connections (Huang 2014f).

Geographic distribution: Asia (China, Japan, Korea, Russia, Taiwan); Pacific Is (Central Polynesia).

Uses: Used for foodstuff and agar production in Korea (Kang 1968, Okazaki 1971, Bonotto 1976, Tseng 1983, Hotta et al. 1989). In eastern Guangdong (China) people use fresh *H. japonica* as a gelling material, which is combined with pork, fish, or vegetables and stewed together. On cooling, the gelled dish is called "Tsai ron dong" and is considered to be a treat (Bangmei and Abbott 1987).

Hypnea pannosa J. Agardh

Common names: English: Blue hypnea (Santhanam 2015); Filipino: Gulot, Lumi cevata, Tattered Sea moss, Kulot (Agngarayngay et al. 2005); Fijian: Lum icevata, Lumivakalolo (Ostraff 2003).

Description: The thalli are distictly intricate cespitode, forming low, prostrate clumps on the substratum, reddish, purplish or greenish when dried; main axes are not percurrent and are compressed or subteret measuring 0.3–1.7 mm broad in dried material; branching is repeatedly subdichotomous to alternate in wide angles forming rounded axil; branchlets bifurcate into short stubby spines at the terminal portions; short ultimate branchlets are not dense and are arranged alternately, pinnately or secundly; these are characteristically stout, stubby and spinose (Paula 2014c).

Geographic distribution: Widely distributed throughout the tropics and subtropics.

Uses: It is a carrageenan yielding plant. In Bangladesh, the Philippines and Fiji, this seaweed is considered edible and the freshly gathered seaweed is commonly prepared as a salad, or eaten cooked in coconut milk with condiments added to make use of their gelling properties (South 1993, Islam 1998, Trono 1998, N'Yeurt 1999, Zemke-White and Ohno 1999, Ostraff 2003, Agngarayngay et al. 2005, Roo et al. 2007, Harrison 2013, Paula 2014c).

Hypnea spinella (C. Agardh) Kützing

Synonym: *H. cervicornis*

Common names: Chinese: Sa ts'ai (Tseng 1935); Hawaiian: Limu huna (Fortner 1978); Indonesian: Boeloeng (Chapman and Chapman 1980).

Description: Thalli forming bushy mats, attached to the substratum by a discoid holdfast, or sometimes epiphytic on other algae, branches dense, lateral ultimate branchlets numerous, short, tapering or forked (Domingo and Corrales 2001).

Geographic distribution: Atlantic Is (Ascension, Bermuda, Canary Is, Cape Verde Is, Selvage Is); Tropical and subtropical W and E Atlantic; Mediterranean; Persian Gulf; Indian Ocean; SW Asia (Arabian Gulf, Cyprus, India, Iran, Oman, Sri Lanka, Turkey); Asia (China, Japan, Korea,

Taiwan); SE Asia (Indonesia, Myanmar, Singapore, Thailand, Vietnam); Australia; Pacific Is (Samoa, Polynesia, Micronesia, Hawaiian Is, Mariana Is, Marshall Is).

Uses: Edible species, usually eaten (boiled in coconut milk) in the Pacific and Asia (Zaneveld 1959, Johnston 1966, Lewmanomont 1978, Arasaki and Arasaki 1983, Bangmei and Abbott 1987, Simoons 1990, Payri et al. 2000, Sidik et al. 2012, Irianto and Syamdidi 2015). In Hawaii this species can be prepared fresh or cooked by methods closely following those for "ogo". Both can be used to make tempura or to thicken a soup or stew; in addition, can be used fresh in a "limu" salad (Fortner 1978). In the Persian Gulf, this species is considered a potential edible alga (Mohammadi et al. 2013).

H. spinella is considered an economically important seaweed because it produces carrageenan, a binding and smoothing agent used in many commercial products such as toothpaste, ice cream, pet foods, etc. In addition, *H. spinella* is important to the medical and pharmaceutical industries, because agar, an important derivative of carrageenan, is a principal component of bacterial culture media (Payri et al. 2000, Hill 2001, Sidik et al. 2012, Pereira 2015a).

Hypnea saidana Holmes

Common name: Filipino: Kulot (Agngarayngay et al. 2005).

Description: Larger diam. of axes (> 500 μ), axes subcomplanate, axial cell smaller than periaxial cells in cross section; absence of lenticular cell wall thickenings in periaxial cells, presence of hooked axes and stichidia-like tetrasporangial sori (Tsiamis and Verlaque 2011).

Geographic distribution: Indian Ocean; Asia (Japan, Korea, Taiwan, the Philippines); Australia; Pacific Is (Polynesia, New Caledonia, Hawaiian Is).

Uses: Used as food in salads, and mixed with meat (Agngarayngay et al. 2005).

Hypnea valentiae (Turner) Montagne

Common names: Filipino: Culot ti pusa (Domingo and Corrales 2001); Tamil: Sem pasi (Kaliaperumal et al. 1995).

Description: Thalli forming tufts of fine entangled branches attached to the substratum by a discoid holdfast, sometimes on other algae. The main axis is cylindrical, giving rise to straight indeterminate branches, which in turn give rise to slender determinate branchlets (Domingo and Corrales 2001).

Geographic distribution: Atlantic Is (Bermuda, Canary Is, Cape Verde Is, Selvage Is); Tropical and subtropical W and E Atlantic; Mediterranean Sea; Indian Ocean; SE Asia (Indonesia, the Philippines, Vietnam); Australia and

New Zealand; Pacific Is (Polyneisa, Micronesia; Fiji, Hawaiian Is, Mariana Is).

Uses: Used as food and for carrageenan production (Rao 1977); in India, Bangladesh and Vietnam this species is used as food (Kaliaperumal et al. 1995, Islam 1998, Zemke-White and Ohno 1999, Nang 2006, Hong et al. 2007, Roo et al. 2007, Harrison 2013).

Kappaphycus alvarezii (Doty) Doty ex P.C. Silva

Common names: Adik goma, Adik kalas, Agal agal, Agal agal besar, Agar agar besar, Agar agar palau, Agar agar seru laut, Agar-agar, Chilin-t' sai, Cottonii, Elkhorn sea moss, Eucheuma, Eucheuman, Guso, Kab kab, Kappa, Kirinsai, Purdoy, Tambalang, Tambalang milo, Vanguarda (SIA 2014); Filipino: Goso (Leyte 2015).

Description: The thalli are smooth cylindrical cartilaginous and bushy with multiaxial prostrate or erect branches. Branching not truly opposite arise often from undersurface of dense thalli. Three different morphotypes are found which are red green and brown in color. The plant shows existence of different morphotypes or color varieties in the cultivation field in response to the environment. Three different color strains such as deep reddish brown, dark green and yellowish green thalli are found. The thalli are smooth cylindrical cartilaginous, bushy multiaxial, prostrate or erect in habit and consist of irregular indeterminate branches that are rigid. Sometimes branching is irregularly pinnate, opposite or falsely dichotomous. The indeterminate primary branches arise one at a time from near the tip of non-damaged axis. Secondary branches are common on the older segments. Branches generally curved upward directed towards the right. Tips of the branches are either pointed or bifurcated. Irregularly arranged wart-like protuberances are found all along the thalli. Each thallus weighs between 500 to 1000 g, and measures 20–40 cm in length (Shaoo 2010).

Geographic distribution: Native to the Indo-pacific (Malaysia and the Philippines) but widely introduced and cultivated in the W Pacific, SW Atlantic, and Indian Oceans.

Uses: *K. alvarezii* is predominantly utilized as raw material for the phycocolloid known as carrageenan, although it is also used directly as a fresh, whole food source (eaten in seaweed salads and used in other recipes) in the tropical areas where it is harvested, namely in Indonesia, the Philippines, Malaysia and Vietnam (Trono 1998, Zemke-White and Ohno 1999, Hong et al. 2007, Roo et al. 2007, Lee 2008, Ahmad et al. 2012, Gray 2015, Irianto and Syamdidi 2015, Leyte 2015).

Kappaphycus cottonii (Weber-van Bosse) Doty ex P.C. Silva

Synonyms: *Eucheuma cottonii, E. okamurae*

Common names: English: Guso (Santhanam 2015); Filipino: Kanot-kanot, Kanutkanot (Domingo and Corrales 2001, Agngarayngay et al. 2005).

Description: Thalli generally prostrate, branching irregular, attached to each other to form cartilaginous frond covering the substratum. The tipper surface of the frond wrinkled, covered by numerous warts, the ventral surface smoother, with numerous hapters for attachment (Domingo and Corrales 2001).

Geographic distribution: Indian Ocean; Asia (China, Japan, Taiwan); SE Asia (Indonesia, Malaysia, the Philippines, Singapore, Vietnam); Pacific Is (Micronesia, Fiji, Guam, Mariana Is).

Uses: Used as food in China, the Philippines and Malaysia (Bangmei and Abbott 1987, Agngarayngay et al. 2005, Matanjun et al. 2009, Foon et al. 2014). *K. cottonii* is used for making jellies and cakes in Vietnam (Nang and Dinh 1998, Zemke-White and Ohno 1999, Roo et al. 2007).

Kappaphycus striatus (F. Schmitz) Doty ex P.C. Silva
(formerly *Eucheuma striatum, K. striatum*)

Common names: Filipino: Kanot-kanot, Kanutkanot (Domingo and Corrales 2001, Agngarayngay et al. 2005), Kapkap (Leyte 2015).

Description: Thalli consist of prostrate and erect branches, covered with spinose branchlets irregularly arranged, attached to rocky substratum by well-developed hapters arising from the ventral portions of the prostrate branches; branching is irregular, alternate to second, the branches forming acute angles, terete attenuated to acute or spinose tips which vary in length from 1–17 mm and 1–2 mm thick near the base (Abbott 1999, Domingo and Corrales 2001).

Geographic distribution: Asia (China, Japan); SE Asia (Indonesia, Malaysia, the Philippines, Singapore); Pacific Is (Micronesia, Fiji, Hawaiian Is, Palau).

Uses: In Malaysia and the Philippines is used in salads and mixed with meat (Agngarayngay et al. 2005, Ahmad et al. 2012, Gray 2015, Leyte 2015). This species is also used for carrageenan production (Pereira 2004, Pereira et al. 2009a, Ahmad et al. 2012).

Laurencia botryoides (C. Agardh) Gaillon (Fig. 85)

Description: Dark red to red-brown thallus, robust, firm, drying cartilaginous, 5–17 cm high, pyramidal in outline, with percurrent axes bearing distichous laterals similarly branched, all branches terete, 1–10 mm

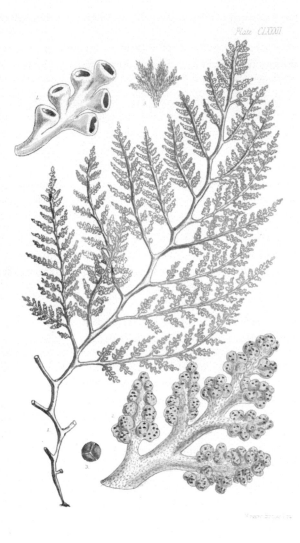

Figure 85. *Laurencia botryoides* (C. Agardh) Gaillon (Rhodophyta). Illustration from William H. Harvey (Plate CLXXXII, Phycologia Australica, 1858–1863).

apart, with reproductive ramuli forming botryoidal clusters; axes 1.5–3 mm in diam., decreasing to 0.5–1 mm in lesser branches; holdfast stoloniferous; epilithic (Womersley 2003).

Geographic distribution: Indian Ocean; SE Asia (Malaysia, Singapore); Australia; Pacific Is (Fiji).

Uses: Used as food in Asia (Zaneveld 1959, Johnston 1966, Sidik et al. 2012).

Laurencia coronopus J. Agardh

Description: Thallus brushy, 5–15 cm high, 0.5–3 mm thick, color varies from yellow to greenish-brown and reddish-brown, attached by basal holdfast; main axis slightly compressed in upper part, bears short tubular or club-shaped branchlets; branches with short proliferations, arranged diversely, in different planes, some regularly, alternately or oppositely, others are whorled, close to apex in same plane (Milchakova 2011).

Geographic distribution: Black Sea; SE Asia.

Uses: *L. coronopus* is a valuable commercial species, the source of antimicrobial, fungicidal and antitumor BASs, PUFA, agaroids; contain rare microelements (strontium, rubidium, zirconium, etc.); used as foodstuff in some countries of SE Asia (Milchakova 2011).

Laurencia flexilis Setchell

Description: Erect thalli, up to 40 cm high, attached by discoidal to rhizoidal to encrusting holdfasts; thalli sparingly to highly branched in all directions or bilateral; branches cylindrical to flat; ultimate branchlets blunt to truncate, often claviform, not much narrower than other branches.

Geographic distribution: Atlantic Is (Canary Is, Madeira, Selvage Is); Indian Ocean; Asia (China, Japan); SE Asia (Indonesia, the Philippines); Australia; Pacific Is (Polynesia, Micronesia, Tahiti).

Uses: Used as food in China (Bangmei and Abbott 1987).

Lemanea fluviatilis (Linnaeus) C. Agardh

Description: Freshwater algae, with plants in clods, in the form of a needle, to 12 cm height, unbranched or branched. The thalli get narrower to apex in thin thread and in base in thin, cylindrical stems. They do not have a cortex around the central axis cell. Their color is from olive green to violet green (Simić 2007).

Geographic distribution: Europe, India, Asia (Russia).

Uses: Used as food in India; as the alga grows in the rivers for only few months, particularly in winter, the local people collect the alga from the river and sundry it for use throughout the year. The sun dried algae are eaten regularly with major meals. It is cooked with vegetables primarily for its characteristic fishy smell. The dried filaments are also added to "Singju" (a local preparation) to make a "Manipuri" delicacy. These plants are served in chutney (a sour preparation) or as a vegetable in dried, fried or roasted form (Bhosale et al. 2012).

Meristotheca papulosa (Montagne) J. Agardh

Synonym: *Eucheuma papulosa*

Common names: English: Rosy pudding plant (Novaczek 2001); Japanese: Tosakanori, Keikansai (Arasaki and Arasaki 1983, Mouritsen 2013), Tosaka-nori (Chapman and Chapman 1980); Lumu mie'ta (Rotuma) (Novaczek 2001).

Description: Thalli are purple colored, cartilaginous, blade-like with irregular branching, erect to prostrate and imbricating with numerous secondary holdfast disks. Thalli are multiaxial. The medulla is broad and filamentous, and is composed mainly of unbranched rhizoids. Inner cortical cells are large and often stellate.

Geographic distribution: Asia (China, Japan, Korea, Taiwan); SE Asia (Indonesia, the Philippines); Australia; Pacific Is (Polynesia).

Uses: Common seaweed used as food in Japan, as seafood and salad (Arasaki and Arasaki 1983, Lembi and Waaland 1988, Zemke-White and Ohno 1999, Lideman et al. 2011, Harrison 2013); used also for foodstuff in China and Korea (Kang 1968, Bonotto 1976, Chapman and Chapman 1980, Tseng 1983, Bangmei and Abbott 1987, Hotta et al. 1989, Hu 2005, Ohno and Largo 2006, Roo et al. 2007, Mouritsen 2013), and as food stabilizer (Okazaki 1971). *Meristotheca* has a nutty flavour and a texture like soft pudding when cooked; boiled in coconut milk with lemon juice, onion, vegetables and fish (Novaczek 2001).

Laurencia pinnata Yamada

Common name: Filipino: Kulot (Agngarayngay et al. 2005).

Description: The fronds are caespitose, erect, cartilaginous and rigid in texture and generally dark purple or purplish brown in color. They stand on a discoid base and never adhere to paper when dried. Stoloniferous basal branches are present but not abundant. The erect main branches are cylindrical, measuring 1100–1800 µ in diam. near the base but becoming complanate upward (Saito 1967).

Geographic distribution: Asia (China, Japan, Korea, Russia, Indonesia, Malaysia, the Philippines); Pacific Is (Polynesia).

Uses: Used in the USA as a seasoning (Chapman and Chapman 1980); this species is also used as food in salads, and mixed with meat in the Philippines (Agngarayngay et al. 2005).

Palisada perforata (Bory de Saint-Vincent) K.W. Nam

Synonym: *Laurencia papillosa*

Common names: Filipino: Culot, Kulot, Culot-tumeng (Domingo and Corrales 2001, Agngarayngay et al. 2005); Hawaiian: Limu maneoneo, Limu lipeepee (Reed 1906).

Description: Thalli erect, cylindrical and cartilaginous, attached to the substratum by a discoid holdfast, forming low dense clusters, irregularly branched, main axis, naked at lower portions. Plants are attached to the substratum by a discoid holdfast, the plant appearing dirty due to the dense branchlets filled with sand and other corals (Domingo and Corrales 2001, Taşkın and Sukatar 2013).

Geographic distribution: Atlantic Is (Ascension, Canary Is, Cape Verde Is); Tropical and subtropical W and E Atlantic; Adriatic Sea; Mediterranean; Persian Gulf; Indian Ocean; Asia (Indonesia, Japan, Taiwan, Korea, the Philippines, Thailand, Vietnam); Australia; Pacific Is (Fiji, Micronesia, Mariana Is, New Caledonia, Hawaiian Is).

Uses: Used as food in jellies, salads and mixed with meat in the Philippines and Thailand (Johnston 1966, Michanek 1975, Lewmanomont 1978, Lee 2008, Agngarayngay et al. 2005, Trono and Montano 2006, Sidik et al. 2012). *P. perforata* is liked by both Hawaiians and Filipinos; the seaweeds are eaten raw with tomatoes or, after hot water has been poured over them, are combined with mashed tomatoes and eaten cold (Reed 1906, Zaneveld 1959, Trono and Montano 2006); also eaten in India (Rao 1970, Kaliaperumal et al. 1995); consumed also in Caribbean and Central America (Harrison 2013). In the Persian Gulf, this species is considered a potential edible alga (Mohammadi et al. 2013).

P. perforata is the source of antimicrobial, fungicidal and antitumor BASs, PUFA, phycocolloids, agaroids and phenol-bromides, rare microelements; also provide natural fertilizer (Milchakova 2011).

Parviphycus pannosus (Feldmann) G. Furnari

Synonym: *Gelidiella tenuissima*

Description: This species forms dense tufts at the upper intertidal level on rocky substrata, exposed to intense wave action. Distinctive morphological characters are compressed upright axes closely arising from short stolons, numerous first-order branches and clavate tetrasporangial sori at main and lateral apices (Bottalico et al. 2014).

Geographic distribution: Adriatic Sea; Tropical and subtropical W Atlantic; Atlantic Is (Azores, Canary Is, Cape Verde Is, Madeira, Selvage Is); SW Mediterranean (Balearic Is, Corsica, France, Greece, Italy, Malta,

Sardinia, Spain, Tunisia); Indian Ocean; SW Asia (Bangladesh, India, Israel, Turkey); SE Asia (Indonesia); Australia; Pacific Is (Polynesia, Micronesia, Fiji, Mariana Is, Samoa).

Uses: In Bangladesh and the Philippines this species is used as food (Islam 1998, Zemke-White and Ohno 1999, Roo et al. 2007, Harrison 2013).

Porphyra atropurpurea (Olivi) De Toni

Common names: Hawaiian: Limu luau, Lipahee; Filipino: Garnet (Zaneveld 1959).

Description: Foliate plants consist of a single layer of cells, 30–40 mm thick, higher than wide in cross-section (7 x 21 µ); reach 15 cm long and 10 cm wide; purple-brown blades or distinctly green, more or less whole or often lobed, with lacerated margins in adults (Joly 1965).

Geographic distribution: Mediterranean (Italy, Turkey); SW Atlantic (Brazil); SE Asia (Indonesia, the Philippines).

Uses: Used for making jellies with coconut milk and vegetable soup in Indonesia; used also in goiter treatments, urinary diseases and dropsy (Istini et al. 1998, Zemke-White and Ohno 1999, Roo et al. 2007, Harrison 2013, Irianto and Syamdidi 2015).

"Limu luau" is considered a great delicacy in the few localities where it occurs in Hawaii. It is found on bold exposed rocks constantly dashed by waves; so it is difficult and dangerous to collect it, especially as it is extremely slippery and has to be scraped forcibly from the rocks in small bunches while the collector clings to his support and avoids the heavy waves (Zaneveld 1959).

Porphyra indica V. Krishnamurthy & M. Baluswami

Common names: English: Laver; Japanese: Nori (Kaliaperumal et al. 1995).

Description: Plants 18 cm high and up to 11 cm broad through its broadest portion; thallus reddish pink with smooth margin without any spinulose processes; attached to the substratum by rhizines (Kaliaperumal et al. 1995).

Geographic distribution: Indian Ocean.

Uses: In India is used as snacks, salad or added to soups and various dishes (Kaliaperumal et al. 1995).

Porphyra kanyakumariensis V. Krishnamurthy & M. Baluswami

Description: The thallus is broad, membranous, undulate, lanceolate in shape and pale violet in color. The blades are 15–33 cm long and 2.0–5.6 cm broad and grow luxuriantly in upper intertidal region. The blades are attached to the substratum by discoid holdfast through a large number of rhizoids. The entire blade starting from the holdfast region to the upper fertile part is covered with thick mucilaginous layers. The margins of the blade are serrated with single or multicellular spines which are 6–8 μ long. The holdfast region has large round, oval, elliptical cells which have long clear, thin protuberances oriented towards the base. These cells are embedded in thick intercellular matrix (Sahoo 2010).

Geographic distribution: Indian Ocean (India).

Uses: Used as food in India (Mathew 1991).

Porphyra marcosii P.A. Cordero

Common name: Gamet (Agngarayngay et al. 2005).

Description: It is characterized as light purplish or brownish red monostromatic blades, up to 14 cm in height. The blades are linear-lanceolate, laterally or very rarely basally branched thallus blades, attached by a small disk. It is monoecious with both male and female gametes interspersed with the archespores and vegetative cells (Monotilla and Notoya 2010).

Geographic distribution: SE Asia (the Philippines); Australia.

Uses: Used as food in salads and mixed with meat in the Philippines (Agngarayngay et al. 2005).

Porphyra monosporangia S. Wang & J. Zhang

Description: Plant is a single membranous blade composed of monostromatic cells, generally round, half-round and irregularly undulated like flowers in shape; measuring about 7–22 cm high and 7–27 cm width; orange yellow in color at the margin of the blade, and green or grass green in the middle of the vegetative portion; with a cordate or umbilicalic base; sessile with a small discoid holdfast near the base; margin of blade with microscopic teeth consists of 1–3 dentate cells, more dense at the base (Sujuan and Jingrong 1980).

Geographic distribution: Asia (China).

Uses: Used as food in China (Bangmei and Abbott 1987).

Porphyra oligospermatangia C.K. Tseng & B.F. Zheng

Description: It has a membranous thallus, obovate or elongated-obovate in shape, generally 6–30 cm reaching 39 cm high, 5–24 cm broad. Apex not undulated, marginal portions sometimes slightly undulated, with cuneate, rotund or cordate base, rhizoidal cells orbiculate, 30–35 μ in diam., with a small discoid holdfast near the base; blade monostromatic, margin microscopically edentate, vegetative cells, irregularly subquadrate or rectangular in surface view (Zheng 1981).

Geographic distribution: Asia (China).

Uses: Used as food in China (Bangmei and Abbott 1987).

Portieria hornemannii (Lyngbye) P.C. Silva

Synonym: *Chondrococcus hornemannii*

Common name: Japanese: Hosoba-naminohana (Oshie-Stark et al. 2003).

Description: Thalli are orange-red colored, gelatinous, 3–12 cm in height, overlapping flattened branches with discoid holdfast, irregularly pinnate-alternate branching in one plane. The terminal branches at the distal portion of the thallus have slightly expanded curved or enrolled tops. Gland cells are scattered in both nemathecial and normal cortical tissue (Huang 2015b).

Geographic distribution: Indian Ocean; Asia (China, Japan, Korea, Taiwan); SE Asia (Indonesia, the Philippines, Singapore, Sri Lanka, Vietnam); Australia; Papua New Guinea; Pacific Is (Fiji, Hawaiian Is, Mariana Is, Micronesia, Polynesia).

Uses: Potentially edible seaweed (Bhuvaneswaris and Murugesan 2013).

Pterocladia lucida (R. Brown ex Turner) J. Agardh (Fig. 86)

Common name: English: Agar weed.

Description: Thallus red-brown to dark red, 8–40 cm long, cartilaginous, with few to several erect, bi- to tripinnate, complanate fronds from a fibrous to hapteroid, stoloniferous base. Axes flat, 3–4 mm broad centrally and above, with a thicker midrib below, with broad apices when actively growing but tapering in some plants, alternately distichously branched at intervals of 1–10 mm, pinnae with rounded axils, not or partly denuded below; pinnae flat, 2–10 cm long and 1–4 mm broad, thicker centrally, slightly basally constricted, usually with broad apices; pinnules alternate, flat, 2–5 mm long and 0.3–1 mm broad, basally slightly constricted; holdfast discoid, becoming stoloniferous; epilithic (Womersley 1994).

Geographic distribution: Atlantic Is (St. Helena); SE Asia (Indonesia); Indian Ocean; Australia and New Zealand.

Uses: Used as food in New Zealand (Wassilieff 2012a).

Figure 86. *Pterocladia lucida* (R. Brown ex Turner) J. Agardh (Rhodophyta). Illustration from William H. Harvey (Plate CCXLVIII, Phycologia Australica, 1858–1863).

Pyropia dentata (Kjellman) N. Kikuchi & M. Miyata (formerly *Porphyra dentata*)

Common name: Japanese: Oni-amonari (Arasaki and Arasaki 1983).

Description: Thallus membranous, orbicular, 2–6 cm high, to 10 cm broad, sessile, commonly aggregate, often overlapping each other, with smooth or slightly undulate margins, purple-red to dark brownish-red, often with lighter marginal stripe, cordate or sometimes cuneate at base with small discoid attachment; blade monostromatic, thin, (40)-50-70 μ; margins inrolled, edges dentate (Guiry and Guiry 2014).

Geographic distribution: Asia (China, Japan, Korea, Taiwan).

Uses: Used as food in China (Bangmei and Abbott 1987), Korea (Kang 1968, Bonotto 1976, Tseng 1983), and Japan (Arasaki and Arasaki 1983).

Pyropia suborbiculata (Kjellman) J.E. Sutherland, H.G. Choi, M.S. Hwang & W.A. Nelson (formerly *Porphyra suborbiculata*)

Common names: Chinese: Tsu choy, Tsz Tsai, Tsu Tsoi, Tsu Tsai, Chi Tsai, Hung Tsoi, Hung Tsai, Chi Choy, Hai Tsai, Hai Tso (Madlener 1977), Zi-cai (Arasaki and Arasaki 1983); English: Red Laver (Madlener 1977); Japanese: Mambiama, Maruba-amanori (Madlener 1977), Iwanori (Arasaki and Arasaki 1983); Korean: Kim (Madlener 1977).

Description: Thallus membranous, orbicular, 2–6 cm high, to 10 cm broad, sessile, commonly aggregate, often overlapping each other, with smooth or slightly undulate margins, purple-red to dark brownish-red, often with lighter marginal stripe, cordate or sometimes cuneate at base with small discoid attachment; monostromatic blade, thin, 40–70 μ; margins inrolled, edges dentate.

Geographic distribution: NW Atlantic; Tropical and subtropical W Atlantic; SW Asia (Sri Lanka); Asia (China, Japan, Korea, Taiwan); SE Asia (the Philippines, Vietnam); Australia and New Zealand.

Uses: Used as food in salads, in Vietnam, Korea and China (Kang 1968, Bonotto 1976, Madlener 1977, Tseng 1983, Bangmei and Abbott 1987, Hong and Hien 2004, Lee 2008, Harrison 2013).

Sarcodia montagneana (J.D. Hooker & Harvey) J. Agardh

Synonym: *Sarcodia ceylanica*

Common names: Indonesian: Bebiroe, Bibiru (Zaneveld 1959); Japanese: Atsuba-nori (Chapman and Chapman 1980).

Description: Thalli are greenish to purplish-red, flattened; reach 8–15 cm in length, irregular dichotomously branched, with greater or lesser numbers of marginal proliferations. Constructions are multiaxial. The medulla is a broadly filamentous mix of primary and rhizoidal filaments, and the cortex is composed of isodiametric cells of progressively decreasing size. Cystocarps are scattered across the blades or confined to the margins (Huang 2014e).

Geographic distribution: Indian Ocean; SW Asia (India, Oman, Sri Lanka, Yemen); Asia (China, Japan, Taiwan); SE Asia (Indonesia, the Philippines); Australia; Antarctica and Subantarctic Is.

Uses: Used as food in Asia (Johnston 1966, Chapman and Chapman 1980, Arasaki and Arasaki 1983, Irianto and Syamdidi 2015). Used frequently as food in the Moluccas (Zaneveld 1955, 1959).

Sarconema filiforme (Sonder) Kylin

Synonym: *S. furcellatum*

Description: Plants tufted, up to 15 cm high, 1–2 mm broad; repeatedly dichotomously branched, forming dense broad intricate tufts; color brick-red or yellowish-red, fleshy consistency, plants breaking quickly when handled (Kaliaperumal et al. 1995).

Geographic distribution: Indian Ocean; Asia (China); SE Asia (the Philippines); Australia; Pacific Is (Polynesia, Samoa).

Uses: It can be used for production of iota-carrageenan (Hoppe 1979). It can also be used for human consumption in India as it controls goiter disease (Kaliaperumal et al. 1995).

Sarcothalia radula (Esper) Edyvane & Womersley
(formerly *Irideae radula*)

Common name: English: Tongue-weed (Braune 2008).

Description: Plants red-brown, of a single or several, large, tough blades 150–400 mm long (over 1 m long in Tasmanian specimens!), lance-shaped when young, broad (≈200 mm) or narrow (≈50 mm) when mature, with small, pimply (papillose) growths on the surface and edges of fertile blades; attached by a minute, inconspicuous stalk (2–8 mm long); some plants form dark, narrow, side blades, and edges of blades may be wavy or have small teeth (Baldock 2014).

Geographic distribution: Falkland Is/Malvinas Is; Antarctic Is; SE Asia; Australia (Tasmania, Victoria).

Uses: Used as food in SE Asia (Chapman and Chapman 1980).

Scinaia hatei Børgesen

Description: Gametophytes erect from discoid holdfasts, dichotomously branched. Axes soft or cartilaginous, noncalcified, terete, complanate or flat, constricted at the nodes or unconstricted, multiaxial (Braune and Guiry 2011).

Geographic distribution: SE Asia (India, Oman, Pakistan, Yemen).

Uses: Considered an edible species (Usov and Zelinsky 2013, Gray 2015); extracts of this species have anti-herpetic (Mandal et al. 2008), antiviral (Pujol et al. 2012), and antileishmanial (Sabina et al. 2005) activity.

Scinaia hormoides Setchell

Common name: Gargarnatis (Agngarayngay et al. 2005).

Description: Erect thalli, repeatedly forking, tufted, axes cylindrical, with continuous constrictions, soft, not calcified; discoid holdfast (Braune and Guiry 2011).

Geographic distribution: Asia (Japan, Indonesia, the Philippines); Australia (Papua New Guinea); Pacific Is (Hawaiian Is).

Uses: Used as food in salads and mixed with meat in the Philippines (Domingo and Corrales 2001, Agngarayngay et al. 2005, Trono Jr. and Montano 2006).

Scinaia moniliformis J. Agardh

Description: Medium brown-red, 5–22 cm high, subdichotomously branched usually from each segment, regularly constricted into ovoid segments 6–17 mm long and 3–5 mm in diam.; holdfast discoid, 1–2 mm across; epilithic (Womersley 1994).

Geographic distribution: Indian Ocean; Asia (China, Japan, Taiwan, the Philippines); Australia.

Uses: Commonly used in vegetable salads in the Philippines (Trono 1998, Zemke-White and Ohno 1999, Roo et al. 2007, Harrison 2013).

Spyridia fusiformis Børgesen

Description: Plants gregarious, forming isolated tufts, erect, 5–10 cm long, bright pinkish red; attachment by well-developed stolonoidal structures; all axes cylindrical, the side branchlets markedly fusiform; basal parts of the main axes (almost) bare, about 1 mm in diam., the upper parts more densely branched, in some specimens becoming even densely intricated, resulting in a bushy form, gradually tapering towards the apices. Main axes and indeterminate side branches completely corticated, but the segments still clearly visible in transparency with the naked eye in the field, cortical cells markedly elongated. All axes bearing relatively stiff, straight, uniseriate filaments, 600–750 μ long, 20 μ in diam., composed of cells about 60 μ long, presenting a single tier of small cortical cells, 10 μ high at their transverse walls; apices of these filaments rounded, without a terminal spine. Tetrasporangia formed at the basal nodes of the uniseriate filaments, singly or in groups of 3–4, oval, 100 μ long and 70 μ broad (Coppejans et al. 2009).

Geographic distribution: Indian Ocean.

Uses: Potentially edible seaweed (Bhuvaneswaris and Murugesan 2013).

Titanophora weberae Børgesen

Common name: Filipino: Aragan elik (Agngarayngay et al. 2005).

Description: Thalli are up to 13 cm tall and 0.8–0.9 mm thick. The holdfast is a fleshy disk about 1 mm in diam., giving rise to a short terete stipe. The stipe measure up to 4 mm long and 0.8 mm in diam., and expand directly into the leading axes which branch repeatedly. The leading axes are complanate and 6–11 mm wide, while the terminal branches are narrower and frequently subterete (Itono and Tsuda 1981).

Geographic distribution: Indian Ocean; Asia (Japan, Taiwan, Indonesia, the Philippines, Vietnam); Australia; Pacific Is (Samoa, Polynesia, Micronesia, Fiji, Palau, Solomon Is).

Uses: Used as food in salads and mixed with meat (Agngarayngay et al. 2005).

Trichogloea requienii (Montagne) Kützing

Common name: Filipino: Barisbaris (Agngarayngay et al. 2005).

Description: Thallus to 25 cm high, dark red-purple to greenish, paniculate, extremely mucilaginous, lightly calcified, generally with several axes arising from an ill-defined holdfast. Primary axes terete, to 7 mm in diam., tapering to the apices; lateral branches 1–2 mm in diam. Medullary cells elongate 15–45 μ in diam.; assimilatory filaments sparsely dichotomously (rarely trichotomously) branched, near periphery unbranched for 10–12 cells; lower cells elongate, 5–12 μ in diam.; distal cells becoming broader and shorter, 15–25 μ in diam.; adventitious rhizoids common, arising from lower cells of assimilatory filaments (Huisman and Parker 2011c).

Geographic distribution: Atlantic Is (Cape Verde Is); Tropical and subtropical W Atlantic; Indian Ocean; Asia (China, Japan, Korea, Taiwan, Indonesia, the Philippines); Australia; Pacific Is (Fiji, Polynesia, Hawaiian Is, Mariana Is).

Uses: Used as food in salads and mixed with meat in the Philippines (Agngarayngay et al. 2005).

2.5 Pacific Islands (Micronesia, Polynesia, Melanesia, and Hawaiian Is)

For the production of food, people living on continents and large island masses have mostly been accustomed to what is regarded as "conventional agriculture" with its cereal crops, pastures and grazing animals. Consequently, the use of seaweeds as food would usually be foreign to such communities except for those living on a seacoast—and they would use algae only to allay hunger in times of dire necessity such as crop-failure, to follow a fad, or to perpetuate the practice of an older indigenous population. It so happens that most if not all of the textbooks on algae have been written by people from societies living on land masses pedologically and climatically suited to conventional agriculture. Thus in these books the use of seaweeds as food finds little mention, and one might form the erroneous opinion that they are rather insignificant as articles of diet (Johnston 1966).

But outside the European, Eurasian, African, Australian and American continents and the larger islands capable of supporting agriculture as one knows it, one finds many millions of people living on islands where conventional agriculture can never be adopted. For this reason, those people inhabiting myriads of islands in the Pacific and Indian Oceans, the South China Sea and the seas around Borneo, New Guinea and Indonesia have traditionally and of necessity relied on the sea as their major source of food. For this reason, many of these island people have come to rely on seaweeds as an important element in their diet. In certain areas such as Japan this reliance has been so great and the demand so constant that methods of seaweed cultivation have been evolved and put into practice. Acquired knowledge and skill have finally produced specialists in this technique, and given rise throughout time to a thriving industry employing considerable labor. Large outside demands have led over the years to the development of an export market (Johnston 1966).

For example, among Polynesians, Hawaiians are unique in their regular use of seaweeds. In the olden times, seaweed was the third component of a nutritionally balanced diet consisting of fish and poi. While seaweed primarily supplied variety and interest, they also added significant amounts of vitamins and other mineral elements to the diet. A common part of the traditional Hawaiian diet, seaweeds are still a common ingredient in foods enjoyed by all. Not only are seaweeds sold in supermarkets, but original Hawaiian methods of preparation are still used (Abbott 1984).

2.5.1 Chlorophyta

Caulerpa bikinensis W.R. Taylor

Description: Thallus grows up to 15 cm high, takes a light green almost tending to yellowish in the best conditions. This *Caulerpa* is composed of creeping stolons up to 1 m long and 1 cm wide, bearing erect fronds 2–3 mm wide provided with pinnules elongated club-shaped, flat or slightly convex. The pinnae are up to 10 mm long and 4 mm wide are arranged alternately or couplet. This species also thrives in the lagoons on the outer slope of the atolls to 70 m depth. It can form a dense and continuous belt on the outer slopes of some atolls (Fortier 2010).

Geographic distribution: Pacific Is (Marshall Is, Micronesia, Palau, and Polynesia).

Uses: In French Polynesia this species is eaten as a salad in the Austral Is (N'Yeurt 1999, Conte and Payri 2006), is also eaten with coconut milk in several Pacific Is (Payri et al. 2000).

Caulerpa chemnitzia (Esper) J.V. Lamououx

Synonyms: *Caulerpa laetevirens, C. racemosa* var. *occidentalis, C. racemosa* var. *turbinata, C. racemosa* var. *peltata*

Common names: Filipino: Ararusip (Agngarayngay et al. 2005); Indonesian: Lata, Boelong (Chapman and Chapman 1980); Fijian: Nama, Nama wawa, Numa balavu (Ostraff 2003); Polynesian: Imu tupua (flower algae), Imu pokupuku (pellet algae), Konini (Conte and Payri 2006).

Description: Similar with other *Caulerpa* (i.e., *C. racemosa*), the arrangement of their branchlets and in shape of the branchlets which are expanded into trumpet-shape ends (Chiang 1960).

Geographic distribution: Pantropical.

Uses: Species used as food in the Fiji Is (South 1993a), French Polynesia (Conte and Payri 2006), Pakistan (Rahman 2002), Indonesia (Arasaki and Arasaki 1983), Vietnam (Hong et al. 2007), and the Philippines (Agngarayngay et al. 2005).

Extracts of this species have antibacterial (Chandrasekaran et al. 2014), and biostimulating (Sivasankari et al. 2006) activity.

Caulerpa cupressoides var. lycopodium Weber-van Bosse

Common name: Mamanga (Conte and Payri 2006).

Description: Plants 2–14 cm in height, stolon glabrous, cylindrical, branched, often with about 1.5 mm in diam.; rhizoids branched covered with substratum material; assimilatory branches upright, cylindrical, with 2.0–4.0 mm in diam., naked in its basal portion and above this coated several rows of ramuli distributed compactly and around the central axis.

Geographic distribution: Atlantic Is (Bermuda); Tropical and subtropical W Atlantic; Indian Ocean; Asia (Japan, Taiwan); Australia; Papua New Guinea; Pacific Is (Samoa, Fiji, Polynesia).

Uses: Used as food in French Polynesia (Conte and Payri 2006).

Caulerpa racemosa (Forsskål) J. Agardh (Fig. 87)

Common name: English: Sea grapes, Mouse plant (Madlener 1977); Hawaiian: Limu fuafua (Chapman and Chapman 1980); Fijian: Nama, Nama levulevu (South 1995, Novaczek 2001, South et al. 2012); Ilocano: Ar-arucep (Lembi and Waaland 1988); Indonesian: Lai lai (Chapman and Chapman 1980); Japanese: Surikogidzuta (Madlener 1977), Sennarizuta (Oshie-Stark et al. 2003); Pacific name (Tonga Is): Toke (Ostraff 2003); Polynesian: Konini (Conte and Payri 2006); Samoan: Limu, Limu fuafua (Ostraff 2003, Skelton 2003); Spanish: Uva de mar, Caviar verde (Radulovich et al. 2013); Tagalog: Lato (Lembi and Waaland 1988); Tahitian: Rimu (Lembi and Waaland 1988); Tamil: Mookkuthi pasi (Kaliaperumal et al. 1995).

Figure 87. *Caulerpa racemosa* (Forsskål) J. Agardh (Chlorophyta). Illustration from Ernst Haeckel (Kunstformen der Natur, 1904).

Description: This plant has erect branches arising from a horizontal stolon attached to the sediment at intervals by descending rhizomes. The erect branches arise every few cm, reaching as much as 30 cm in height. A large number of branchlets, resembling ovate or spherical bodies on stalks, arise from each erect branch. Where branches and stolons are close together, the branchlets form a dense mat of seemingly spherical structures. The plants are coenocytic, i.e., the plant is multinucleate and non-septate.

Geographic distribution: From Bermuda and Florida to Brazil, including all of the Caribbean; NW Pacific (Japan); NE Pacific (Mexico); W Pacific and Indo-Pacific (the Philippines, Indonesia, Vietnam); Pacific Is; Indian Ocean; Australia; W Africa.

Uses: Consumed fresh in salads or as an appetizer in the Caribbean Is (Radulovich et al. 2015); in India, it is eaten raw or as a salad or boiled with other foods (Kaliaperumal et al. 1995).

The edible seaweed *C. racemosa* is widely consumed in the Pacific Is and has been the subject of several reports on its harvesting and consumption in Fiji, Samoa, Tonga and French Polynesia (South 1993a-c, Ostraff 2003, Skelton 2003, Conte and Payri 2006, South and Pickering 2006, South et al. 2012, Morris et al. 2014). This is one of the best known edible seaweeds in the Pacific and is a delicacy of the Samoans; it is eaten fresh or with coconut meat and green bananas. It is easy to recognize from other *Caulerpa* species by its small rounded or mushroom shaped ramuli. There are over 20 different types of this species, and some people can easily distinguished the forms by the taste. One of the forms of *C. racemosa* shown above has firm compact ramuli that is very bitter. It leaves a strong bile-like after-taste and is less favored as an edible seaweed. The longer erect thallus bearing softer ramuli is preferred due to its juicy content and slightly sweeter taste that doesn't leave a strong after-taste (Skelton 2003, Morris et al. 2014).

When escaping to the Islands during winter not only will one find sunshine and friendly faces but also a pace of life that is conducive to enjoying the simpler pleasures. There is an abundance of topical crops and wherever one goes, it is possible to experience the fresh produce of the country: taro, cassava, papaya, coconut, banana and loads of fish. Although seaweed is not as abundant and luxurious in the warm waters near the equator, there are some delicious varieties well worth noting. Sea grapes, in particular, named after their grape-like shape, grow well in Fiji and are an integral part of some of the local dishes (Fawcett 2015).

Caulerpa scalpelliformis (R. Brown ex Turner) C. Agardh (Fig. 70)

Common names: Pacific name (Tonga Is): Palalafa (Ostraff 2003).

Description: Stolon slender (0.5–1 mm in diam.) in small forms of young plants, robust (3 mm in diam.) in large, rough-water plants, cartilaginous, naked, epilithic; erect fronds medium to dark green, simple to occasionally branched, from 4–10 cm high and 3–6 mm broad in slender forms, to 20 cm high and 2–3 cm broad in robust plants, terete for the basal 1–3 cm, then strongly compressed with an axis 2–3 mm broad in slender plants to 4–10 mm broad in robust plants, bearing alternately distichous, closely adjacent ramuli. Ramuli 2–4 mm long and 1–2 mm broad in slender plants, to 1–1.5 cm long and 3–5 mm broad in robust plants, strongly compressed, basally broadest but often slightly constricted, usually curved slightly upwards and with an acute angle between the adjacent ramuli, tapering gradually over their lower half to three quarters, then abruptly to a distinct spinous tip (Womersley 1984).

Geographic distribution: Atlantic Is (Canary Is); Caribbean Is (Barbados, Lesser Antilles); Tropical and subtropical Atlantic; Indian Ocean; Asia (Japan); SE Asia (Indonesia, Singapore, Vietnam); Australia; Pacific Is (Tonga Is).

Uses: Used as food in Bangladesh (Aziz 2009), and Tonga Is (Ostraff 2003).

Chaetomorpha crassa (C. Agardh) Kützing

Common names: Chinese: Tai tyau (Madlener 1977); English: Hair-shaped green algae, Big green algae (Madlener 1977), Curly fishing line (Novaczek 2001); Pacific name: lumot (the Philippines) (Novaczek 2001).

Description: This bright green or dark green plant is a simple unbranched thread less than 1 mm wide but can be 10–50 cm long. It is springy and curls around itself and other plants, forming a tangled ball that looks like a green fishing line. It may be attached to a small stone or shell but is often free floating or entangled with other plants (Novaczek 2001).

Geographic distribution: NE Atlantic; Adriatic Sea; Atlantic Is (Azores, Bermuda); Caribbean Is; Tropical and subtropical Atlantic; Indian Ocean; Asia (China, Japan, Korea, Taiwan); SE Asia (Indonesia, the Philippines, Singapore, Thailand, Vietnam); Australia; Pacific Is (Micronesia, Fiji, Mariana Is, Solomon Is).

Uses: *Chaetomorpha* is eaten raw, either salted or in salads, in the same way as *Ulva* (formerly *Enteromorpha*) spp. (Novaczek 2001, Sidik et al. 2012).

Cladophora patentiramea (Montagne) Kützing

Common name: Polynesian: Imu ouoho (Hair algae) (Conte and Payri 2006).

Description: Filaments fine, dark, dirty green, with branches arising at distal end of cells just below septa; cells 10 times as long as broad, length 0.78–1.55 mm and breath 77–110 µ (Wei and Chin 1983).

Geographic distribution: Indian Ocean; Australia; Pacific Is (Polynesia, Micronesia, Hawaiian Is, Palau, Samoa).

Uses: Used as food in French Polynesia (Conte and Payri 2006).

Cladophora sericea (Hudson) Kützing

Synonym: *C. nitida*

Common names: English: Graceful green hair; Hawaiian: Limu hulu-ilio (dog's hair) (Reed 1906, MacCaughey 1918); Japanese: Haiiro-shiwogusa (Tokida 1954).

Description: Plants bushy or spreading not tufted. Filaments slightly or profusely branched; plants usually in soft to stiff clumps, often floating; the main axis branches dichotomously and has lateral branches that are shorter, narrower (20–40 µ diam.) and arranged alternately, opposite or on one side. Lateral branches near the tip often have shorter branches between longer branches. Attachment is by rhizoids from basal poles of mid to lower cells, or by cluster of basal rhizoids; pale green to grass green (Russel et al. 2000).

Geographic distribution: NE Atlantic (Iceland to Canary Is); North Sea and Baltic; Mediterranean; S Africa; Caribbean; Indian Ocean; NE Pacific (Alaska to California); SW Pacific (the Philippines); S Australia, New Zealand; Pacific Is (Polynesia, Micronesia, Hawaiian Is, Solomon Is).

Uses: Edible, used as a source of food (Reed 1906, MacCaughey 1918, Braune and Guiry 2011).

C. sericea is a potential source of BASs for pharmaceuticals, and PUFA (Milchakova 2011).

Codium adhaerens C. Agardh (Fig. 88)

Common name: Hawaiian: Limu aalaula (Reed 1906).

Description: Spongy thallus, green light, prostrate, irregularly shaped and with the appearance of a plane carpet firmly fixed to the substratum. Consists of entangled coenocytic filaments, ending on the surface by narrow and elongated utricles that are difficult to separate. Firm, gelatinous texture and smooth to the touch (Pereira 2015c).

Geographic distribution: NE Atlantic; Adriatic Sea; Atlantic Is (Azores, Canary Is, Cape Verde Is, Madeira, Selvage Is); Mediterranean; Asia (Japan, Taiwan); SE Asia (Indonesia, Vietnam); Australia; Pacific Is (Polynesia); Antarctica Is.

Uses: Used as food in Hawaii (Reed 1906).

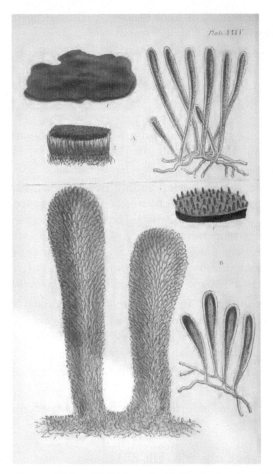

Figure 88. *Codium adhaerens* C. Agardh (Chlorophyta). Illustration from William H. Harvey (Plate XXXV, Phycologia Britannica, 1846–1951).

Codium arabicum Kützing

Common names: Pacific name: Imu tutae kioe (= Rat-faces algae) (Conte and Payri 2006); Filipino: Bagaba (Leyte 2015).

Description: Plant firm, compressed or recumbent, adherent to substratum, often convoluted or with rounded superficial knobs 1–3 cm in diam. on the upper surface or along margins; utricles cylindrical or clavate, developed in clusters joined near bases by plugged connections, variable in size among plants and from margin to center of individual plants, 400–1440 μ in length, 40–250 μ in diam., with sub-truncate or slightly rounded apices; secondary utricles arising as buds from lower parts of primary utricles and forming large groups of closely related utricles. Hairs and hair scars abundant on older utricles (Chang 2015a).

Geographic distribution: Cosmopolitan in the Pacific area.

Uses: Used as food in the Philippines (Leyte 2015), and French Polynesia (Conte and Payri 2006).

Codium bartlettii C.K. Tseng & W.J. Gilbert

Common name: English: Green sea antler.

Description: Thalli erect, green to greenish brown in color. Branching repeatedly sub-dichotomous—divaricate, forming broad rounded axils (especially at the basal portion of the thallus); one branch of the dichotomy more developed than the other and continuous with the main branch below it; branches cylindrical-compressed, adhering to each other at some points by cushion-like rhizoidal structures; medullary filaments cylindrical and branched, giving rise to inflated club-shaped utricles forming the cortex; thallus up to 6 cm in height (Trono 2001).

Geographic distribution: Asia (China, Indonesia, the Philippines).

Uses: Used as food in salads in the Philippines (Trono 1998, Zemke-White and Ohno 1999, Roo et al. 2007, Harrison 2013).

Codium dwarkense Børgesen

Description: Thallus mostly upright but with lower branches sometimes repent, dark green, attached to the substratum by a spongy base, with terete, dichotomously branched axes. Utricles often slightly unevenly constricted below the apex, medullary siphons with a plug closely adjacent to the point of departure from the utricle (Huisman and Parker 2011a).

Geographic distribution: Indian Ocean; Australia.

Uses: Considered an edible species (Rao 1979).

Codium edule P.C. Silva

Common names: Filipino: Pukpuklo (Agngarayngay et al. 2005); Hawaiian: Limu, Wawae'iole, Ala'ula (Fortner 1978, Arasaki and Arasaki 1983), A'ala (Abbott 1984); Japanese: Miru (Arasaki and Arasaki 1983).

Description: Plants are dark green with felt-like surface; form large mats. Found throughout the Hawaiian Is from low intertidal to subtidal, 2–4 m depth. Requires careful cleaning as this is attached in several places to rubble and coral. Thalli erect with decumbent portions, branching dichotomously or sub-dichotomously; branches cylindrical, dark green at middle and light green to almost transparent at the periphery.

Geographic distribution: Indian Ocean; SE Asia (Indonesia, the Philippines; Australia (Papua New Guinea); Pacific Is (Polynesia, Micronesia, Hawaiian Is, Mariana Is, Marshall Is).

Uses: In Hawaii usually chopped or pounded and mixed with salt. Under refrigeration may be kept indefinitely, but best if eaten within 10 days. Served alone with fish, seafood, or stew, or may be mixed with other seaweeds (Arasaki and Arasaki 1983, Abbott 1984, Novaczek 2001, Hart et al. 2014); used as food in the French Polynesia (Conte and Payri 2006), and used in salads in the Philippines (Lembi and Waaland 1988, Trono 1998, Zemke-White and Ohno 1999, Agngarayngay et al. 2005, Roo et al. 2007, Harrison 2013).

Codium intricatum Okamura

Common names: English: Finger algae; Filipino: Pokpoklo, Pukpuklo, Pupu-lo; Pakistani: Mosure-miru (Velasquez 1972, Rizvi and Shameel 2005).

Description: Plants creeping, compressed with broader segment, irregularly and divaricately dichotomous, deep-green, intricated by attaching to each other with root-fibers. Branches little flattened, 1.5–4 mm in diam.; utricles cylindrical to clavate, 520–700 µ in length, 80–250 µ in diam. Hairs or hair scars 1–2 per utricle (Chang 2014).

Geographic distribution: Indian Ocean; Asia (China, Japan, Korea, Taiwan, the Philippines); Pacific Is (Micronesia).

Uses: Used as food in salads in Pakistan and the Philippines (Levring et al. 1969, Velasquez 1972, Trono 1998, Rizvi and Shameel 2005).

Codium muelleri Kützing

Common names: Filipino: Pokpoklo; Siling siling; Hawaiian: Limu aala-ula, Limu wawae-iole, Limu wawae-moa (Reed 1906, MacCaughey 1918, Zaneveld 1969).

Description: Thallus moderately firm, medium green, erect, repeatedly dichotomous, to 25 cm high, branches terete, 2.5–5 mm in diam., decreasing slightly to about 1.5 mm in diam. near apices. Utricles relatively short and often broadest somewhat below the apex, 130–520 µ in diam. and 290–960 µ long, apices broadly rounded to sub-truncate; hairs (or scars) common (Womersley 1994).

Geographic distribution: SE Asia (the Philippines); Australia and New Zealand; Pacific Is (Hawaiian Is).

Uses: In the Hawaiian Is the species is eaten raw, usually with tomatoes, after being thoroughly washed in fresh water (Reed 1906, Zaneveld 1969, Chapman and Chapman 1980, Zemke-White and Ohno 1999, Roo et al. 2007, Harrison 2013).

Codium reediae P.C. Silva

Common names: English: Dead man's fingers, Antler seaweed; Filipino: Pokpoklo; Hawaiian: Limu a'ala'ula (Kent 1986), Limu wawae'iole; Japanese: Miru (Fortner 1978).

Description: Thallus is dark green, spongy to the touch, fleshy, erect, and composed of somewhat flattened fronds arising from a single discoid holdfast. Fronds are from 1–2 cm wide, slightly flattened in the lower half, and markedly flattened in the upper half. Branching is primarily irregularly dichotomous, expanding above to cuneate, flattened dichotomies at ends of fronds (Abbott 1999).

Geographic distribution: *C. reediae* is native to the Hawaiian Is and the S Pacific, but is also found in the Indian Ocean and around Taiwan.

Uses: In Hawaii markets this seaweed is labeled as "Pokpoklo" or "Limu wawae'iole" (Arasaki and Arasaki 1983); used as food in French Polynesia (Conte and Payri 2006). It is particularly prized by the Filipinos and most of the recipes for *Codium* are of Filipino origin. The old Hawaiians used to prepare the seaweed by soaking it in a brine solution. In the brine the branches would shrink, become limp, and exude a red liquid. From this it acquired the name "Limu 'a'ala'ula" or "the seaweed making a red fragrance". The wilted plant was prepared with sea cucumber, sea urchin gonads, or raw octopus (Fortner 1978, Kent 1986).

Codium repens P.L. Crouan & H.M. Crouan

Common name: Filipino: Pukpuklo (Agngarayngay et al. 2005).

Description: Thallus spongy, consisting of cylindrical branches, mostly creeping, loosely following substratum contour, attached at random points (branches occasionally fusing together); branching mainly pseudo-dichotomous, branches 1-3(-5) mm in diam. Hair common, generally one (occasionally two) per utricle (Fikes 2008).

Geographic distribution: Atlantic Is (Bermuda, Canary Is, Cape Verde Is); Tropical and subtropical W Atlantic; India Ocean; Asia (China, Japan, Korea, Taiwan, the Philippines, Vietnam); Pacific Is (Micronesia).

Uses: Used as food in salads in the Philippines (Agngarayngay et al. 2005).

Pithophora Wittrock

Common name: Hawaiian: Lipala'o, Limu palawai, Lipalawai (Kent 1986).

Description: Freshwater algae; thallus structure similar to *Cladophora* but always with large terminal and intercalary akinetes-like densely packed with food reserves.

Geographic distribution: *Pithophora* is common in tropical and temperate regions throughout the world.

Uses: Considered an edible alga in the Hawaiian Is (Kent 1986).

Pithophora roettleri (Roth) Wittrock

Synonyms: *Pithophora affinis, Pithophora polymorpha*

Common names: Hawaiian: Limu palawai, Limu lipalawai (Reed 1906).

Description: Freshwater species; filaments slender, branching mostly solitary, rarely opposite; cells long cylindrical, 45–80 µ wide, up to 20 times their diam. in length; akinetes intercalary and terminal, cylindrical or swollen, conical or acuminate when terminal, 60–150 µ wide, 81–380 µ long.

Geographic distribution: NE Atlantic; Tropical and subtropical W Atlantic; Asia (Japan); SE Asia (Singapore, Vietnam); Pacific Is (Hawaiian Is).

Uses: Used as food in Hawaiian Is (Reed 1906).

Ulva flexuosa Wulfen

Synonyms: *Enteromorpha flexuosa, E. tubulosa, E. prolifera* var. *tubulosa*

Common names: Arabic: Tahalib (FAO 1997); Chinese: hu-t'ai (Tseng 1983); English: Winding Nori (FAO 1997); Hawaiian: Limu ele-ele (MacCaughey 1918, Chapman and Chapman 1980), Limu pipilani (Reed 1906); Polynesian: Imu vai (freshwater algae), Imutapaa (ripe algae), Imu ketaha (encroaching algae), Imu ouohu (hair algae) (Conte and Payri 2006); Swedish: Gallertarmalg, Tarmalg (Tolstoy and Österlund 2003).

Description: Plants soft, green, fading to a browner color, 30–60 mm tall branching near the base, branches hollow, about 8 mm broad (Baldock 2009a).

Geographic distribution: N and S Atlantic; SW Asia (Abu Dhabi, Arabian Gulf, Cyprus, Israel, Kuwait, Turkey); Asia (China, Japan, Korea, Taiwan); SE Asia (Indonesia, Singapore, Vietnam); Australia; Pacific Is (Samoa, Polynesia, Hawaiian Is).

Uses: Commonly eaten as food in the Caribbean, central America, and the Hawaiian Is (Reed 1906, MacCaughey 1918, Chapman and Chapman 1980, McDermid and Stuercke 2003, Harrison 2013, Hart et al. 2014); used as food in French Polynesia (Conte and Payri 2006), and Cook Is (N'Yeurt 1999); is often collected as food by people in the Yellow Sea and Gulf of Chihli areas of northeastern China and by the aboriginal Yemei tribe in Taiwan (Tseng 1983, Bangmei and Abbott 1987). This species is also eaten on the shores of the Malay Peninsula (Zaneveld 1959).

Extracts of this species have antioxidant, antibacterial, antivirus and fungicide (Farasat et al. 2014) activity; is used also in cosmetics (Milchakova 2011).

Ulva flexuosa subsp. *paradoxa* (C. Agardh) M.J. Wynne

Synonyms: *Enteromorpha hopkirkii, E. plumosa*

Common names: Filipino: Lumot (Galutira and Velasquez 1963); Hawaiian: Limu ele'ele (MacCaughey 1918), Limu pipilani (Reed 1906); Japanese: Watage-awonori (Tokida 1954).

Description: These green algae preferring mainly salt-wedge saline waters.

Geographic distribution: NE Atlantic; Tropical and subtropical Atlantic; Caribbean Sea; Atlantic Is (Bermuda, Canary Is); Adriatic Sea; Baltic Sea; Black Sea; Indian Ocean; Asia (Japan); SE Asia (Indonesia, the Philippines, Singapore); Australia; Pacific Is (Fiji, Samoa, Hawaiian Is).

Uses: Used as food in Hawaii (Reed 1906, MacCaughey 1918), and the Philippines (Galutira and Velasquez 1963).

Ulva lactuca Linnaeus (Fig. 89)

Synonym: *U. fasciata*

Common names: English: Sea lettuce, Lettuce laver, Green laver, Sea grass, Thin stone brick (Madlener 1977), Chicory sea lettuce (FAO 1997);

Figure 89. *Ulva lactuca* Linnaeus (Chlorophyta). Illustration from William H. Harvey (Plate CCXLII, Phycologia Britannica, 1846–1951).

French: Laitue de mer (Boisvert 1984); German: Meersalat (Madlener 1977); Hawaiian: Limu papahapapa (Chapman and Chapman 1980), Pahapaha, Limu pahapaha (Arasaki and Arasaki 1983), Limu palahalaha, Pakaiea, Papahapaha, Limu paha-paha, Limu pala-haloha (Reed 1906, MacCaughey 1918, Abbott 1984, Kent 1986); Japanese: Aosa (Smith 1904, Madlener 1977); Polynesian: Imu kokuu, Kokuu, Imu sarata (salad or lettuce algae) (Conte and Payri 2006).

Description: Thalli thin, sheet-like, up to 50 cm long, consisting of wide blades, 10 to 15 cm wide at base, tapering upward to less that 2–5 cm wide at tip. Basally broadened, but the upper portions divided deeply into many ribbon-like segments; margins smooth, often undulate. Holdfast is small without dark rhizoids. Bright grass-green to dark green, gold at margins when reproductive, may be colorless when stressed (Braune and Guiry 2011, Pereira 2015a).

Geographic distribution: Worldwide distribution in temperate and tropical waters.

Uses: This species, the "sea lettuce" is eaten widely in the East, particularly in Hawaii, French Polynesia and Japan, in soup, as salad and is used for garnishing (Reed 1906, MacCaughey 1918, Zemke-White and Ohno 1999). Easily collected, remove small black snails that feed on blades; wash well and chopp into little pieces; mix with other seaweeds and serve with raw fish, or add to light soups (Zaneveld 1959, Arasaki and Arasaki 1983, Kent 1986, Conte and Payri 2006, Hart et al. 2014, Pereira 2015a); is also consumed in Caribbean, Central America, and Alaska (Harrison 2013, Kellogg and Lila 2013).

Ulva prolifera O.F. Müller

Synonyms: *Enteromorpha prolifera*, *E. comporessa* var. *trichodes*

Common names: Chinese: Tai-tiao (Arasaki and Arasaki 1983); Hawaiian: Limu ele-ele (MacCaughey 1918), Hulu'ilio (Abbott 1984, Kent 1986); Japanese: Suji-awnori (Tokida 1954), Suji-aonori (Arasaki and Arasaki 1983); Korean: Parae (Sohn 1998); Swedish: Spretig tarmalg (Tolstoy and Österlund 2003).

Description: The fronds are tubular, though often more or less flattened, few to many branches. The arrangement of the cells, in longitudinal and transverse rows in the central part of the frond, is characteristic of this species, as are the cylindrical chloroplasts seeming to fill the cell and the usually single, central pyrenoids (Pereira 2015a).

Geographic distribution: Arctic (Canada, Svalbard, White Sea); Adriatic Sea; Black Sea; NE Atlantic; Atlantic Is (Azores, Bermuda, Canary Is, Greenland, Iceland, Madeira); Tropical and subtropical Atlantic; W Mediterranean; Asia (China, Japan, Korea, Taiwan, Indonesia, the Philippines); New Zealand; Pacific Is (Samoa, Polynesia, Hawaiian Is); NE Pacific (Alaska).

Uses: Used for preparation of miso soup, nori-jam, salads, and seaweed powder for various foods in Japanese cuisine (Ohno et al. 1998, Zemke-White and Ohno 1999, Mori et al. 2004). *U. prolifera*, the species most commonly used in Fujian, is particularly favored as a condiment with "chun bing" or "spring cake." This alga is also exported to Indonesia and Singapore (Bangmei and Abbott 1987, Istini et al. 1998, Harrison 2013, Irianto and Syamdidi 2015). In Korean cuisine, this species is prepared with sesame oil, and sometimes with vinegar (Kang 1968, Bonotto 1976, Sohn 1998, Roo et al. 2007). In Malaysia and the Hawaiian Is, *U. prolifera* is used for the preparation of salads (MacCaughey 1918, Abbott 1978, Arasaki and Arasaki 1983, Kent 1986, Sidik et al. 2012); considered also an edible species in India (Naidu et al. 1993), and Pakistan (Rahman 2002).

Its extracts have hypocholesterolemic action (Tsuchiya 1969); used in the treatment of aphthae, back pain, paronychia, lymphatic swellings, goiter, cough, bronchitis, antipyretics, sunstroke treatment, tonsillitis, asthma, nosebleeds *fulvescens* and sore-hand (Tseng and Zhang 1984, Tokuda et al. 1986).

Valonia utricularis (Roth) C. Agardh

Common name: Hawaiian: Limu lipuu-puu (Reed 1906, MacCaughey 1918, Kent 1986).

Description: Thallus, translucent light- to dark green, primarily consisting of a large (up to 5 mm thick and 20 mm long) bladder- or club- to hose-like cell, branching at the base rhizoidally. Later due to outgrowths of this cell cylindrical-clavate branches, often contorted and almost gapless densely packed, thus forming intertwined erect stands (Braune and Guiry 2011, Pereira 2015a).

Geographic distribution: Warm NE Atlantic (Portugal to Canary Is); Mediterranean; Caribbean; Indian Ocean; Asia (Japan, China); SE Asia (the Philippines, Vietnam); Pacific Is; Australia.

Uses: Considered and edible alga in the Hawaiian Is (Reed 1906, MacCaughey 1918, Kent 1986).

2.5.2 Ochrophyta – Phaeophyceae

Dictyota acutiloba J. Agardh

Synonym: *Dictyota acutiloba* var. *distorta*

Common names: English: Brown ribbon weed; Fiji: Vutua (Novaczek 2001); Hawaiian: Limu alani, Alani (Reed 1906, MacCaughey 1918, Kent 1986), False Lipoa (Reed 1906).

Description: There are several species of *Dictyota*. Some are tiny (2–5 mm wide) others are more than 10 mm wide and 20 cm long. All are golden to dark brown with flat blades, and have a Y-shaped form of branching (that is, dividing into two). Sometimes the blades are also spirally twisted. Some species shine with iridescent colors when seen underwater. Smaller species often grow in clumps, while the larger ones may grow as separate, scattered plants (Novaczek 2001).

Geographic distribution: NE Pacific (Chile); Australia and New Zealand; Pacific Is (Polynesia, Easter Is, Micronesia, Hawaiian Is).

Uses: The thalli of several *Dictyota* are edible and used in Indonesia (Sulawesi), Malaysia and Thailand. In the Hawaiian Archipelago, *D. acutiloba* is cultivated in "algal gardens" and sold in local markets. *Dictyota* is either eaten raw or cooked with coconut milk, pickled or preserved by smoke-drying and is very nutritious. Some *Dictyota* are known to have a somewhat bitter taste. If you like the bitter flavor you can eat *Dictyota* raw with vinegar or lemon juice dressing, or cook the fresh plants in stir fry, soup or stew. To reduce bitterness, the sea plants can be soaked overnight in fresh water. Remove from the water, squeeze dry and sprinkle with salt. Store in the fridge until you want to use it as a spice (Reed 1906, MacCaughey 1918, Levring et al. 1969, Chapman and Chapman 1980, Kent 1986, Novaczek 2001).

Dictyota bartayresiana J.V. Lamouroux

Common name: Hawaiian: Alani (Kent 1986).

Description: Thallus erect, iridescent blue and green in the water, or light brown, often with dark olive-brown bands, 9–14 cm high, erect, not entangled, a little harsh to the touch, attached to the substratum by irregularly shaped holdfast with rhizoids, thallus branched dichotomously; segments without midrib, 1–1.5 cm long, 2–4 mm broad above a fork, broadening to 6–10 mm below the next fork; margin entire, tips are pointed except in young branches (Pereira 2014).

Geographic distribution: Atlantic Is, Pacific Mexico, Caribbean Is, Brazil, Venezuela, Africa, Indian Ocean Is, SW Asia, Japan, Taiwan, SE Asia, Australia and New Zealand, Pacific Is.

Uses: Used as food in the Hawaiian Is (Kent 1986).

Dictyota sandvicensis Sonder

Description: Iridescent yellow-green with banding and wide Y-shaped branch tips, margins with many branchlets.

Geographic distribution: Australia and New Zealand; Pacific Is (Samoa, Polynesia, Hawaiian Is).

Uses: Commonly eaten as food in the Hawaiian Is (McDermid and Stuercke 2003).

Dictyopteris australis (Sonder) Askenasy (Fig. 90)

Synonym: *Haliseris pardalis*

Common name: Hawaiian: Limu lipoa (Reed 1906, Fortner 1978, Kent 1986).

Description: Thallus medium brown, usually 10–30 cm long, with one to several complanate fronds arising from a matted rhizoidal holdfast 0.2–1.5 cm across and 0.2–1.5 cm long; epilithic. Growth from several apical cells in a rounded apex; fronds sub-dichotomously branched at intervals of 3–10 cm, with or without proliferous branchlets from adjacent to the midrib, of fairly uniform width (usually 0.8–1.5 cm) throughout, with a central midrib and usually with faint (microscopic) lateral veins, less than 1.5 mm apart, running upwards from midrib to margin; branches often denuded below (Womersley 1987).

Geographic distribution: Indian Ocean; Australia; Pacific Is (Easter Is, Hawaiian Is).

Uses: In the Hawaiian Is, traditionally, the collected "Limu" is carefully cleaned and washed, removing all coralline algae, then salted in a process known as "Pa'akai". The salted "Limu" will keep indefinitely under refrigeration. Recipes calling for "Lipoa" refer to the salted "Limu". "Limu lipoa" took the place of sage and pepper in the old Hawaiian diet and today serves as a spice, usually accompanying raw fish or *Octopus*. It can be used in a meat stew or as a substitute for an olive in a Martini (Reed 1906, Michanek 1975, Fortner 1978, Kent 1986); considered an edible species in India (Rao 1970).

Figure 90. *Dictyopteris australis* (Sonder) Askenasy (Ochrophyta, Phaeophyceae). Illustration from William H. Harvey (Plate XXIX, Phycologia Australica, 1858–1863).

Dictyopteris plagiogramma (Montagne) Vickers

Synonym: *Haliseris plagiogramma*

Common names: Hawaiian: Limu-lipoa (Arasaki and Arasaki 1983), Limu lipoa (MacCaughey 1918, Abbott 1984, Kent 1986).

Description: Thalli dark olivaceous green to light tan or almost yellowish, occurring as dense often intertwined tufts of multiple axes arising from a common matted holdfast; fronds 2–8 cm long, 1–8 mm wide, usually crisp in texture and pungently aromatic when fresh; branching of juvenile thalli basically dichotomous, becoming pseudo-monopodial at maturity and then highly irregular as wing tissue wears away to leave midribs as several orders of stalks. Blades from apex to base traversed from midrib to margin by regularly spaced microscopic to faintly visible veins that arise at angles of 10–20° and form broad arches bending away from the apices (Pereira 2015a).

Geographic distribution: Atlantic Is (Bermuda, Canary Is); Tropical and subtropical W Atlantic; Indian Ocean; Asia (Japan); Australia; Pacific Is (Polynesia, Hawaiian Is).

Uses: In Hawaii, it is a highly regarded foodstuff (Reed 1906, MacCaughey 1918, Levring et al. 1969, Arasaki and Arasaki 1983, Kent 1986, Hart et al. 2014); is used as a relish for penetrating spicy flavor (Chapman and Chapman 1980). All Hawaiians like the odor and flavor of this alga, especially with raw fish. It is considered particularly delicious with raw flying fish, if simply broken and salted slightly (MacCaughey 1918). Leafy branches are washed, and heavily salted for indefinite storage. Young plants can be chopped or pounded, lightly salted, and refrigerated for future use. Spicy flavor good with fish and meat dishes, especially stews (MacCaughey 1918, Preskitt 2002).

Dictyopteris repens (Okamura) Børgesen

Common name: Auke (Easter Is) (Ohni 1968).

Description: The plants are 2–3 cm long and dark brown in color. The branches are strap-shaped, dichotomously divided, 2–3 mm wide, with a distinct midrib that appears throughout the entire length. It grows with sand bottom at deeper places along moderately wave exposed shorelines (JIRCAS 2012b).

Geographic distribution: Indian Ocean; Asia (China, Japan, Taiwan); SE Asia (Indonesia, the Philippines, Singapore, Thailand); Australia and New Zealand; Pacific Is (Samoa, Polynesia, Easter Is, Micronesia, Fiji, Hawaiian Is, Mariana Is, Marshall Is, Solomon Is).

Uses: Used as food in Easter Is (Ohni 1968).

Feldmannia indica (Sonder) Womersley & A. Bailey

Synonym: *Ectocarpus indicus*

Common names: Hawaiian: Limu hulu-ilio (MacCaughey 1918), Limu 'àka, Limu 'ako'a (MacCaughey 1918, Kent 1986), Limu akaakoa, Limu huluilio (Reed 1906).

Description: Soft filamentous tufts, to 5 cm high, brown-green; branching irregular, forming as lateral projections. Filaments 20–34 µ diam., tips tapering to 10 μ diam.; cells 0.5–5.0 diam. long, most growth intercalary (cell division not at branch tips but elsewhere in the filament); shorter branchlets tapered, darkly pigmented; stolons filamentous, with fine rhizoids. Plurilocular sporangia cylindrical, 20–50 μ diam., 100–260 μ long, rarely stalked, with blunt apices, forming laterally on filaments, scattered, rarely in linear series (Womersley and Bailey 1970, Kraft 2009).

Geographic distribution: Tropical and subtropical Atlantic; Caribbean Sea; Arabian Gulf; Indian Ocean; Asia (China, Japan, Korea, Taiwan); SE Asia (the Philippines, Singapore, Vietnam); Australia and New Zealand; Pacific Is (Micronesia, Polynesia, Samoa, Solomon Is, Fiji, Easter Is, Marshall Is, Mariana Is, Hawaiian Is).

Uses: Used as food in the Hawaiian Is (Reed 1906, MacCaughey 1918, Kent 1986).

Padina Adanson

Common name: English: Sea fan (Novaczek 2001).

Description: The golden to dark brown, fan-shaped blades of *Padina* grow in clusters; one side of the blade are concentric, or curved, rows of white lines, where the plant produces chalk (Novaczek 2001).

Geographic distribution: Common throughout the Pacific region; reported from New Caledonia, Fiji, Tahiti, Cook Islands, Samoa, Tonga, Solomon Is, Marshall Is, Palau, Kiribati and elsewhere in Micronesia.

Uses: *Padina* contains algin. The chalk on its surface makes it crunchy and the blades are rather tough when fresh, but it is edible and is a good source of calcium. Vinegar or lemon juice can be used to remove the chalk from the surface of the blade. Add the sea plant to soups, stews, fritters and salads, or stir-fry together with other vegetables. Dried *Padina* flakes can be sprinkled on salads, omelets, potatoes or any other dish to add calcium, other minerals and flavor (Novaczek 2001).

Rosenvingea intricata (J. Agardh) Børgesen

Common names: English: Slippery cushion (Novaczek 2001); Filipino: Samsamit (Agngarayngay et al. 2005).

Description: Light golden-brown, prostrate (lying flat along the seabed), branched and hollow. It has many branches that are somewhat flattened, entangled and stuck to one another to form a cushion. Branches are 1–9 mm wide, and are often more slender at the tips. The cushions may be 5–20 cm across, and have a distinctive strong smell. *Rosenvingea* has a soft, papery texture and is more smooth and gooey than *Colpomenia* (Novaczek 2001).

Geographic distribution: N and S Atlantic (temperate and tropical Atlantic); Indian Ocean Is; SW Asia (Arabian Gulf, Bangladesh, India, Yemen); Asia (China, Japan); SE Asia (Indonesia, the Philippines, Vietnam); Australia; Pacific Is (Polynesia, Micronesia, Fiji, Hawaiian Is, Mariana Is, Marshall Is, New Caledonia; Samoan Archipelago).

Uses: *Rosenvingea* has the same edible and medicinal value as *Colpomenia* and *Hydroclathrus*, making it a good preventative medicine for heart disease and a healthy food for pregnant women (Rao 1970, Novaczek 2001). Add fresh chopped *Rosenvingea* to any salad, soup or stew (Agngarayngay et al. 2005). Crisp the dried plants to make flavor flakes (Novaczek 2001).

Sargassum aquifolium (Turner) C. Agardh

Synonyms: *S. echinocarpum, S. crassifolium*

Common names: Arabic: Qunbar al-ma (FAO 1997); English: Binder sargassum weed (FAO 1997), Sea oak (Novaczek 2001); Hawaiian: Limu kala (Reed 1906, MacCaughey 1918, Fortner 1978, Chapman and Chapman 1980, Novaczek 2001), Limu-Kala (Arasaki and Arasaki 1983), Limu Honu, Holly Limu, Kala-launui, Kala-lauli'i (Abbott 1984); Pacific name: Rimu akau (New Zealand) (Novaczek 2001, Ostraff 2003); Indonesian: Arien wari (Zaneveld 1955); Samoan: Limu vaova (Novaczek 2001, Ostraff 2003).

Description: This large, dark brown plant with golden-brown tips has a long central stem and branches with blades that resemble oak tree leaves. The plant may be a meter or more in length. Blades have toothed edges and the larger ones are rather tough. Leaves at the bottom are large, compared with the ones near the top. The stem also has many small grapelike bladders attached to it, which are filled with air and hold the plant up in the water (Novaczek 2001).

Geographic distribution: Asia (China, Japan, Taiwan); SE Asia (Indonesia, the Philippines, Singapore, Vietnam, Malaysia); SW Asia (Arabian Gulf, India, Iran, Kuwait); Indian Ocean Is; Australia; Pacific Is (Samoa, Polynesia, Fiji, Hawaiian Is, Solomon Is).

Uses: In the Hawaiian Is and SE Asia, *Sargassum* is edible raw or cooked, and has many medicinal properties (Reed 1906, MacCaughey 1918, Arasaki and Arasaki 1983, Zemke-White and Ohno 1999, Hart et al. 2014). The tops of the weeds are eaten raw or cooked with coconut milk (Zaneveld 1959, Istini et al. 1998); is also eaten in New Zealand, Indonesia, Malaysia and Pakistan (Zaneveld 1955, Zemke-White and Ohno 1999, Rahman 2002, Ostraff 2003, Harrison 2013, Irianto and Syamdidi 2015).

It contains high levels of iodine, which prevents goiter. It also has algin, fucoidan and laminarin, substances which act as a preventative medicine for heart disease and stroke. Algin can help remove poisonous metals such as lead and radioactivity from one's body. *Sargassum* can be made into a tea that promotes weight loss. Basal parts rich in algin can be dried for use as a dressing for cuts and burns (Novaczek 2001, Spalding 2009).

In Samoa, fresh young tips are used in soups or eaten fresh, dressed with soya sauce. Fresh or dried plants can be cooked with garlic, spices, green onion (scallion), and hot red peppers, and then served as a main dish. *Sargassum* can also be added to tomato sauce and served with spaghetti noodles. Use young leaves to stuff baked fish. Add to sea vegetable fritters or sundry the tips and eat as a crispy snack. Sprinkle dried *Sargassum* flakes on salads, omelets, soups and cooked vegetables (Novaczek 2001, Ostraff 2003, Spalding 2009).

When used as food by the Hawaiians the leaves were broken, soaked in fresh water until they turned dark, and used as a stuffing for baked fish, or chopped with fish heads and salt. The "limu" was also eaten fresh at the beach with raw fish or *Octopus* (Fortner 1978).

Sargassum obtusifolium J. Agardh

Common name: English: Ribbon sargassum (Stender and Stender 2014d).

Description: Abundant on reefs exposed to wave action during winter. Blades are thin, narrow, and elongated; endemic to Hawaii (Stender and Stender 2014d).

Geographic distribution: E Pacific (Chile, Ecuador); Pacific Is (Easter Is, Hawaiian Is, Mariana Is, Micronesia, Polynesia).

Uses: Commonly eaten as food in the Hawaiian Is (McDermid et al. 2005).

Sargassum polyphyllum J. Agardh

Description: Tough, bushy, erect thallus, 4–70 cm tall. Plant consists of a primary axes, rounded or slightly compressed below first or second blade, 0.5–4 mm diam., with spines. Plants usually highly branched, with secondary branches variable, some short compared to the primary branch on some plants, and as long as the primary branch on others. Blades narrow, oblong, and either flat or spirally twisted, 1–6 cm long, 0.1–0.8 cm wide, with short flattened petiole, petiole usually with spines. Blade margins are often spiny, occasionally smooth, wavy or straight, with spines or wings developing on upper or lower surface of blade midrib (Abbott 2001).

Geographic distribution: Indian Ocean; Australia; Pacific Is (Polynesia, Fiji, Hawaiian Is, New Caledonia).

Uses: Used as food in the Hawaiian Is (Reed 1906).

2.5.3 Rhodophyta

Acanthophora spicifera (M. Vahl) Børgesen

Common names: English: Spiny sea plant (Novaczek 2001), Spiny seaweed (Santhanam 2015); Filipino: Kulot (Trono 1998, Agngarayngay et al. 2005), Culot (Velasquez 1972); Fijiana: Lumikaro, Lumi karokaro or Lumi karo (Novaczek 2001, Ostraff 2003); Indonesian: Boeloeng, Bideng (Chapman and Chapman 1980); Pacific names: Kirokiro (Vanuatu) (Novaczek 2001).

Description: Thalli erect, loosely branched, greenish brown to purple, with a small discoid holdfast. Branches terete throughout slightly attenuated towards the acute tips. Spinous projections, a characteristic of this genus, are densely borne on the spirally arranged determinate branchlets; thalli to about 15 cm in height (Trono 2001).

Geographic distribution: E Atlantic (tropical W Africa), W Atlantic (Brazil), Caribbean, Indian Ocean (India, E Africa), NW Pacific (Japan, China), Pacific Is, Indo-Pacific (Indo-China); Australia (Braune and Guiry 2011).

Uses: Used as food in the Caribbean and Central America, India, China, Indonesia, the Philippines, Vietnam, Thailand, Tahiti and Fiji (Velasquez 1972, Lewmanomont 1978, Arasaki and Arasaki 1983, Bangmei and Abbott 1987, South 1993a, Kaliaperumal et al. 1995, Trono 1998, Zemke-White and Ohno 1999, Agngarayngay et al. 2005, Roo et al. 2007, Braune and Guiry 2011, Harrison 2013, Irianto and Syamdidi 2015). Eat fresh with vinegar or lemon dressing; boil briefly in lemon water or coconut milk; makes sweet or savory puddings; add to soup or stew as a thickener (Novaczek 2001, Lee 2008). *A. spicifera* is also a source of lambda-carrageenan and antibiotics (Kaliaperumal et al. 1995, Trono 2001).

Ahnfeltiopsis concinna (J. Agardh) P.C. Silva & DeCew
(formerly *Ahnfeltia concinna*)

Common names: English: Tufted seaweed; Hawaiian: Limu aki'aki (MacCaughey 1918, Fortner 1978, Abbott 1984, Kent 1986), Limu lko'ele'ele (Chapman and Chapman 1980), Limu elau (MacCaughey 1918); Japanese: Saimi (Chapman and Chapman 1980).

Description: Golden-brown, dense erect cartilaginous tufts of terete, branching axes; branching denser in the upper section, irregular, in parts forking; holdfast crustose (Braune and Guiry 2011).

Geographic distribution: Asia (Japan, the Philippines); Pacific Is (Hawaiian Is).

Uses: Used as food in Hawaii, it can be used as a base for a gelled salad or as a thickener in soups. The young, tender tips of the plant can be eaten fresh in a salad (MacCaughey 1918, Fortner 1978, Arasaki and Arasaki 1983, Kent 1986, Braune and Guiry 2011).

Ahnfeltiopsis gigartinoides (J. Agardh) P.C. Silva & DeCew

Common names: English: Sea nibbles (Chapman and Chapman 1980); Hawaiian: Limu aki aki, Limu koeleele (Madlener 1977); Japanese: Itanigusa saimi (Madlener 1977).

Description: The thalli of reproductively mature female gametophyte are erect multiaxial fronds. The erect fronds grow from well developed basal crusts, strongly attached to the substratum. The fronds vary in size from 1.5–2.6 cm and have a stipe 8–12 mm long. The branching is generally irregular, with 2 to 4 dichotomies, at 25–35° angles. Upper branch segments (1.9 mm) are shorter than the lower ones (4.2 mm). The branch apices are apiculate (Leon-Alvarez et al. 1997).

Geographic distribution: Atlantic Is (Cape Verde Is); NE Pacific (Alaska to California); SE Pacific (Chile), Asia (Japan, Russia); Pacific Is (Hawaiian Is).

Uses: Used as food in Japan, Russia and Hawaii (Arasaki and Arasaki 1983).

Ahnfeltiopsis vermicularis (C. Agardh) P.C. Silva & DeCew

Synonym: *Gymnogongrus vermicularis*

Common names: Hawaiian: Limu vavaloli (Schönfeld-Leber 1979), Limu ko'ele'ele (Chapman and Chapman 1980).

Description: Plants of the largest species reach 25 cm in length and occur singly or in dense aggregates from extensive (to 40 cm diam.) crustose bases; thalli compressed, dichotomous, and have varying numbers of proliferous marginal laterals.

Geographic distribution: Auckland Is; New Zealand; Galápagos Is; Indian Ocean.

Uses: Used as food in the Hawaiian Is (Schönfeld-Leber 1979, Chapman and Chapman 1980).

Asparagopsis taxiformis (Delile) Trevisan de Saint-Léon (Fig. 91)

Synonym: *Asparagopsis sanfordiana*

Common names: English: Supreme Limu (Chapman and Chapman 1980, Novaczek 2001); Hawaiian: Limu kohu, Limu lipaakai, Kohu lipeche, Kohu koko, Limu kohu, Limu koko, Limu nipaakai (Reed 1906, MacCaughey 1918, Madlener 1977, Fortner 1978, Chapman and Chapman 1980, Novaczek 2001), Kohu lipehe (Abbott 1984); Japanese: Kagikenori (Madlener 1977).

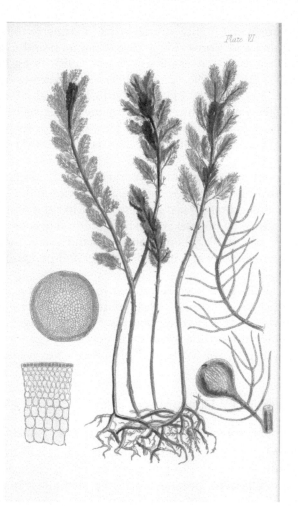

Figure 91. *Asparagopsis taxiformis* (Delile) Trevisan de Saint-Léon (Rhodophyta). Illustration from William H. Harvey (Plate VI, Phycologia Australica, 1858–1863).

Description: In water this sea vegetable looks like a forest of tiny pink trees, about 10–20 cm tall. Sometimes the color can be yellowish red or dark red. There is a creeping base with rigid upright branches. Each upright branch is covered in many fuzzy, soft, branchlets that get shorter towards the top (Novaczek 2001).

Geographic distribution: Native to the Pacific (Australia, New Zealand, Hawaii), but has been introduced elsewhere and is now found throughout the Gulf of Mexico, Indian Ocean as well in the Mediterranean Sea; America (California, Ecuador); Asia (Japan, Taiwan).

Uses: *A. taxiformis* is a highly valued edible marine red alga in Hawaii, Indonesia, China and Korea (Reed 1906, Chapman and Chapman 1980, Abbott and Cheney 1982, Arasaki and Arasaki 1983, Tseng 1983, Bangmei and Abbott 1987, Abbott 1996, N'Yeurt 1999, Zemke-White and Ohno 1999, Novaczek 2001, Roo et al. 2007, Harrison 2013, Hart et al. 2014). This species is used fresh in Hawaii as "limu" (generic name for seaweed) to flavor meat and fish dishes. It is also prized by the Hawaiians for its peppery flavor that only ali`i (Hawaiian royalty) were allowed to eat it (MacCaughey 1918, Abbott 1978).

This alga is found throughout the world in tropical marine environments and more recently, is considered an invasive species in the Mediterranean Sea. In China, this is first soaked in fresh water to rid it of its bitter iodine taste; then it is mixed with pork; the "shao" method of cooking is followed (Bangmei and Abbott 1987, Lembi and Waaland 1988). In India is eaten after cooking in beef stew or with fruit powder and pepper (Kaliaperumal et al. 1995).

The species has also been examined for its potential as a source of pharmaceutical agents since its extracts contain anti-*Leishmania* compounds and antibacterial halogenated compounds. Found on edges of reef in areas of constant water motion, the plant has largely disappeared from the main Hawaiian Is where it was once very abundant (Novaczek 2001, Pereira 2014b).

Bostrychia tenella (J.V. Lamouroux) J. Agardh

Common name: Filipino: Pakpako (Agngarayngay et al. 2005).

Description: Matt forming prostrate thallus is black to purple; pinnate branching with incurred apices. Several layers of cortication surround 6–8 pericentral cells. Tetrasporangia formed in stichidia at branchlets apices (Taylor 1960).

Geographic distribution: Tropical and subtropical W and E Atlantic; Indian Ocean; Asia (China, Taiwan, the Philippines, Singapore, Vietnam); Australia; Pacific Is (Samoa, Polynesia, Micronesia, New Caledonia, Fiji, Mariana Is, Solomon Is, Vanuatu).

Uses: Considered edible in Pakistan and the Philippines, used in salads, and mixed with meat (Rahman 2002, Agngarayngay et al. 2005).

Callophycus serratus (Harvey ex Kützing) P.C. Silva

Common name: English: Large wire weed (Novaczek 2001).

Description: This plant is named by the author as large wire weed because it is stiff and rubbery. The branching is opposite, like on a feather and the dark red branches are flattened. It grows to be quite large (10–20 cm tall) (Novaczek 2001).

Geographic distribution: Indian Ocean; Asia (Philippines); Australia; Papua New Guinea; Pacific Is (Fiji, Solomon Is).

Uses and recipes: Dry and bleach before rehydrating with boiling water. Use in salad mixed with vinegar, onion, salt and tomato; or boil into jelly with fruit juice and sugar, strain it, and let it set. Fresh plants can also be baked with meat in a ground oven (lovo) (Novaczek 2001).

Champia compressa Harvey (Fig. 92)

Common names: Limu oolu (Reed 1906), Limu o-olu (MacCaughey 1918).

Description: Small red algae, highly branched, erect forming clusters, 3–5 cm in height; thalli are highly mucilaginous and in deeply dark-red to brown red color; fronds are articulated, compressed to flattened and vary in width from 3–5 mm; branching is more or less bi-pinnate in the lower portion, bearing irregular to verticillate in the upper region; branches are distinctly attenuated at the point of origin and expanded above with obtuse tips.

Geographic distribution: Tropical and subtropical Atlantic; Indian Ocean; Asia (Korea); SE Asia (Indonesia, Malaysia, the Philippines, Thailand); Australia; Papua New Guinea; Pacific Is (Samoa, Polynesia, Micronesia, Fiji, Mariana Is, Hawaiian Is).

Uses: Used as food in Hawaii (Reed 1906, MacCaughey 1918).

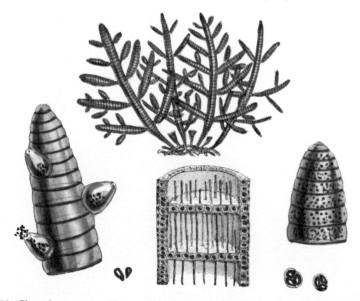

Figure 92. *Champia compressa* Harvey (Rhodophyta). Illustration from William H. Harvey (Phycologia Britannica, 1846–1951).

Champia parvula (C. Agardh) Harvey (Fig. 93)

Common names: English: Little fat sausage weed (Bunker et al. 2010); Hawaiian: Limu'o'olu (Kent 1986).

Description: Soft, gelatinous, pinkish red, much branched fronds, densely matted, with blunt apices, to 100 mm high. Axes segmented, with nodal diaphragms, segments about as broad as long, filled with watery mucilage (Braune and Guiry 2011).

Geographic distribution: World-wide in tropical to subtropical and warmer seas with bordering warm-temperate zones.

Figure 93. *Champia parvula* (C. Agardh) Harvey (Rhodophyta). Illustration from William H. Harvey (Plate CCX, Phycologia Britannica, 1846–1951).

Uses: Considered an edible alga in the Hawaiian Is (Kent 1986), and the Persian Gulf (Mohammadi et al. 2013).

Chondria capillaris (Hudson) M.J. Wynne (Fig. 94)

Synonym: *C. tenuissima*

Common name: Limu 'o'olu (Reed 1906, Kent 1986).

Geographic distribution: NE Atlantic; Adriatic Sea; Atlantic Is (Azores, Bermuda, Canary Is, Madeira, Selvage Is); Tropical and subtropical Atlantic; Mediterranean; Caribbean; Asia (China, Japan); Pacific Is (Hawaiian Is).

Figure 94. *Chondria capillaris* (Hudson) M.J. Wynne (Rhodophyta). Illustration from William H. Harvey (Plate CXCVIII, Phycologia Britannica, 1846–1951).

Description: Thallus bushy, vivid purple to pale yellowish red, upright cylindrical main axes cartilaginous-firm, with numerous long, softer irregularly spirally arranged lateral axes, branching to the fourth degree; the recent branches also stand rather loosely, spindle-shaped tapering at both ends, the apices pointed and crowned by hair-thin filamentous tufts (magnifying glass required); plants with reproductive bodies appear more densely branched peripherally; discoid holdfast, on prostate axes additional rhizoids (Braune and Guiry 2011).

Uses: Used as food in the Hawaiian Is (Reed 1906, Kent 1986).

Some *Chondria* species contain laminin, spasmolytic and hypotensive α-amino acid, and domoic acid; extract has the selective suppressant of parotitis virus; these plants are the source for antihelminthic and antiviral medicine; extracts of this species have antioxidant and vermifuge activity (Milchakova 2011, Pereira 2015a).

Chondria dasyphylla var. *intermedia* (Grunow) P.C. Silva

Synonym: *Chondriopsis intermedia* Grunow

Common name: Hawaiian: Limu o-olu (MacCaughey 1918).

Description: Thalli of cylindrical erect axes or decumbent tufts, brownish-red or yellowish, 8–15 cm high, to 1 mm diam., tubular, coarse-fibrous, and attached by basal holdfast.

Geographic distribution: Pacific Is (Polynesia, Samoa, and Hawaiian Is).

Uses: Considered an edible species in Hawaii, is common in fish-markets (MacCaughey 1918).

Chondrophycus dotyi (Y. Saito) K.W. Nam (formerly *Laurencia dotyi*)

Common name: Hawaiian: Limu lipe'epe'e (Abbott 1984).

Description: Forming clumps which have but a few erect axes standing on a discoid holdfast, without stoloniferous basal branches, up to about 5 cm high, cartilaginous; the erect axes slightly but constantly compressed, usually percurrent, broadest in the middle part, and there up to 1.8 mm broad and 1 mm thick; branching alternately or oppositely distichous. The branches also slightly compressed, longest at the middle portions, becoming prominently shorter toward the base and gradually shorter toward the upper portions. The ultimate stichidial branchlets up to 1 mm long, never compressed, clavate, truncate, or rounded at the apices; color brownish purple or somewhat greenish when fresh, changing to black on drying (Saito 1969).

Geographic distribution: Indian Ocean; Asia (Taiwan, Indonesia); Pacific Is (Polynesia; Hawaiian Is).

Uses: Used as food in the Hawaiian Is (Snakenberg 1987).

Chondrophycus succisus (A.B. Cribb) K.W. Nam (formerly *Laurencia succisa*)

Common name: Hawaiian: Limu Lipe`epe`e (Abbott 1984).

Description: Thalli 3–7 cm in length, dark reddish in color, compressed, 0.8–1.5 mm broad by 0.6–0.8 mm thick, cartilaginous in texture, not adhering to paper on being pressed, attached to the substratum by a discoid holdfast, without stoloniferous branches. Primary branches distichously, oppositely or rarely alternately arranged, bearing second order branches unbranched or sometimes bearing short simple branchlets. Branches and branchlets often restricted at the base (Wynne et al. 2005).

Geographic distribution: SW Asia (Oman); Australia; Pacific Is (Micronesia, Polynesia, Samoa, Tonga, Hawaiian Is, Mariana Is).

Uses: Used as food in the Hawaiian Is (Snakenberg 1987).

Chylocladia rigens (C. Agardh) J. Agardh

Common names: Hawaiian: Limu akuila, Limu kihe (Reed 1906, MacCaughey 1918).

Description: Thallus with erect fronds segmented by inner septa, the segments usually longer than wide and also externally visible as constrictions; mall discoid holdfast (Braune and Guiry 2011).

Geographic distribution: Indian Ocean; Pacific Is (Polynesia, Micronesia, Hawaiian Is).

Uses: Considered an edible species in Hawaii (Reed 1906, MacCaughey 1918).

Gelidium attenuatum (Turner) Thuret

Common name: Limu loloa (Reed 1906).

Description: Thallus cartilaginous, somewhat crispate, 2 to 40 cm tall, composed by one or several erect axes, terete or compressed, distichously, plumose or irregularly branched, red to deep purple, although in some species it can be blackish. Erect axes arise from cylindrical to compressed, branched (Wynne and Schneider 2010).

Geographic distribution: NE Atlantic; Mediterranean Sea; Pacific Is (Hawaiian Is).

Uses: Used as food in Hawaii (Reed 1906).

Gelidium filicinum Bory de Saint-Vincent

Common names: Limu loloa, Limu ekahakaha (Reed 1906).

Description: These algae are composed of several erect, flat axes similar to a sword blade but with rounded and somewhat widened apices. Attachment is by a cylindrical creeping axis with stolons at irregular intervals. Branches can be found along most of the axis, although they are scarce in the basal third. Branching can be of up to three orders. First-order branches are similar to the main axis, frequently disposed in an irregular manner and generally producing second-order or terminal, ovate-lanceolate branchlets. Often first order branches end in a truncate (grazed) apex whence a group of ovate-lanceolate or heart like branchlets originate.

Geographic distribution: SE Pacific; Pacific Is (Hawaiian Is).

Uses: Used as food in Hawaii (Reed 1906).

Gelidium micropterum Kützing

Description: Copiously branched plants, about 1 cm high. The diam. of the cells vary from 30–40 µ near the base to 8–10 µ at the tip, growth is diffuse; thin rhizoidal filaments may develop from the upper part of the main branches.

Geographic distribution: SE Atlantic; Indian Ocean; Pacific Is (Hawaiian Is).

Uses: Algae eaten in the Hawaiian Is (Reed 1906).
    This species is used as source of agar (Chapman and Chapman 1980).

Gelidium pulvinatum (Kützing) Thuret ex Bornet

Description: These algae appear as dark red, almost blackish cushions of entangled filaments. The thallus is cylindrical, up to 200 µ in diam., slender and creeping through most of its extension, producing short, erect axes and holdfasts at short intervals. The erect axes are cylindrical, lanceolate, or broadly ovoid, up to 100 µ broad, with small proliferations from the margins (Santelices 1977).

Geographic distribution: Atlantic Is (Madeira); Mediterranean; Pacific Is (Hawaiian Is).

Uses: This species is used as food in the Hawaiian Is (Reed 1906).

Gigartina polycarpa (Kützing) Setchell & N.L. Gardner

Common name: Tongue weed (Heyden 2013, Maneveldt and Frans 2015).

Description: Fleshy seaweed with oval blades. Individuals of this species are characteristically rough and bear numerous papillae (sometimes the blades are even rippled) giving the appearance of a rough tongue. The color varies from yellow-brown to reddish-brown to almost black (Maneveldt and Frans 2015).

Geographic distribution: Indian Ocean (S Africa).

Uses: In S Africa is fried in coconut oil, then salted. It is a source of carrageenan, used as a gel (Heyden 2013).

Gracilaria coronopifolia J. Agardh

Synonym: *Gracilaria lichenoides* f. *coronopifolia*

Common names: Hawaiian: Limu-manauea (MacCaughey 1918, Fortner 1978, Arasaki and Arasaki 1983), Limu Manauea (Reed 1906, Abbott 1984, Kent 1986); Filipino: Cao-caoyan, Gargararao (Domingo and Corrales 2001), Kawkawayan (Agngarayngay et al. 2005).

Description: Thalli bushy, erect to decumbent, greenish brown to reddish purple in habitat, cartilaginous, attached to the substratum by a disk-like holdfast. Main axis cylindrical, branching frequent, irregularly alternate, with slight constriction at the base (Domingo and Corrales 2001).

Geographic distribution: Asia (China, Japan, Taiwan, Indonesia, Malaysia, the Philippines, Singapore, Vietnam); Australia; Pacific Is (Micronesia, Fiji, Hawaiian Is).

Uses: In SE Asia is used as food in salads, soups, and for agar extraction (Fortner 1978, Chapman and Chapman 1980, Lembi and Waaland 1988, Nang and Dinh 1998, Roo et al. 2007, Lee 2008, Agngarayngay et al. 2005, Sidik et al. 2012, Harrison 2013, Kılın et al. 2013). The plants of this variety are eaten both raw and cooked. It is eaten with salt and tomatoes, either fresh or after dipping it in boiling water; sometimes also in combination with onions and vinegar (Zaneveld 1959). Extensively used for food in the Hawaiian Is; a specialty is cooking it with octopus (Reed 1906, MacCaughey 1918, Abbott 1978, Kent 1986, Zemke-White and Ohno 1999, Harrison 2013).

Gracilaria debilis (Forsskål) Børgesen
(formerly *Gymnogongrus javanicus*)

Common name: Hawaiian: Limu koelen (Zaneveld 1959).

Description: Thalli purple to dark red, repeatedly forked, bushy and coriaceous in texture; height up to 10 cm. The branches are compressed or flat, consisting of a central axis or colorless cells and a peripheral layer of smaller, closely packed assimilative cells, arranged perpendicular to the surface (Zaneveld 1959).

Geographic distribution: Tropical and subtropical W and E Atlantic; Indian Ocean; Asia (Indonesia, the Philippines, Singapore); Pacific Is (Hawaiian Is).

Uses: In Hawaii this "dry or hard limu" is eaten raw as a salad or as a relish (Zaneveld 1959); is also used as food in Indonesia (Arasaki and Arasaki 1983).

Gracilaria ephemera Skelton, G.R. South & A.J.K. Millar

Common name: Samoan: Limu aau (Ostraff 2003).

Description: A flattened species from Samoa (Skelton et al. 2004).

Geographic distribution: Pacific Is (Polynesia, Samoa).

Uses: Considered an edible species from Samoa (Ostraff 2003).

Gracilaria maramae G.R. South

Description: Thallus succulent, erect, solitary or caespitose, arising from a small disk 1.5–2.0 mm in diam.; plant terete throughout, 25 cm tall (average, 40 cm), axes 1.0–2.5 mm in diam. (average, 1.5 mm), with two to four orders of branching, the branches flagelliform; main axis persistent, or limited to the lowermost part; in very sheltered locations, lower laterals showing a tendency to become decumbent, or becoming the lead branches. First-order branches alternate, second-order branches predominantly second, unilateral branches arranged predominantly abaxially. Branches markedly fine tip at the base, tapering terminally to a fine tip (Abbot 1995).

Geographic distribution: Pacific Is (Polynesia, Fiji).

Uses: This species is an important food species in the Fiji Is, where it is gathered and marketed by Fijian women (South 1993a).

Gracilaria parvispora I.A. Abbott

Common name: Hawaiian: Limu ogo (Lembi and Waaland 1988), Long ogo; Ogo (Hart et al. 2014).

Description: Solid, commonly compressed branches, 1–4 mm in diam., with long narrow, pointed tips. The thalli grow tall, to 30 cm or more, with a single dominant axis, 0.8–3.5 mm diam., usually with three orders of branching or, if more, the last order is short, slender and spine-like. The plant is often red, but can become light brown, light green, or almost white in areas of bright sunlight; can become very dark brown to almost black in habitats of low water motion or in mariculture (Abbott 1995).

Geographic distribution: E Pacific (Baja California); Asia (Japan, Korea); Pacific Is (Polynesia, Hawaiian Is).

Uses: *G. parvispora*, or long ogo, was the most common seaweed in Hawaiian markets until the 1970s, sold as a fresh vegetable in Honolulu, usually mixed with raw fish (Lembi and Waaland 1988, Zemke-White and Ohno 1999, Roo et al. 2007, Rao 2010, Harrison 2013). Due to its large popularity as an ingredient in many types of cuisine, a long time ago was overharvested and the government restricted its growth (Kılınç et al. 2013).

Grateloupia filicina (J.V. Lamouroux) C. Agardh (Fig. 95)

Common names: English: Chop-chop (Abbott 1984); Hawaiian: Limu paka-ele-awa'a, Limu hulu-hulu-waen (MacCaughey 1918), Limu hula hula

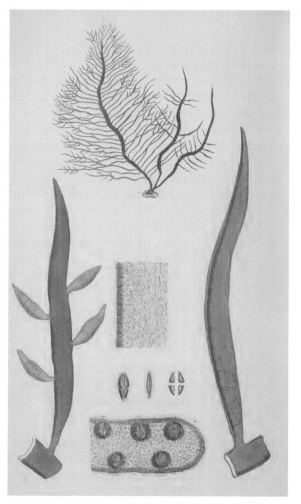

Figure 95. *Grateloupia filicina* (J.V. Lamouroux) C. Agardh (Rhodophyta). Illustration from William H. Harvey (Phycologia Britannica, 1846–1951).

waena (Fortner 1978, Chapman and Chapman 1980), Limu Huluhu-Iuwaena (Arasaki and Arasaki 1983), Pakeleaw'a, Ake limu (Abbott 1984); Japanese: Mukade-nori (Yendo 1902, Smith 1904, Subba Rao 1965, Arasaki and Arasaki 1983, Ohno et al. 1998); Portuguese: Ratanho, Ratenho (INII 1966).

Description: Bushy, pyramidal plants with flattened stems and fine branches in one plane. Grow in small groups on rocks covered with fine sand (Pereira 2015a).

Geographic distribution: NE Atlantic; Atlantic Is (Azores, Canary Is, Cape Verde Is, Madeira); Adriatic Sea; Mediterranean Sea; Tropical and subtropical W and E Atlantic; Indian Ocean; Asia (China, Japan, Korea, Taiwan, Indonesia, the Philippines, Singapore, Vietnam); Australia; Pacific Is (Polynesia, Micronesia, Fiji, Hawaiian Is, Mariana Is, Marshall Is); Antarctic Is (Fuegia).

Uses: In Japan, after washing, plants are usually finely chopped and lightly salted. Eaten with raw liver, raw fish; added to cooked beef at serving time; or eaten with dried or broiled *Octopus* (Subba Rao 1965, Kang 1968, Okazaki 1971, Bonotto 1976, Abbott 1984, Hotta et al. 1989, Ohno and Largo 2006, Roo et al. 2007, Harrison 2013). In the Philippines, *G. filicina* is used in vegetable salads (Trono 1998, Zemke-White and Ohno 1999). The indigenous of the Batan Is (Indonesia) market the seaweeds in Ilocos province, where they are eaten (Zemke-White and Ohno 1999). In the Hawaiian Is, is the best known edible species of *Grateloupia* in Hawaii; its many names reflect this. In ancient Hawaii it was reserved for the "ali'I" and was called "the queen limu" (Abbott 1978, Fortner 1978), used in beef stew, with octopus, especially beef liver (Kent 1986), and with limpets (Zaneveld 1959). In Indonesia is used to make salad, vegetable soup, pickle, and sweetened jellies (Istini et al. 1998, Irianto and Syamdidi 2015); used also as food in Korea (Kang 1968, Bonotto 1976, Hotta et al. 1989). This species is also eaten in India (Rao 1970).

Extracts of this species have antiangiogenic (Yu et al. 2012), and vermifuge (Istini et al. 1998) activity.

Grateloupia hawaiiana E.Y. Dawson

Description: Displays a wide variability in form, from deeply lobed, broad, and flattened to narrow, twisted, and wiry. The texture of the fresh plant is rubbery and the color is a deep and shiny reddish-purple (Fortner 1978).

Geographic distribution: Pacific Is (Hawaiiana Is).

Uses: Like the other edible species of *Grateloupia*, *G. hawaiiensis* has a pleasant taste reminiscent of the ocean and combines well in a vinegar dressing or pickling sauce. The blades can be pressed together and shredded or sliced, then eaten fresh in a salad or cooked in a soup or "tsukudani" (Fortner 1978).

Griffithsia ovalis Harvey (Fig. 96)

Common names: Hawaiian: Limu moopunaa-ka-lipoa, Limu moo-puna, Limu ka-lipoa, Limu au-pupu (Reed 1906, MacCaughey 1918, Abbott 1946, Kent 1986).

Description: Thallus light red, 3–6.5 cm high, branching spreading, three-four times sub-dichotomous, ecorticate; attached by rhizoids from the basal cell; epiphytic on sea grasses or algae (Abbott 1946, Womersley 1998).

Figure 96. *Griffithsia ovalis* Harvey (Rhodophyta). Illustration from William H. Harvey (Plate CCIII, Phycologia Australica, 1858–1863).

Geographic distribution: S Atlantic; Indian Ocean; SE Asia (the Philippines, Singapore); Australia; Pacific Is (Polynesia, Micronesia, Fiji, Mariana Is, Marshall Is, Hawaiian Is).

Uses: Sometimes used for food on Maui and S Hawaii (Reed 1906, MacCaughey 1918, Abbott 1946, Kent 1986).

Gymnogongrus disciplinalis (Bory de Saint-Vincent) J. Agardh

Common names: Limu awikiwiki, Limu nei (Reed 1906), Limu vavaloli (Subba Rao 1965).

Description: Plants of the largest species reach 25 cm in length and occur singly or in dense aggregates from extensive (to 40 cm diam.) crustose bases. Gametophytes are erect, generally compressed, dichotomous, and either lack or have varying numbers of proliferous marginal laterals.

Geographic distribution: SE Pacific; Pacific Is (Hawaiian Is).

Uses: This species is used as food in the Hawaiian Is (Reed 1906, Subba Rao 1965).

Halymenia formosa Harvey ex Kützing

Common names: Hawaiian: Limu Lepe-o-hina, Limu lepe'ula'ula (Fortner 1978, Abbott 1984), Limu lepeahina (Reed 1906).

Description: Blades are up to 1 m wide, commonly 5–50 cm, with toothed or lobed margins; dark magenta or purple color. Shape highly variable, may have slender branches coming from a broad blade, or be branched throughout its length (Novaczek 2001).

Geographic distribution: Indian Ocean; SW Asia (Oman); Asia (Japan, Korea, Taiwan); SE Asia (Indonesia, Malaysia, the Philippines, Singapore); Australia; Pacific Is (Hawaiian Is, Samoan Is).

Uses: In Hawaii, the classical method of preparation is as a thickener in a stew or soup. The tasty dried fronds can be served as a snack, somewhat like potato chips (Reed 1906, Fortner 1978, Lembi and Waaland 1988, McDermid and Stuercke 2003). In Samoa, *Halymenia* is included with fish in a ground oven, or cooked as a vegetable with coconut milk. The fronds may also be cooked in coconut milk with local herbs and eaten as a vegetable. In Vanuatu is eaten fresh or sundried and used as a thickener and flavoring in seafood soup or stew. Dry fronds can be eaten as a snack. Fresh *Halymenia* is also sliced thin and eaten as a salad vegetable (Novaczek 2001). This species is also eaten in the Philippines and Malaysia (Zaneveld 1959, Sidik et al. 2012).

Hypnea cenomyce J. Agardh

Description: Plants with persistent axes with densely branched, cushion like basal part and erect fronds (Chiang 1997).

Geographic distribution: Atlantic Is (Cape Verde Is); Tropical and subtropical W Atlantic; Indian Ocean; SW Asia (Bangladesh, Sri Lanka); Asia (Japan, Taiwan); SE Asia (Indonesia, the Philippines, Vietnam); Australia; Pacific Is (Easter Is).

Uses: Used as food in Indonesia, Borneo, Hawaii, the Philippines (Zaneveld 1955, 1959, Johnston 1966, Irianto and Syamdidi 2015), and also used for agar extraction (Sidik et al. 2012).

Hypnea musciformis (Wulfen) J.V. Lamouroux

Common names: Chinese: Su-wei-tung (Tseng 1935); English: Crozier weeds (Santhanam 2015), Hooked red seaweed (Stender and Stender 2014); Hawai'i: Maidenhair, Limu huna (Fortner 1978, Novaczek 2001); Fiji: Lumt cevata, Lumi tabia, Lumi vakalolo (Novaczek 2001); Wormroos (used by the Netherlanders in Indonesia) (Zaneveld 1959); Spanish: Alga con gancho (Radulovich et al. 2013); Tamil: Sem pasi (Kaliaperumal et al. 1995).

Description: This pale pink, greenish, brownish or yellowish, highly branched plant creeps along the bottom. It forms tangled mats that are often iridescent in the sun. The plant is generally floppy, entangled and loose lying. The cylindrical branches are only about 0.5–1 mm wide; branches alternate and taper at the tips. Branchlets are short, stubby spines. The plant feels fluffy and light like cotton threads. *Hypnea* is tolerant of brackish water (Novaczek 2001, Stender and Stender 2014b).

Geographic distribution: Tropical and subtropical Atlantic; Mediterranean; Indian Ocean; Asia (China, Indonesia, Myanmar, the Philippines, Singapore); Australia; Pacific Is (Hawaiian Is, Mariana Is).

Uses: This edible plant is used to make jellies. It has a mild flavor and delicate texture. Make jelly with coconut and fish or use to make sweet puddings and jellies. Cook in a batter to make fritters. Chop fresh for use in salads. Use as a thickener for gravy, soup or stew (Zaneveld 1959, Johnston 1966, Simoons 1990, Novaczek 2001). Some populations of Panama use it in the preparation of soups (Radulovich et al. 2013).

Hypnea is high in carrageenan, which can thin the blood and lower cholesterol, helping to prevent strokes. It is also used against diarrhea (Novaczek 2001).

Extracts of this species have also antibacterial (Ravikumar et al. 2002, Selvin and Lipton 2004), antiviral (Mendes et al. 2012), anti-inflammatory (de Brito 2013), psychotropic, anxiolytic and hypocholesterolemic (Rahila et al. 2010) activity.

Hypnea nidifica J. Agardh

Common names: Hawaiian: Limu huna (Reed 1906), Wane'one'o (Ostraff 2003).

Description: *Hypnea* is characterized by a terete to compressed, much branched thallus with short lateral branchlets, globular cystocarps, and zonate tetrasporangia on short laterals (Womersley 1994).

Geographic distribution: Tropical and subtropical Atlantic; Indian Ocean; SE Asia (Indonesia); New Zealand; NE Pacific (Alaska to California); Pacific Is (Hawaiian Is).

Uses: Is boiled with squid or *Octopus* in Hawaii, and used for gastritis treatment (Reed 1906, Chapman and Chapman 1980, Zemke-White and Ohno 1999, Roo et al. 2007, Harrison 2013).

Hypnea spicifera (Suhr) Harvey

Synonym: *H. armata*

Common names: African: agar agar; Hawaiian: Limu huna (MacCaughey 1918).

Description: Elongate, corymbosely branched (MacCaughey 1918).

Geographic distribution: SE Atlantic; Indian Ocean; Pacific Is (Hawaiian Is); SE Pacific.

Uses: This species and *H. nidifica* are especially relished by the Hawaiian natives when boiled with octopus (Reed 1906, MacCaughey 1918).

Laurencia spp.

Common names: Hawaiian: Limu mane`one`o (Novaczek 2001), Limu pe'epe'e (Fortner 1978); English: Flower limu (Novaczek 2001), Mustard limu (Abbott 1984).

Description: There over 16 species of *Laurencia* in Hawai`i. Plants are usually erect and fleshy with variable branching patterns, each order of branching shorter than the preceding. Most *Laurencia* species have cylindrical branches, but a few intertidal species are compressed. Branch tips club-shaped with pits in the tips. Plants usually range in size from a few cm to 20 cm, and are pink-purple to red but can have yellow and even green portions.

Laurencia spp. are found in clumps or as components of turfs attached to eroded coral or basalt rocks on intertidal to subtidal zones, and are often associated with *Acanthophora spicifera*.

Geographic distribution: Various edible species are found throughout the region; reported from New Caledonia, Fiji, Solomon Is, Tahiti, Samoa, Tonga, Cook Is, Hawai'i, Palau, Marshall Is, Kiribati and elsewhere in Micronesia (Novaczek 2001).

Uses: Some types have a very peppery flavor and are chopped and salted for use as a spice; its spicy flavor combines well with raw fish. It should only be eaten fresh (Fortner 1978, Novaczek 2001). Others are sweet and used as vegetables in salads or cooked dishes (Novaczek 2001).

Laurencia composita Yamada

Common name: Filipino: Kulot (Agngarayngay et al. 2005).

Description: Erect thalli, up to 40 cm high, attached by discoidal to rhizoidal to encrusting holdfasts; thalli sparingly to highly branched in all directions or bilateral; branches cylindrical to flat; ultimate branchlets blunt to truncate, often claviform, not much narrower than other branches (Papenfuss 1947).

Geographic distribution: Tropical and subtropical W Atlantic; Asia (China, Japan, Korea, the Philippines).

Uses: Used as food in salads, and mixed with meat (Agngarayngay et al. 2005).

Laurencia glomerata (Kützing) Kützing

Synonym: *L. virgata*

Common names: Limu maneoneo, Limu lipuupuu (Reed 1906).

Description: Thallus with cylindrical branches densely crowded with rounded, purplish-brown tips (Lubke and Moor 1998).

Geographic distribution: SE Atlantic; Indian Ocean; SE Asia (Indonesia); Pacific Is (Hawaiian Is).

Uses: Used as food in the Hawaiian Is (Reed 1906).

Laurencia nidifica J. Agardh

Common names: English: Mustard limu (Abbott 1984); Hawaiian: Limu Mane'one'o (Abbott 1984), Limu Ape'ape'e (Kent 1986).

Description: Firm, erect plant, to 10 cm tall, arising singly or in tufts from an entangled base. Terete axes are relatively thin, 0.5–1 mm in diam., branching rarely more than three orders with the main divisions sub-dichotomous.

Next orders are varied: alternate, opposite, or occasionally whorled. Branchlets are short, with blunt, indented tips. Because of the high variation in color, branching pattern and texture, it is not simple to identify *Laurencia* species in the field (Abbott 1999).

Geographic distribution: Atlantic Is (Cape Verde Is, Madeira); SE Atlantic; Indian Ocean; Asia (Japan, Korea); SE Asia (Indonesia, the Philippines, Thailand, Vietnam); Australia; Pacific Is (Polynesia, Micronesia, Hawaiian Is, Samoan Is).

Uses: *L. nidifica* is used as a condiment by Hawaiians because of its peppery taste (Abbott 1978, Arasaki and Arasaki 1983, Kent 1986).

Liagora albicans J.V. Lamouroux

Synonym: *L. decussata*

Common names: Hawaiian: Limu pu-aki (MacCaughey 1918), Limu puaki (Reed 1906, Zaneveld 1959), Limu puak (Scönfeld-Leber 1979).

Description: Fronds filiform, calcareous, pinnate or radial branching, segments distinctively constricted at base, deep pink, green or white depending on depth (MacCaughey 1918, Stender and Stender 2014c).

Geographic distribution: Warm seas worldwide.

Uses: Is eaten in Hawaii (Reed 1906, MacCaughey 1918, Zaneveld 1959, Scönfeld-Leber 1979).

Melanamansia glomerata (C. Agardh) R.E. Norris
(formerly *Amansia glomerata*)

Common names: Hawaiian: Limu li-pepe-iao, Limu pepe-iao (Reed 1906, MacCaughey 1918), Amansia, Lipepeiao, Limu ha'ula (Kent 1986).

Description: The thallus is rose-red, up to 100 (on average 40) mm high. Branching is lax to profuse and usually forms deep-red leafy rosettes of secondary branchlets when fully mature. Secondary blades develop as successive pairs on the dorsal side of the midrib, and are crisp and lanceolate, up to 35 mm long and 6 mm broad, 85–110 µ thick, with more or less pronounced marginal serrations 1–4 mm long (which represent endogenous branches) always present. A distinct midrib up to 1 mm wide is present in the lower half of ultimate branchlets, and the apex is characteristically in-rolled (N'Yeurt 2002).

Geographic distribution: Tropical Indian Ocean; SE Asia; Japan; Tropical Pacific Ocean (N'Yeurt 2002).

Uses: Used as food in the Hawaiian Is (Reed 1906, MacCaughey 1918, Kent 1986).

Mastocarpus papillatus (C. Agardh) Kützing
(formerly *Gigartina papillata*)

Common names: English: Grapestone (Madlener 1977), Tar spot, Turkish washcloth (Gutschmidt 2010b).

Description: It has a long-lived tetrasporophyte phase, which looks like a crust or dried tar spot on rocks, with colors ranging from reddish-brown to black. Each year, a new gametophyte phase grows from the "tar spot", known in some areas as "Turkish washcloth", consisting of short, upright, rubbery, deeply divided and curled blades, growing up to 15 cm. The ruffly blades taper to a small stipe and holdfast. Actually, even this gametophyte stage of *M. papillatus* can have two different appearances. There are male and female gametophytes, which are produced on separate plants. The female plants are covered with papillae—short, bumpy or nipple-like outgrowths which are the sites of fertilization. Colors can range from very dark purple-red to dark red-brown dark brown to reddish black to almost black. The male gametophyte has no papillae, is yellowish to light pink or light rose, and has a thinner blade. The spore producing tar spot phase of *M. papillatus* is a slow growing, long lived perennial (one source estimates up to 90 years). The female and male bladed gametophyte phase dies back every year leaving just the encrusting "tar spot" (Gutschmidt 2010b).

Geographic distribution: N Atlantic (Iceland); NE Pacific (Alaska to California, Mexico).

Uses: Used as food in Iceland and NE Pacific (Madlener 1977, Gutschmidt 2010b).

Meristotheca procumbens P.W. Gabrielson & Kraft

Common name: Lum mi'a (N'Yeurt 1999).

Description: Plant deep-pink and turgid when fresh, procumbent, up to 10 cm in diam., branched and lobed. Thallus attached at various points to support coral via terete haptera up to 2 mm long (N'Yeurt 1995).

Geographic distribution: Australia; Pacific Is (Cook Is, Fiji, Polynesia, Samoa).

Uses: This species is a favorite food item in the Rotuman (Fiji Is) and Cook Is diet (N'Yeurt 1995, 1999, Zemke-White and Ohno 1999, Roo et al. 2007, Harrison 2013).

Nitophyllum adhaerens M.J. Wynne

Common name: Hawaiian: Limu haula (Reed 1906).

Description: Thalli consisting of extensive tangled mats composed of prostrate and overlapping, decumbent, lobed blades lacking stipe and holdfast; lobed blades each up to 2 cm in length by 0.5–6 mm in width, mats reaching up to 13 cm in width, attached to the substratum by rhizoidal haptera borne underneath the prostrate blades and by marginal rhizoidal filaments along the thallus margin; blades membranous and monostromatic throughout; macro- and microscopic veins absent.

Geographic distribution: Tropical and subtropical W Atlantic; Caribbean Sea; SE Asia (Vietnam); Pacific Is (Polynesia, Micronesia, Guam, Hawaiian Is).

Uses: Used as food in the Hawaiian Is (Redd 1906).

Palisada intermedia (Yamada) K.W. Nam

Synonyms: *Laurencia intermedia, Chondrophycus intermedius*

Common name: Filipino: Kulot (Agngarayngay et al. 2005).

Description: Erect thalli, up to 40 cm high, attached by discoidal to rhizoidal to encrusting holdfasts; thalli sparingly to highly branched in all directions or bilateral; branches cylindrical to flat; ultimate branchlets blunt to truncate, often claviform, not much narrower than other branches (Papenfuss 1947).

Geographic distribution: Atlantic Is (Cape Verde Is); Tropical and subtropical W and E Atlantic; Asia (China, Korea, Taiwan, Indonesia, the Philippines); Australia (Papua New Guinea).

Uses: Used as food in salads, and mixed with meat (Agngarayngay et al. 2005).

Polyopes J. Agardh

Common names: Hawaiian: Limu luau (Reed 1906); Japanese: Matsunori (Chapman and Chapman 1980).

Description: Sub-cylindrical to flat, of uniform width (relatively narrow), irregular constrictions sometimes present. Dichotomously to irregularly branched; margins entire or with pinnate club-shaped or bifurcate branchlets.

Geographic distribution: Indian Ocean; Asia (China, Japan, Korea); SE Asia (Indonesia, the Philippines); Pacific Is (Fiji, Hawaiian Is).

Uses: Used as food in the Hawaiian Is (Redd 1906).

Porphyra vietnamensis T. Tanaka & Pham-Hoàng Ho

Common name: Hawaiian: Limu pahe'e (Abbott 1984).

Description: This species is membranous, monostromatic, attached to the rocky substratum with the help of slightly circular disk shaped holdfast like attachment organ (Ghosh and Keshri 2009).

Geographic distribution: Indian Ocean; Asia (China, Indonesia, Taiwan, Thailand, Vietnam); Pacific Is (Hawaiian Is).

Uses: Commonly eaten as food in the Hawaiian Is (McDermid and Stuercke 2003); used in India (Rao 1970), Pakistan (Rahman 2002), and Thailand as foodstuff (Lewmanomont 1978, Lewmanomont 1998, Zemke-White and Ohno 1999, Roo et al. 2007, Harrison 2013).

Scinaia aborealis Huisman

Common name: English: Glassweed (Novaczek 2001).

Description: Thallus medium gray-red to dark red, sometimes brownish, 5–38 cm high, sub-dichotomously branched every 1–11 cm usually with constrictions at points of branching, segments 1.5–3 mm in diam. below, 3–4 mm in diam. above, ultimate segments tapering to a point; holdfast discoid, 1.5–2 mm across; epilithic (Womersley 1994).

Geographic distribution: Tropical and subtropical W Atlantic; SE Asia (the Philippines); Australia; Pacific Is (Fiji, New Caledonia).

Uses: *Scinaia* is eaten as salads or cooked with coconut milk (Novaczek 2001, Rao et al. 2010).

Solieria robusta (Greville) Kylin

Common names: English: Blubber weed (Santhanam 2015), Tender golden weed (Novaczek 2001); Filipino: Tajuk bau'no (Tito and Liao 2000); Fijian: Lumitamana (N'Yeurt 1999).

Description: Thallus gray-red to brown-red, often bleached to yellow-gray, erect, 10–30 cm high, moderately and irregularly branched at intervals of 1–4 cm, branches relatively soft, terete to slightly compressed, 2–5 mm in diam., basally constricted and tapering above to rounded or sub-acute tips; holdfast fibrous, branched, 1–2 cm across, with several fronds; epilithic (Womersley 1994).

Geographic distribution: Asia (China, Japan, Taiwan); Indian Ocean; SE Asia (Indonesia, the Philippines, Singapore); SW Asia (Arabian Golf, India, Iran, Kuwait, Pakistan, Sri Lanka, Yemen); Australia; Pacific Is (Fiji).

Uses: Commonly used as food in Fiji, Cook Is and the Philippines, eaten cooked in coconut milk with condiments added to make use of their gelling properties (South 1993a, N'Yeurt 1999, Tito and Liao 2000). Make into pickles or relish, use in salads. Prepare as a vegetable, like *Gracilaria*, *Hypnea* or *Acanthophora*. Chop into 0.5–1 cm bits and mix with salted raw fish (Novaczek 2001).

Spyridia filamentosa (Wulfen) Harvey (formerly *Spyridia spinella*)

Common names: English: Hairy basket weed (Bunker et al. 2010, Santhanam 2015); Hawaiian: Limu hulupuaa (Reed 1906), Hulu pua'a (Kent 1986).

Description: Principal axes with 0.5–2 mm in diam., covered with a large number of fine and delicate branches which give a fluffy appearance; red color to almost white; the fact that it often traps small particles of sediment between their branches can make it appear to have other colors.

Geographic distribution: Adriatic Sea; NE Atlantic; Mediterranean; Atlantic Is (Azores, Bermuda, Canary Is, Cape Verde Is, Madeira, Selvage Is); Tropical and subtropical Atlantic; Caribbean Is (Bahamas, Barbados, Cayucos Is, Cuba, Jamaica, Martinique, Puerto Rico, Trinidad); Indian Ocean; Asia (China, Japan, Korea, Taiwan); SE Asia (Indonesia, the Philippines, Vietnam, Singapore); Australia and New Zealand; Pacific Is (Samoa, Polynesia, Micronesia, Fiji, Hawaiian Is, Mariana Is, Marshall Is, Solomon Is); E Pacific (California, Mexico).

Uses: Used as food in the Hawaiian Is (Reed 1906, Kent 1986).

2.6 East Pacific (Alaska to California, Mexico, Peru, Chile, Argentina)

2.6.1 Chlorophyta

Codium dichotomum S.F. Gray

Common names: Japanese: Imose-miru, Yezo-miru (Tokida 1954).

Description: Thallus spongy, dark green, anchored to rocks or shells by a weft of rhizoids, varying in size from 15 to 25 cm long. Dichotomously branched (to nine orders); erect or repent; branches wholly terete or variously flattened, at times anastomosing. Internal structure composed of a colorless medulla of densely intertwined siphons and a green palisade-like layer of vesicles called utricles.

Geographic distribution: NE Atlantic (France, Sweden), Adriatic Sea, E Pacific (California, Mexico, Galápagos Is); Mediterranean Sea (Egypt); Asia (South Korea).

Uses: Direct use as food in Japan and Korea (Tokida 1954, Levring et al. 1969, Bonotto 1976); used also as food in Argentina (Zemke-White and Ohno 1999).

Monostroma quaternarium (Kützing) Desmazières

Description: Frond widely stretched like a membrane, ultimately often freely vessel, consisting of a single seated cells rounded or angular, often ternate or even closely approach and arranged without any order appreciable, towards the base stretched-clubbed or transformed into fibers by radical agglutinated mucilaginous substance.

Geographic distribution: NE Atlantic (Britain, Spain); NE Pacific (Alaska to Baja California, Costa Rica, Peru).

Uses: Used as food in Peru (Aldave-Pajares 1985).

Protomonostroma undulatum (Wittrock) K.L. Vinogradova

Synonym: *Monostroma undulatum*

Common name: Japanese: Hida-hitoe (Tokida 1954).

Description: Thalli light green, thin, with delicate blades which are one cell layer thick (as seen in cross section here). This alga is common from winter to late spring, and growing attached to rocks or on other algae.

Geographic distribution: N Atlantic; Atlantic Is (Greenland, Iceland); Tropical and subtropical W Atlantic, Baltic Sea; Asia (China, Japan, Korea, Russia); Argentina; Antarctic Is; NE Pacific (Alaska).

Uses: Considered an edible species in Argentina (Riss et al. 2003).

Ulva lactuca Linnaeus (Fig. 97)

Synonym: *U. fenestrata*

Common names: English: Sea lettuce, Green laver (Garza 2005); Japanese: Aosa (Smith 1904); Portuguese: Alface-do-mar (Pereira 2010a).

Description: Sea lettuce is vivid green and cellophane thin. One can see the outline of one's hand even through larger mature fronds. The frond is connected to rocks with a small, almost invisible holdfast. The frond grows as a single, irregular, but somewhat round shaped blade with slightly ruffled edges. The young fronds are small, thin, and very fragile, while the large fronds can grow to 6–8 inches and look slightly thicker. There may be randomly spaced holes in the frond. Sea lettuce can be confused with other green seaweeds which may not have the same flavor as *Ulva* (Garza 2005).

Geographic distribution: This species is ubiquitous, common to most shorelines around the world.

Uses: Sea lettuce is usually picked to dry and used as a seasoning. It is a delicate seaweed that dries to a small amount (Zemke-White and Ohno 1999, Garza 2005, Braune and Guiry 2011, Harrison 2013); is used as food in Vietnam and Indonesia (Roo et al. 2007).

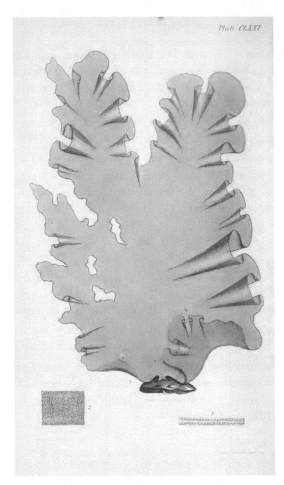

Figure 97. *Ulva lactuca* Linnaeus (Chlorophyta). Illustration from William H. Harvey (Plate CLXXI, Phycologia Britannica, 1846–1951).

2.6.2 Ochrophyta – Phaeophyceae

Alaria marginata Postels & Ruprecht

Common names: English: Winged kelp (McConnaughey 1985), Pacific wakame, Atlantic wakame, Wild wakame (Mouritsen 2013), Strap kelp, Winged kelp, Honey ware; Japanese: Wakame (McConnaughey 1985).

Description: *Alaria*, an olive-brown colored seaweed, can grow up to more than 60 cm long and 5–20 cm wide in Alaska. It is anchored to the ocean floor with a visible holdfast. A short but noticeable stipe runs from the holdfast to the frond. Oval shaped blades are attached to the stipe; these are sporophylls. The frond is long and narrow. The edges of the frond are ruffled and tend to fray near the tip, likely due to wave actions. A thick midrib runs the length of the frond. Small *Alaria* fronds look similar to older ones (Garza 2005).

Geographic distribution: NE Pacific (Alaska to Oregon, Aleutian Is); Asia (Japan, Russia).

Uses: Used as food in British Columbia (Canada) (Levring et al. 1969, Kuhnlein and Turner 1996, Zemke-White and Ohno 1999, Lindstrom 2006, Harrison 2013, Kellogg and Lila 2013, Edibleseaweed 2014). In Alaska, when fresh the midrib of *Alaria* is often cut out, creating two sheets and a thick celery-like stalk. The midrib can be chopped fresh and quickly stir-fried. The two side sheets can be used like a tortilla. Rice and fish are piled in the center and the seaweed sheet is rolled. Dried fronds or strips may be broken into pieces. Store the pieces in airtight containers, in a cool, dry place (Garza 2005, Roo et al. 2007).

Costaria costata (C. Agardh) De A. Saunders

Common names: English: Ribbed kelp, Seersucker kelp (Holmes 2012), Short kelp (Kuhnlein and Turner 1996); Japanese: Sujime (Tokida 1954).

Description: Stipe cylindrical, unbranched, with longitudinal stripes, flattened towards the top and merging into a single, undivided, broad, elongated to broadly oval, dark chocolate brown, blade. This has five characteristic longitudinal ribs running the entire length and alternately protrudes from both sides of thallus. Thallus wrinkled, blistered between ribs. Stipe basally attached by a strong holdfast of branching haptera (Braune and Guiry 2011).

Geographic distribution: Found commonly along the Pacific Coast of the Americas from Alaska through N Mexico, also occurring in Chile, Argentina, and Japan.

Uses: Used for foodstuff in NE Pacific, especially for the manufacture of potash salts (Tokida 1954).

C. costata is sometimes eaten by coastal people (Korea) when the plants are young and tender. Lacking the muciliage of other kelp species makes *C. costata* much more palatable. The tissues of this plant, like all kelps, are high in potassium, making it useful in a variety of applications from folk medicine to cosmetics to fertilizer (Tokida 1954, Bonotto 1976, Kuhnlein and Turner 1996).

Recent studies have also found that kelps like *C. costata* contain sterols which are useful in the lowering of cholesterol levels as well as fucoidans which are being studied as antitumor agents for human melanoma and colon cancer. The antithrombotic serine protease (CCP), purified from this species, could have therapeutic potential for the treatment of thrombosis (Kim et al. 2013).

Cymathaere triplicata (Postels & Ruprecht) J. Agardh

Common name: Three-ribbed Kelp (Lindeberg and Lindstrom 2015a).

Description: Thallus of this light to sometimes reddish brown kelp has a discoidal holdfast, a stipe up to 25 cm long, a linear blade up to 4 m long and 18 cm wide, and three riblike folds. No other kelp has this configuration. This species has a distinctive cucumber-like aroma that can often be smelled before the kelp is seen (Lindeberg and Lindstrom 2015a).

Geographic distribution: NE Pacific (Alaska, British Columbia); Asia (Japan, Russia).

Uses: Considered an edible species (Lembi and Waaland 1988).

Durvillaea antarctica (Chamisso) Hariot

Common names: Chilean: Cochajugo (Lembi and Waaland 1988), Cochayuyo (Astorga-España and Mansilla 2014); English: Bull kelp (Ohmi 1968), Cape, Thonged (Akatsuka 1994); New Zealand (Maori): Rimurapi (Brooker and Cooper 1961), Rimuroa (Novaczek 2001, Ostraff 2003); Spanish: Cochayuyo, Huilte, Ulte, Coyofe (Akatsuka 1994).

Description: Is said to be the strongest kelp in the world, able to withstand the rough seas that are characteristic of the S Ocean; a massive, discoid holdfast supports a thick, flexible, cylindrical stipe (up to 15 m long) which gives rise to several narrow strap-like blades. The number, length, and width of these blades are variable and are influenced by growing conditions with more divisions occurring in more exposed sites.

Geographic distribution: Atlantic Is (Gough Is); Antarctic and the sub-Antarctic Is; Circumpolar distribution between Chile and Macquarie Is.

Uses: Is eaten in Chile and Peru, after washing, drying and roasting (Ohni 1968, Levring et al. 1969, Chapman and Chapman 1980, Masuda 1981a, Zemke-White and Ohno 1999, Alveal 2006, Roo et al. 2007, Harrison 2013, Astorga-España and Mansilla 2014); stipe used mainly on Is of Chiloe in soups and as vegetable. Also used by aborigines in Australia (Chapman and Chapman 1980), and in New Zealand (Ostraff 2003, Smith et al. 2010).

Ecklonia arborea (Areschoug) M.D. Rothman, Mattio & J.J. Bolton

Synonym: *Eisenia arborea*
Common names: English: Southern sea palm, Stiped kelp; Japanese: Arame (Arasaki and Arasaki 1983, Santhanam 2015).

Description: It has a relatively short (up to 1 m), broad stipe running from the holdfast to the cluster of strap-like blades that grow off at a split at the top of the plant. The entire plant reaches about 1.5 m tall and the two clusters of blades at the top, make the plant resemble a mini palm tree or a cheerleader's pom-pom.

Geographic distribution: E Pacific (British Columbia to Mexico); Asia (Japan).

Uses: *E. arborea* is often a part of Asian cuisine and is sold as dried whole or crushed leaves under the name "Arame" and is also used as feed for aquaculture of abalone. Used as food in N Pacific (Arasaki and Arasaki 1983, Santhanam 2015). This seaweed contain up to 20 times the levels of elements found in land plants. Their mineral content can include calcium, copper, iron, magnesium, potassium and zinc. They are also rich in vitamins. They are highly nutritious; containing beta-carotene (a potent antioxidant) and "Arame" contains particularly high levels of iodine (Gupta and Abu-Ghannam 2011).

In Peru it has been used as a folk medicine with anti-allergic properties. Because of this and the naturally moisturizing properties of *E. arborea*, it is often used in cosmetics, soaps, and skin care products (Gupta and Abu-Ghannam 2011).

Egregia menziesii (Turner) Areschoug (formerly *Egregia laevigata*)

Common names: English: Feather boa, Boa kelp (McConnaughey 1985), Feather boa kelp (Abbott and Hollenberg 1976, Braune 2008).

Description: Thick, flattened strap-like axes (each referred to as a rachis), with numerous small, lateral blades (can be flat, broad, narrow, filiform to cylindrical) and floats along the margin; thallus arises from dense hapterous holdfast, which can become fleshy and cone-like in large specimens (Abbott and Hollenberg 1976).

Geographic distribution: NE Pacific (Alaska to Baja California, Mexico).

Uses: Used as food in British Columbia, Canada (Arasaki and Arasaki 1983, Lindstrom 1998, Zemke-White and Ohno 1999, Lindstrom 2006, Roo et al. 2007, Harrison 2013).

Eualaria fistulosa (Postels & Ruprecht) (formerly *Alaria fistulosa*)

Common names: English: Dragon kelp (Rao 2010); Japanese: Kausam, Kauan, Oni-wakame (Tokida 1954).

Description: Thallus of this canopy-forming kelp is brown with a large branching holdfast (haptera), a stipe 25 cm long, and a blade with midrib up to 25 m long and 1 m wide. The midrib is 2–3 cm wide with gas-filled chambers (fistulae) that hold the blade in the water column. Reproductive sporophylls develop on the upper portion of the stipe.

Geographic distribution: NE Pacific (Aleutian Is, Alaska); Asia (Japan, Russia).

Uses: Is eaten freshly or dried in Japan and Alaska (Tokida 1954, Zemke-White and Ohno 1999, Roo et al. 2007, Rao 2010, Harrison 2013).

Fucus evanescens C. Agardh

Synonym: *Fucus distichus* ssp. *evanescens*

Common names: English: Popweed, Rockweed (Garza 2012), Arctic wrack (Pereira 2015a); Japanese: Hibatsunomata (Tokida 1954), Hibamata (Arasaki and Arasaki 1983).

Description: Usually grows to 10–25 cm in length, but may reach up to 40 cm. It is dichotomously branched, and can vary in color from olive brown to dark brown to green. In spring the species is distinguished by the long, swollen, yellowish green reproductive bodies on the tips of its branches (Pereira 2015a).

Geographic distribution: Widely distributed throughout the Arctic and temperate seas from the Arctic Ocean through N Europe, NW Atlantic, NE Pacific (Alaska through Central California), through Japan and Russia.

Uses: Used as food in Japan (Tokida 1954) and Canada (Edibleseaweed 2014), commonly used in Europe as a nutritional additive to livestock feed. It can also be used as food by humans, though it is not sold commercially for this application. Many coastal people enjoy collecting small amounts of *F. evanescens* for personal use. This species is eaten whole, blanched or cooked into dishes such as stir-frys. Younger plants that have not developed enlarged bulbs at their tips are preferred to avoid large amounts of somewhat bitter mucus. Fronds turn bright neon green when blanched. Like all brown seaweeds, *F. evanescens* is also high in alginic acid, making it useful as a raw material for the production of alginates (compounds used in a wide range of industries from textiles to pharmaceuticals to food stabilization) (Pereira 2015a).

Laminaria setchellii P.C. Silva

Common names: English: Southern stiff-stiped kelp, Split blade kelp, Split kelp, Stiff-stiped kelp (Klinkenberg 2015), Oar weed (Yuan and Walsh 2006).

Description: The sporophyte is medium to dark brown. It has a well branched holdfast that gives rise to a single, erect, unbranched and somewhat rigid stipe up to 80 cm long and about 2 cm in diam. This stipe supports a flat blade that is up to 80 cm long and 25 cm wide. The blade is divided into several to many distinctive, narrow straps and is basally rounded to wedge-shaped (Klinkenberg 2015).

Geographic distribution: *L. setchelli* ranges from the Aleutian Is in Alaska to Baja California, Mexico.

Uses: In N America, this species is sold as a similar alternative to the more widely known Japanese Kombu (*Saccharina japonica*). It is available in many forms, such as whole leaf, granules, powder, and in blends with other types of seaweed. It is an ingredient in certain health and energy bars, and also packaged as a snack food (Zemke-White and Ohno 1999, Yuan and Walsh 2006, Harrison 2013).

Lessoniopsis littoralis (Farlow & Setchell ex Tilden) Reinke

Common names: English: Short kelp (Kuhnlein and Turner 1996), Ocean ribbons, Flat pompom kelp, Ocean ribbon, Strap kelp (OHSVC 2015).

Description: Woody kelp and sturdy seaweed that is found only in the low intertidal zone in wave exposed environments. It is usually dark brown in color, and may grow up to 2 m in length. The holdfast of *L. littoralis* is stout and woody, and looks like a miniature tree trunk. The holdfast is highly branched and massive, build to withstand high energy waves. The stipe system is also woody and branches repeatedly. The holdfast and stipe base can be up to 20 cm thick and 40 cm long. At the end of each branch is a long, flattened blade with a 2 to 3 mm wide midrib. The slim blades are 7 to 12 mm wide, and up to 1 m long.

Geographic distribution: Pacific Coast of N America, from Kodiak Is, Alaska to central California.

Uses: Eaten by the indigenous people of Canada (Kuhnlein and Turner 1996).

Lobophora variegata (J.V. Lamouroux) Womersley ex E.C. Oliveira

Description: Brown to orange prostrate thallus. Blades fan shaped and overlapping; comprised of a medullary layer (1 cell thick) surrounded by a cortex (2–3 cells thick). Number of outer cortical cells may be different on two different blade surfaces; surface hairs random or in concentric lines (Kelly et al. 2014).

Geographic distribution: Atlantic Is (Canary Is, Cape Verde Is); Pacific Is (Easter Is); S America (Brazil); Africa (Mauritius); Asia (Sri Lanka, Taiwan).

Uses: Used for foodstuff (Gray 2015); extracts have antiviral activity (Kremb et al. 2005).

Macrocystis pyrifera (Linnaeus) C. Agardh (Fig. 98)

Synonym: *Macrocystis integrifolia*

Common names: Chilean: Huiro (Astorga-España and Mansilla 2014); English: Giant Kelp, Sea Ivy (Madlener 1977), Giant Pacific Kelp (McConnaughey 1985), Small perennial kelp, Giant perennial kelp (McConnaughey 1985); Spanish: Kelp gigante (Radulovich et al. 2013).

Description: Is one of the fastest-growing organisms on Earth, commonly growing to 30 m, and in ideal conditions can grow over 50 cm per day in one season and reach over 50 or 60 m and 100 Kg in mass. Like most kelp species, *M. pyrifera* is perennial, and has a lifespan of five to eight years. This species will often die back to its holdfast each winter, as much as three quarters of an individual can be lost during the winter season. The impressive holdfasts of old plants are conical and may reach 1 m in diam. and 1 m in height,

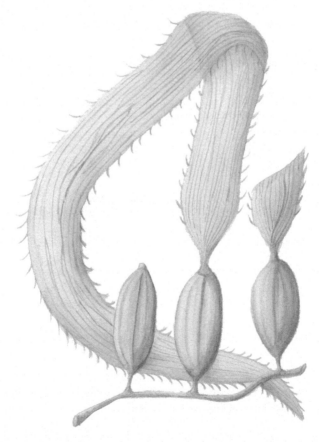

Figure 98. *Macrocystis pyrifera* (Linnaeus) C. Agardh (Ochrophyta, Phaeophyceae), adapted from Pereira and Correia 2015.

consisting of branching extensions called haptera. Numerous stem-like stipes, often four or five times dichotomously branched near the base, arise from the holdfast and are tough but flexible, allowing the kelp to sway in ocean currents. Long, tapered blades develop at irregular intervals along each stipe, with a single oval-elongated pneumatocyst (gas-filled bladder) at the base of each blade. These help to hold the kelp upright, maximizing the amount of sunlight the blades receive. The flattened blades, which develop via splitting of a small terminal blade, are distinctly corrugated or wrinkled with marginal teeth, and can grow up to 70 cm in length. The blades grow towards and spread out along the surface in a dense floating golden-brown canopy, whose thickness varies greatly with season and location (John 2014).

Geographic distribution: From the Kodiak Archipelago, Alaska, to northern Baja California, Mexico, with the primary distribution south of Monterey, California. In the Southern Hemisphere, *M. pyrifera* can be found along the coast of S America, where it grows from Valparaíso, Chile southward to Tierra del Fuego, Argentina. Elsewhere in the Southern Hemisphere, *M. pyrifera* grows in the waters of S Africa, S Australia, Tasmania, New Zealand, and several sub-Antarctic Is.

Uses: *M. pyrifera* is a commercially important seaweed; this species is harvested for its rich iodine, potassium, polysaccharides, and other minerals, but the primary product obtained from giant kelp is alginic acid for alginates; is used as food and for alginic acid extraction, for example, in Canada, California, Chile and New Zealand (Madelener 1977, Lembi and Waaland 1988, Kuhnlein and Turner 1996, Zemke-White and Ohno 1999, Smith et al. 2010, Harrison 2013, Astorga-España and Mansilla 2014).

Nereocystis luetkeana (K. Mertens) Postels & Ruprecht

Common names: English: Ribbon Kelp, Giant Kelp, Bull Whip Kelp, Bull Kelp, Sea Whip, Horsetail Kelp, Bladder Kelp, Sea Otter's Cabbage (Madlener 1977), Serpent kelp (Rhoads and Zunic 1978), Giant bull kelp (Arasaki and Arasaki 1983), Bullwhip kelp (McConnaughey 1985).

Description: Sporophyte large, differentiated into a single, very long stipe in the shape of an inverted whip, anchored to the substratum by a frequently branched, tough holdfast at its base. It carries a head of broad, ribbon-shaped blades (phylloids) at the apex. The stipe is continuously cylindrical and solid, its diam. gradually widening towards the top, inflating into a hollow, club-shaped, slightly oval apophysis; attached to this is a terminal spherical to oval gas-filled pneumatocyst about the size of a fist (Braune and Guiry 2011).

Geographic distribution: NE Pacific (Alaska to Central California), NW Pacific (Commander Is).

Uses: This species is harvested in British Columbia (Canada) and California for food (Kuhnlein and Turner 1996, Zemke-White and Ohno 1999, Braune and Guiry 2011, Edibleseaweed 2014). A variety of pickles or relishes can be made from the thicker portion of the stipe/stem. You can experiment by using pickle or relish recipes in a standard cookbook and replacing the cucumber with bull kelp (Chapman and Chapman 1980, Garza 2005, Gutschmidt 2010d, Harrison 2013).

See recipes with *Nereocystis* in Annex I.

Padina crispata Thivy

Description: Thallus in leaf-like clusters, to 4 cm high, brown. Blades fan-shaped, split or lobed, to 4 cm wide, banded, lightly calcified on lower surface or uncalcified, 130–360 µ thick, 2–4 cells thick near margin, 6–10 cells thick at base; outer margins inrolled (Littler and Littler 2015b).

Geographic distribution: Tropical W Atlantic; Caribbean Sea; Tropical E Pacific.

Uses: With food potential (Radulovich et al. 2013).

Padina durvillei Bory Saint-Vincent

Description: Thalli erect or prostrate, attached by a rhizoidal holdfast, up to 20 cm long, complanate, flabellate or becoming lacerate often to the base, segments 1–5 cm broad, calcified on one or both surfaces. Growth initiated by an entire marginal row of apical cells in an involute apical fold, directed towards the upper (= inner) thallus surface.

Geographic distribution: Tropical W and E Atlantic; Tropical E Pacific.

Uses: With food potential (Radulovich et al. 2013).

Padina fernandeziana Skottsberg & Levring

Description: It has leafy, fan-like blades. The leaves are lightly calcified and are about 5 cm in diam. This species lays down the calcium carbonate on the cell wall surface rather than between the cells (intercellularly) (Goecke et al. 2012, Santhanam 2015).

Geographic distribution: SE Asia (the Philippines); Pacific Is; SE Pacific (Chile).

Uses: With food potential (Goecke et al. 2012, Santhanam 2015).

Pleurophycus gardneri Setchell & Saunders ex Tilden

Common names: English: Kelp (Madlener 1977), Tender Kelp (Chapman and Chapman 1980), Sea spatula (McConnaughey 1985).

Description: Thallus of this dark brown kelp has a branched holdfast (haptera), a stipe, cylindrical at the base, flattening above, up to 50 cm long, and an elliptical blade to 150 cm long and 28 cm wide with a distinctive thick, broad midrib 5 cm wide. The blade is puckered along the sides of the midrib. The stipe and holdfast are perennial while the blade is annual and dies back in the fall (Lindeberg and Lindstrom 2015c).

Geographic distribution: NE Pacific (Alaska to California).

Uses: Used steamed and in salads in the US (Madelener 1977, Chapman and Chapman 1980).

Postelsia palmaeformis Ruprecht

Common names: English: Sea palm, Sea palm kelp (Madlener 1977); German: See-palme (Braune 2008); Indian (California): Gaye (Goodrich et al. 1980).

Description: Resembles a miniature palm tree; small hapterous holdfast with thick, flexible stipe; blades with longitudinal grooves; a protected species in California; seasonal (Kendall et al. 2002).

Geographic distribution: NE Pacific (from Hope Is, British Columbia south to the southern central coast of California).

Uses: Used as food in California (Madlener 1977, Goodrich et al. 1980, Zemke-White and Ohno 1999).
    See recipes with *Postelsia* in Annex I.

Pterygophora californica Ruprecht

Common name: English: Pompom kelp (McConnaughey 1985).

Description: *P. californica* is large brown seaweed and may grow to about 3 m in height. It is attached to a rocky substratum by its holdfast and has a single, tough, woody stipe or stem up to 2 m long and 2 cm in diam. It is a long-lived perennial plant which may survive for 25 years, and annual growth rings can be seen in the stipe. There are a number of smooth blades arranged on either side of the top half of the stipe. The terminal blade is linear, has a midrib and is larger than the others, growing up to 90 cm.

Geographic distribution: NE Pacific (Alaska to California, Mexico).

Uses: Eaten by the Inuit of Broughton Is (Canada) (Kuhnlein and Turner 1996).

Saccharina bongardiana (Postels & Ruprecht) Selivanova, Zhigadlova & G.I. Hansen (formerly *Laminaria bongardiana*)

Common name: Split kelp (Gutschmidt 2010e).

Description: This species has large visible holdfast, longer stipe, large blade with smooth surface, split into several larger strips, and dark brown color (Garza 2005).

Geographic distribution: NE Pacific (Alaska to California), NW Pacific (Kamchatka, Japan and Russia).

Uses: Used as food in Alaska, this species dries thicker, and when dried pieces are soaked in freshwater for 30 min they will appear fresh, as if never dried (Chapman and Chapman 1980, Zemke-White and Ohno 1999, Garza 2005, Roo et al. 2007, Gutschmidt 2010e, Harrison 2013).

See Annex I for kelp chips, kelp seasoning, and other dishes.

Saccharina dentigera (Kjellman) C.E. Lane, C. Mayes, Druehl & G.W. Saunders (formerly *Laminaria dentigera*)

Common names: English: Mendocino coast Kombu (OHSVC 2015); Japanese: Kumade-kombu (Tokida 1954).

Description: Thallus is dark brown, thick, reaching 1.5 m tall. The holdfast is branched, the stipe somewhat rigid, and the blade often split down to 10 cm above its base; mucilage ducts occur near the surface of stipe (Abbot and Hollenberg 1976).

Geographic distribution: NE Pacific (Alaska to California); Asia (Japan, Russia).

Uses: The native California Kombu may be used interchangeably with the Japanese variety as a healthy and flavorful addition to soup stocks, vegetable dishes, pickles and condiments. Don't forget that of the entire marine alga, Kombu isolates heavy metals and radioactive substances in the body for elimination and contains the amino acid Laminin, which effectively combats hypertension and high blood pressure. It is also a rich source of protein and B vitamins.

At low tide, the arching stems of the rich brown Kombu plants hover above the water's edge, the flat shiny blades immersing themselves in the swirling currents. Each blade is trimmed 5–8 cm above the stipe to ensure the continued growth of Kombu, and then individually hung to be sun-dried in the redwood hills.

Indispensable in soups and beans also makes tasty pickles and nutritious condiments. Restorative and refreshing, mineral rich Kombu has been shown to eliminate heavy metals and radiation from our bodies.

Exceptionally rich in nutrients including protein, vitamins, high in iron, iodine, zinc, essential elements and trace minerals; excellent base ("Kombu Dashi") for soups and sauces; makes a fine pickle—"Shio Kombu" (OHSVC 2015).

Saccharina groenlandica (Rosenvinge) C.E. Lane, C. Mayes, Druehl & G.W. Saunders (formerly *Laminaria groenlandica*)

Common names: English: Short kelp (Kuhnlein and Turner 1996), Kombu, Split kelp, Sugar wrack.

Description: Thallus of this common kelp is medium to dark brown with a branched holdfast (haptera), a stipe up to 60 cm long with microscopic mucilage ducts, and a blade up to 2 m long. The blade is often bullate when young but becomes thicker and smoother with age; it often splits into two-three segments. The stipe is cylindrical at the holdfast but is often flattened at the base of the blade.

Geographic distribution: Arctic (Canada); Atlantic Is (Greenland); NW Atlantic; NE Pacific (Alaska to California); Asia (Russia).

Uses: *S. groenlandica* is a desirable sea vegetable having an appetizing texture and flavor in various cooked dishes, and is sold by companies in France and Canada, as a dried sea vegetable, and in bath soak products (Kuhnlein and Turner 1996, Zemke-White and Ohno 1999, Roo et al. 2007, Harrison 2013, Kellogg and Lila 2013).

Saccharina sessilis (C. Agardh) Kuntze (formerly *Hedophyllum sessile*)

Common names: English: Bubbly kelp (Kuhnlein and Turner 1996), Sea cabbage (McConnaughey 1985), Sweet Kombu (OHSVC 2015).

Description: Common kelp that is unique in the fact that it has no stipe or "stem" (and is therefore sessile). Colors range from light to medium to dark brown, or can also be olive drab. There is a well-developed and sturdy holdfast which gives rise directly to several blades that can be from 30 cm to up to 1.5 m long. The branches are often split into several pieces and depending on growing conditions can be smooth or puckered and vary in thickness. These often form a lush canopy which can obscure the many haptera that make up the holdfast. The blades of *S. sessilis* take on different appearances depending on the amount of wave action and water motion in their environment. Where wave action is greatest the blades have a smooth texture, and are split into slender, strap shaped segments 5 to 10 cm wide. In sheltered environments the blades are usually larger, have a ruffled, wavy form, and few longitudinal splits. The unruffled, split blades are thought to offer less resistance to water motion, allowing these plants to remain attached to the rocks in wave exposed environments. *S. sessilis* is perennial seaweed that lives up to three years if it is not washed up by storms. This seaweed is able to live in both sheltered water and in more exposed sites, altering its morphology to match growing conditions, though *S. sessilis* prefers intertidal to upper subtidal zones where it is firmly attached to rocky substratum (Abbot and Hollenberg 1976).

Geographic distribution: NE Pacific (Alaska, Aleutian Is, California), Asia (Commander Is).

Uses: *S. sessilis* is harvested and sold as tasty edible seaweed, used historically by NW indigenous American tribes. It is also cultivated and sold dried throughout Asia as "Sweet Kombu". *S. sessilis* is considered a tasty kelp that adds flavor, color, and texture to soups, casseroles and many other dishes (Kuhnlein and Turner 1996).

Stephanocystis osmundacea (Turner) Trevisan (formerly *Cystoseira osmundacea*)

Common names: English: Woody chain bladder (Arasaki and Arasaki 1983), Bladder chain kelp (Mondragon and Mondragon 2003).

Description: Is one of the largest of the brown algae species, often termed "kelp" for its size even though it is not a member of the order Laminariales. *S. osmundacea* has quite a distinctive appearance. The discoid holdfast supports a thick, woody stipe that is triangular in cross-section. This produces irregular branches of flattened, leaf-like blades (some have described them as similar to those of a coastal oak) which are darker olive-brown colored with a distinctive midrib that is slightly lighter in color. The apical region of the *S. osmundacea* thallus consists of distinctive chains of five or more cylindrical pneumatocysts (floats) which are lighter in color than the basal leaves. These plants can grow over 8 m in length.

Geographic distribution: *S. osmundacea* is less widely distributed than other species in the genus *Cystoseira*, and is restricted to the Pacific Ocean. This species can be found commonly from N Oregon through N Mexico.

Uses: *S. osmundacea* is an edible seaweed, gathered on small scales by individuals or coastal communities. Some companies are now incorporating this seaweed into health supplements for its high nutrient content. Seaweeds are very high in minerals and trace elements that are so often missing from the modern western diet. These can be available in capsules, powders, etc.

2.6.3 Rhodophyta

Callophyllis variegata (Bory de Saint-Vincent) Kützing

Common names: Spanish: Carola (McHugh 2003); English: Large wire weed (Rao et al. 2010).

Description: Anchored to rocky substrates by a discoid holdfast. The thallus, typically made up of flat, sometimes deeply and irregularly divided branches and glossy or semi-glossy blades (leaf-like structures), extends upward from the holdfast, sometimes supported on a short stipe (stem-like stalk). The blades, which lack midribs and veins, have large cells in the medulla (central region) and ragged or smooth margins. *Callophyllis* have fleshy, flexible branches and blades, though some (particularly intertidal specimens) may be brittle. Most species grow to between 5 and 30 cm in height, though some are slightly smaller or larger.

Geographic distribution: *C. variegata* occurs in Concepción de Chile and other parts of S America such as Peru, the Falkland Is, and Tierra del Fuego; but also in New Guinea, S Africa, Australia, New Zealand, Alaska, St. Paul Is (Indian Ocean), Antarctic and sub-Antarctic Is such as the Graham Land, Kerguelen, Macquarie Is, S Georgia, and the S Orkney Is.

Uses: *C. variegata*, harvested off the southern coast of Chile, is popular edible seaweed; consumed in cooked or dried and rehydrated form (Rao et al. 2010, Astorga-España and Mansilla 2014). In Chile, the demand for edible seaweeds has increased and *C. variegata* (Carola) is one of the most popular (McHugh 2003, Harrison 2013).

Chondracanthus chamissoi (C. Agardh) Kützing (formerly *Gigartina chamissoi*)

Common names: Japanese: Suginori, Yuyo; Spanish: Alga roja, Chicorea de mar, Chicoria de mar, Gigartina, Mococho, Cochayuyo (Polo 1977).

Description: Is benthic red algae that may reach up to 50 cm in length. The holdfast gives rise to one or several erect, cylindrical axes which are sparsely pinnately branched with numerous vegetative or reproductive branchlets. Overall *C. chamissoi* resembles dark reddish-purple tumbleweed and is usually attached by the holdfast to the substratum but may be free-living in deep pools and calmer benthic regions. *C. chamissoi* inhabits a variety of habitats from the lower intertidal to a depth of 15 m.

Geographic distribution: *C. chamissoi* is distributed from Paita, Peru to Ancud, Chile. In northern Chile, the main harvesting areas are Caldera, La Herradura, and Puerto Aldea (Buschmann et al. 2005, Bulboa et al. 2008).

Uses: This species is used to make soup and stew in Peru (Polo 1977); is also an economically important seaweed in Chile where it is the main source of carrageenan. Carrageenan is a very widespread, naturally derived compound which is used extensively as a stabilizer in food and cosmetic products. This seaweed is also used as a whole food throughout S America and is even exported to Asia for use in their food industry. *C. chamissoi* is a highly valued as a delicacy in Japan (Buschmann et al. 2005, Bulboa et al. 2008).

Chondracanthus exasperatus (Harvey & Bailey) Hughey

Common name: English: Turkish towel (Arasaki and Arasaki 1983, McConnaughey 1985).

Description: Thalli 20–70 cm tall, fleshy or crisp, yellowish-green to brownish-red in sheltered habitats, brick red (north of Point Conception and in Baja California) or pink (south of Point Conception to the Mexican border) on exposed coasts, consisting of one to several blades from a discoid holdfast up to 1 cm broad; blades 6–20 cm wide, 0.2–3 mm thick, simple or forked, lanceolate to long-elliptical and tapered to a blunt or acuminate tip, the margins entire or dentate or sometimes bearing blades; stipitate, the stipe usually 2–3 cm long bearing lateral blades and expanding to a short, fleshy and flattened cuneate base; surface and margins of blades, except the base covered with closely to widely spaced, spinose outgrowths (Braune and Guiry 2011; Guiry and Guiry 2015a).

Geographic distribution: NE Pacific (Alaska to California, Mexico).

Uses: Used as food in Pacific Coast of N America (Arasaki and Arasaki 1983); raw material for carrageenan-production; occasionally cultivated in small quantities on ropes for the production of "Turkish towels" for beauty salons (Braune and Guiry 2011).

Chondracanthus glomeratus (Howe) Guiry
(formerly *Gigartina glomerata*)

Common name: Cochayuyo (Polo 1977).

Description: Densely caespitose, 3–5 cm high, rather firm (in preservative fluid), corneous when dry; main axes of cystocarpic plants sub-terete or lightly compressed, 1–3 mm in diam., usually one–three times sub-dichotomous below, otherwise naked or 1–25 cm in basal and median parts, bearing toward their apices at first sub-distichous, later irregularly polystichous, short, simple or compound branches; the simple branches mammiform or sub-ovoid; the compound branches irregularly and densely short-ramose, or longer (5–8 mm) and two–five times closely, often somewhat palmately or pedately sub-dichotomous or cervicornis, sometimes a little inflated towards the apex with a sub-rotate or coronate verticil of short branchlets, or very irregularly fasciculate-ramose, the ultimate ramuli mostly short-digitiform, patent or sub-divaricate, obtuse or acute, the penultimate ramuli commonly narrowed towards the base (Hommersand et al. 1993).

Geographic distribution: SE Pacific (Chile, Peru).

Uses: Used in Peru to make soup and stew (Polo 1977).

Cryptonemia obovata J. Agardh

Description: Foliose, linear-lanceolate to broad and lobed, entire, proliferous or dissected into numerous cuneate, lanceolate or obovate lobes, occasionally laciniate, usually with distinct stipe, continuing as a midrib in some species, branched or unbranched, and with small discoid holdfast.

Geographic distribution: Tropical and subtropical W Atlantic; Australia; SE Pacific.

Uses: Considered edible in Chile (Ortiz et al. 2010).

Gigartina skottsbergii Setchell & N.L. Gardner

Common names: English: Red marine algae; Spanish: Cuero de chancho, Luga gruesa, Luga roja.

Description: Thick, leathery, brick red thallus that is generally circular to ellipsoidal in shape. *G. skottsbergii* attaches to rocky substratum by a short strong stipe and holdfast.

Geographic distribution: Limited to S America and Antarctica.

Uses: Used for foodstuff, is exported from Chile as food and raw for the Asiatic market. Commonly known as "Luga roja", is one of the most important and highly valued carrageenophyte resources growing in cold temperate water (Lembi and Waaland 1988, Alveal 2006, Gray 2015).

Gracilaria chilensis C.J. Bird, McLachlan & E.C. Oliveira

Common name: Spanish: Pelillo.

Description: The cylindrical, filamentous thalli are about 1 to 2 mm in diam., usually purple-red, occasionally yellow-brown, and can grow quite long, up to 5 m in length. Identification of the specific species is difficult in the field however because of a lack of unique diagnostic characteristics and the variability of species morphology. Depending of growing conditions (light availability, depth, water movement, nutrient levels), *G. chilensis* can take on a broad variety of overall shapes from short, bushy, and densely branched to long and sparsely branched. *G. chilensis* is naturally found in the intertidal or subtidal zones (up to 10 m depth), where it can grow in sheltered harbors, bays, or estuaries on sand, mud, or rocky substrates.

Geographic distribution: S America (Chile) as well as Africa, China, and the Indo-Pacific.

Uses: Enjoyed as a food source by humans and is often seen in Japanese, Hawaiian, and Filipino cuisine (McHugh 2003).

Gracilaria pacifica I.A. Abbott

Common names: Red ogo seaweed, California limu, Red spaghetti (Fretwell and Starzoski 2015, Han 2015).

Description: This red alga has long and thin cylindrical branches; it reaches 50 cm tall and only 2 mm wide. It is sparsely branched, often with small spiny branches around the bases of the main axes. Its color is reddish-brown to brown (Fretwell and Starzoski 2015).

Geographic distribution: NE Pacific (Alaska to California).

Uses: This species is edible, and is a source of the jelly-like thickening agent agar (Fretwell and Starzoski 2015).
    See recipes in Annex I.

Gracilaria tikvahiae McLachlan

Common names: English: Graceful red weed (Hill 2001); Ogo (Hart et al. 2014).

Description: *G. tikvahiae* is a highly opportunistic species common in estuaries and bays, especially where nutrient loading leads to either seasonal or year-round eutrophication. Its morphology is highly variable, with colors ranging from dark green to shades of red and brown; with outer branches that can be either somewhat flattened or cylindrical in shape. It can be found in protected, quiescent bays, as well as in high energy coastline habitats. This species grows free or attached to rocks or other substrata, and can reach a height of 30 cm. *G. tikvahiae* grows to depths of approximately 10 m, but is most common at depths less than 1 m (Hill 2001).

Geographic distribution: W Atlantic (Tropical and subtropical W Atlantic); Caribbean Sea; SW Atlantic; Pacific Is (Hawaii).

Uses: Used as food and for production of agar in New England (US) and Hawaii (Hill 2001, Kılınç et al. 2013, Hart et al. 2014, Redmond et al. 2014).

Gracilariopsis andersonii (Grunow) E.Y. Dawson

Synonym: *G. sjoestedtii*

Common name: English: Sea spaghetti (Gutschmidt 2010a, Klinkenberg 2014a); Hawaiian: Ogo (McConnaughey 1985).

Description: Is anchored by a small, discoid holdfast, from which extend several branches up to 2 m tall. The branches are reddish brown to reddish purple early in the season, often fading to yellowish brown later; sparingly branched, with slender, flexible axes up to almost 2 mm in diam. The branches are rather cartilaginous in texture (Klinkenberg 2014a).

Geographic distribution: Caribbean Sea; Tropical W Atlantic; Asia (China); NE Pacific (British Columbia to Mexico).

Uses: Considered edible in the NE Pacific (Gutschmidt 2010a, Harrison 2013).

Halosaccion glandiforme (S.G. Gmelin) Ruprecht (Fig. 99)

Common names: English: Sea sac (Madlener 1977), Dead man's fingers, Salt sacs (Braune and Guiry 2011); Japanese: Benifukuronori (Madlener 1977).

Description: Thallus erect, yellow-brown to crimson, occasionally greenish, hollow, short-stiped, elongated-oval sacs with watery interior, endings rounded, older specimens may be compressed; single or several growing on a common basal disk (Braune and Guiry 2011).

Geographic distribution: Asia (Commander Is, Russia); NE Pacific (Alaska to California).

Uses: *H. glandiforme* is an edible species, and can be made into tasty soups or tempuras; traditionally consumed in Russia and NE Pacific (Tokida 1954, Madlener 1977, Chapman and Chapman 1980, Roos-Collins 1990, Harrison 2013).

Figure 99. *Halosaccion glandiforme* (S.G. Gmelin) Ruprecht (Ochrophyta, Phaeophyceae).

Iridaea cordata (Turner) Bory de Saint-Vincent

Common names: Rainbow kelp, Rainbow seaweed (Edibleseaweed 2015).

Description: *I. cordata* has a broad, smooth, flat blade, with an iridescent sheen when wet; the thalli are undivided, it can be heart-shaped to narrow and tapered shape; is sometimes lobed and with ruffled margins, and is rubbery to the touch (Braune and Guiry 2011, Edibleseaweed 2015).

Geographic distribution: Antarctic Is; NE Pacific (Alaska to California); SE Pacific (Argentina, Chile); Auckland Is.

Uses: It is used as nutritional supplement; it is loaded with vitamins and minerals in maintaining and promoting healthy skin; high in protein, vitamin A, E, C and iodine; it also contains Vitamin B_1, sodium, calcium, magnesium, iron, essential fatty acids, beta-carotene, and zinc (Edibleseaweed 2015).

Mazzaella denticulata (E.Y. Dawson, Acleto & Foldvik) Fredericq

Synonym: *Rhodoglossum denticulatum*

Description: Thallus erect, flat or leaf-like, undivided or irregularly lobed, often proliferating from the margins; very similar to *Iridaea* (Braune and Guiry 2011).

Geographic distribution: SE Pacific (Chile, Peru).

Uses: Is consumed in Peru (Polo 1977).

Mazzaella splendens (Setchell & N.L. Gardner) Fredericq

Common names: Rainbow seaweed, Rainbow leaf seaweed, splendid iridescent seaweed (Gutschmidt 2010c, Edibleseaweed 2014).

Description: Beautiful red seaweed that is named for its iridescent qualities; when wet, the deep red to purple blades shimmer from blue to purple, sometimes gold or green iridescence, giving it the common name of "rainbow leaf". The blades of *M. splendens* are broad and long, up to 20 to 40 cm in length, with an oblong to lanceolate shape, and are somewhat thick and rubbery, almost stretchy, with a short 3 cm stipe. One to several of these blades arises from a single, fleshy, perennial holdfast (Gutschmidt 2010c).

Geographic distribution: NE Pacific (Alaska to California).

Uses: Used as food in British Columbia (Canada) (Zemke-White and Ohno 1999, Lindstrom 2006, Roo et al. 2007, Gutschmidt 2010c, Harrison 2013, Edibleseaweed 2014).

Palmaria hecatensis M.W. Hawkes

Common names: English: Leathery dulse, stiff red ribbon (Lindeberg and Lindstrom 2010).

Description: Thallus is leathery, somewhat shiny, deep red to maroon, to at least 20 cm tall. The blade is strap-like, unbranched or with one or two lobes with rounded tips, tapering to a discoidal holdfast. Habitat: This annual is found on rocks in the mid to low intertidal (usually above *P. mollis*) of semi-exposed habitats. This species is edible, a good source of vitamins and minerals, and similar to the Atlantic species *P. palmata* (Dulse). Similar *taxa*: *P. mollis* is thinner and more glove-shaped than the mitten-shape of *P. hecatensis* (Lindeberg and Lindstrom 2010).

Geographic distribution: NE Pacific (Alaska to Oregon, Aleutian Is).

Uses: In British Columbia (Canada), *P. hecatensis* is consumed as a food, and is high in many vitamins and minerals (Zemke-White and Ohno 1999, Lindstrom 2006, Roo et al. 2007, Harrison 2013).

Palmaria mollis (Setchell & N.L. Gardner) van der Meer & C.J. Bird

Common names: English: Dulse, Ribbon seaweed (Garza 2012), Pacific dulse, Red ribbon, Red kale.

Description: Ribbon seaweed is brick red or reddish-brown in color. Several blades branch off a single small holdfast. There is no stipe or stem. Fronds sometimes appear like lobes while others may appear more irregular in shape. They grow longer than wide, up to 20–25 cm long and several inches wide. The blades are thick and almost feel leathery to the touch (Garza 2005).

Geographic distribution: NE Pacific (Alaska to Oregon, Aleutian Is).

Uses: *P. mollis* has a tough, chewy texture; some people prefer it cooked in a dish, or fried or roasted like a chip (Zemke-White and Ohno 1999, Lindstrom 2006, Roo et al. 2007, Garza 2012, Harrison 2013).

Polyneura latissima (Harvey) Kylin

Common name: Crisscross network (Lindeberg and Lindstrom 2015d).

Description: Thallus is a thin, crinkly, rose-red blade, to 15 cm tall and about half the width. A crisscrossing network of veins permeates the thallus, which often has bumps (cystocarps) and may develop holes (Lindeberg and Lindstrom 2015d).

Geographic distribution: NE Pacific (Alaska to California, Mexico); Asia (Russia, the Philippines).

Uses: Used in soups and salads in California (US), and the Philippines (Madlener 1975, Chapman and Chapman 1980).

Polysiphonia mollis J.D. Hooker & Harve

Common names: Hawaiian: Limu pualu, Limu hawane (Reed 1906); Limu pepe-iao (MacCaughey 1918).

Description: Thallus red-brown, usually 4–20 cm high, with a single, erect, basal axis (occasionally a slight prostrate part) and profusely branched above (often denuded below in older plants) sub-dichotomously to laterally to form dense, fastigiate to spreading, soft tufts; holdfast discoid, small; commonly epiphytic on *Posidonia, Heterozostera, Halophila* (marine vascular plants) or larger algae (Womersley 2003).

Geographic distribution: Indian Ocean; SE Asia (Indonesia, the Philippines); Australia; Pacific Is (Polynesia, Fiji, Hawaiian Is); NE Pacific (Alaska to Baja California); SE Pacific (Chile).

Uses: Considered an edible species in Hawaii (Reed 1906, MacCaughey 1918).

Pyropia abbottiae (V. Krishnamurthy) S.C. Lindstrom
(Formerly *Porphyra abbottiae*)

Common names: English: Black seaweed (Garza 2005), summer seaweed (Kuhnlein and Turner 1996).

Description: Black seaweed begins to grow in early spring. It is recognized by the near black strands hanging down rock faces. The near transparent fronds may be a dark rose-purple or a black-green color. The colors are most apparent when the fronds are wet, and black when drying (Garza 2005).

Geographic distribution: NE Pacific (Alaska to British Columbia), Asia (Commander Is).

Uses: *P. abbottiae* is a prized food in Tlingit, Haida, and Tsimshian (Alaska and British Columbia, Canada). It is an important trade item, because there are many areas where it does not grow, and because many Elders who enjoy eating it have stopped harvesting black seaweed because it is too much work to harvest. Members of non-Native rural communities also enjoy harvesting black seaweed in the spring. Generally, black seaweed is dried and not used fresh. Dried black seaweed is eaten as a snack like popcorn. It can also be added to a meal like the Seaweed Chop Suey (Kuhnlein and Turner 1996, Zemke-White and Ohno 1999, Garza 2005, Lindstrom 2006, Roo et al. 2007, Harrison 2013).

Pyropia columbina (Montagne) W.A. Nelson
(formerly *Porphyra columbina*)

Common names: English: Common porphyra, Purple laver; Maori: Karengo (Brown and Zemke-White 2006); Spanish: Luche, Luche rojo; Peru: Yuyo (Cesar 2006), Cochayuyo (Polo 1977).

Description: It consists of a short holdfast and a broad blade 2 to 4 cm across. Its morphology is somewhat plastic, depending on location and growth conditions. Fully hydrated underwater, *P. columbina* is an expanded, translucent greenish red leaf-like thallus ranging from small rosette-like leaves to longer, ribbon-like growth forms; upon drying, however, *P. columbina* loses up to 80% of its weight and becomes a dark purple.

Geographic distribution: Chile, Argentina, Falkland Is, New Zealand.

Uses: *P. columbina* is an important edible algal species. It contains 25–30% protein, 40% carbohydrates, significant iodine, and is rich in vitamin C and B. It is often dried and sold as food in Argentina, Peru, as "luche" in Chile, and "karengo" in New Zealand. It is used as a supplement to other foods, as a laxative, and even as a chewing gum (Polo 1977, Chapman and Chapman 1980, Zemke-White and Ohno 1999, Alveal 2006, Zaixsoa et al. 2006, Cesar 2006, Zemke-White 2006, Roo et al. 2007, Harrison 2013, Astorga-España and Mansilla 2014, Cian et al. 2014).

Pyropia fallax (S.C. Lindstrom & K.M. Cole) S.C. Lindstrom
(formerly *Porphyra fallax*)

Common name: English: False laver.

Description: Thallus is a reddish brown blade, to about 50 cm long, elongate, with a greenish center, one cell layer thick, and often ruffled along the margin. The margin becomes pale red when female cells are fertilized; male cells are in whitish patches or streaks.

Geographic distribution: NE Pacific (Alaska to Oregon).

Uses: Used as food in British Columbia and Canada (Zemke-White and Ohno 1999, Lindstrom 2006, Roo et al. 2007, Harrison 2013, Kellogg and Lila 2013).

Pyropia nereocystis (C.L. Anderson) S.C. Lindstrom
(formerly *Porphyra nereocystis*)

Common names: Chinese: Chi Choy (Madlener 1977); English: Bull Kelp laver (Lindeberg and Lindstrom 2014e), Red Laver, Purple Laver (Madlener 1977), Rose Nori (Chapman and Chapman 1980), Red nori (McConnaughey 1985); Japanese: Nori (Madlener 1977).

Description: Thallus is a reddish pink strap-shaped blade, one cell layer thick, that often reach 100 cm or more in length and 30 cm broad; epiphytic on bull kelp, *Nereocystis luetkeana* (Lindeberg and Lindstrom 2014e).

Geographic distribution: N America (Alaska, British Columbia, California, Washington).

Uses: *P. nereocystis* is edible and nutritious (Garza 2005).

Pyropia perforata (J. Agardh) S.C. Lindstrom
(formerly *Porphyra perforata*)

Common names: Chinese: Chi Choy (Madlener 1977); English: Black seaweed, Nori, Laver, Wild nori, Red laver (Arasaki and Arasaki 1983, Kuhnlein and Turner 1996).

Description: A translucent leafy alga that occurs in thin ruffled blades in a circular shape about 30 cm in diam., growing in small groups from a discoid holdfast. Each blade of *P. perforata* is composed of only a single layer of cells. The blades can range from gray-green and burgundy red and appear shiny, darkening to a brownish purple when dry. The name *perforata* comes from the fact that the alga is often found with small holes that occur naturally, not as a result of herbivory.

Geographic distribution: NE Pacific (Alaska to Oregon); Asia (Japan, Malaysia); Antarctic and sub Antarctic Is.

Uses: Used as food in Canada and the US (Arasaki and Arasaki 1983, Kuhnlein and Turner 1996, Zemke-White and Ohno 1999, Roo et al. 2007, Harrison 2013), Malaysia (Sidik et al. 2012), and Japan (SIA 2014).

Pyropia pseudolanceolata (V. Krishnamurthy) S.C. Lindstrom
(formerly *Porphyra pseudolanceolata*)

Common name: Laver seaweed (Kuhnlein and Turner 1996).

Description: Thallus is an olive green blade, to about 30 cm long but usually much shorter, one cell layer thick; lanceolate when young, becoming more rotund with age.

Geographic distribution: NE Pacific (Alaska to California); E Pacific (Peru).

Uses: Used as food in British Columbia and Canada (Kuhnlein and Turner 1996, Zemke-White and Ohno 1999, Lindstrom 2006, Roo et al. 2007, Harrison 2013).

Pyropia torta (V. Krishnamurthy) S.C. Lindstrom (formerly *Porphyra torta*)

Common names: English: Teal nori (Klinkenberg 2014b), winter seaweed (Kuhnlein and Turner 1996), winter black seaweed (Garza 2005).

Description: This alga has a distinctive bluish green color that has led to this unusual common name. Blades can be broadly ovate (shaped like fat eggs) to irregularly circular, sometimes lobed, and almost always abundantly ruffled; blades seldom reach 15 cm (6 in) in their longer dimension (Klinkenberg 2014b).

Geographic distribution: N America (Alaska, British Columbia, Washington), Asia (Commander Is).

Uses: Used as food in Alaska and British Columbia (Canada), *P. torta* is edible and nutritious (Kuhnlein and Turner 1996, Zemke-White and Ohno 1999, Garza 2005, Lindstrom 2006, Roo et al. 2007, Harrison 2013).

Rhodymenia corallina (Bory de Saint-Vincent) Greville

Description: Species with stipe cylindrical and flat sheets divided dichotomously or irregularly; wedge-shaped and proliferates distally; sometimes it branches from the surface of the sheet.

Geographic distribution: Falkland Is/Islas Malvinas; Indian Ocean, Australia (Papua New Guinea); SE Asia (Indonesia); Pacific Is (Polynesia); E Pacific (Chile, Peru).

Uses: Considered edible in Chile (Ortiz et al. 2010).

Sarcodiotheca gaudichaudii (Montagne) P.W. Gabrielson

Common name: Red string seaweed (IIMHO 2015).

Description: Medium to large species with cylindrical, rather brittle fronds, laterals tapering both basally and apically. Color varies from straw yellow in summer when bleached to pale pink to deep red or reddish brown; morphologically similar to *Gracilaria bursa-pastoris* but differing anatomically (filamentous medulla), with zonate tetrasporangia, and less obvious cystocarps (Braune and Guiry 2011).

Geographic distribution: NE Atlantic (Britain); NE and E Pacific (Alaska to California; Chile, Galápagos Is; Peru).

Uses: Can be eaten fresh in salads (IIMHO 2015).

Sarcothalia crispata (Bory de Saint-Vincent) Leister

Common names: Spanish: Luga negra (Avila et al. 1996), Negra, Crespa.

Description: Thallus erect, to 30 cm long, strongly compressed, linear band-like, almost dichotomously to irregularly branching, with rounded axillae and terminal sections; with marginal proliferations; crustose base (Braune and Guiry 2011).

Geographic distribution: SE Pacific (Chile).

Uses: Considered edible in Chile (Bucholz et al. 2012); used for extraction of carrageenan (Buschmann et al. 2005, Pereira et al. 2009a).

Schizymenia binderi (J. Agardh ex Kützing) J. Agardh

Description: Upright fronds leaf-like, flattened, often irregularly lobed or split; texture soft-fleshy; small discoid holdfast for anchoring (Braune and Guiry 2011).

Geographic distribution: S America (Chile, Peru); Antarctic and sub-Antarctic Is (Fuegia).

Uses: Considered an edible species (Gray 2015); extracts of this species have antiviral (Matsuhiro et al. 2005, Uzair et al. 2011), and antioxidant (Barahona et al. 2011) activity.

Sebdenia flabellata (J. Agardh) P.G. Parkinson

Synonym: *S. polydactyla*

Description: Thalli are purple-red, texture is soft and flaccid or gelatinous, tubular or cylindrical branches, dichotomous branching, erect, up to 8–20 cm in length. In cross section, plants are multiaxial and composed of a filamentous medulla surrounded by parenchymatous cortical layers with cells decreasing in size towards the exterior. Medullary filaments incorporate "ganglioid" cells (cells with radiating arms filled with highly refractive contents) that can bear "gland" cells. Cruciate tetrasporangia scattered in the outer cortex (Huang 2014h).

Geographic distribution: SW Atlantic; Asia (Japan, Korea, Taiwan); SE Pacific; Australia.

Uses: Used as food in Japan (Gray 2015), and for carrageenan extraction (Doshi et al. 1988). Extracts of this species have antiviral activity (Gosh et al. 2009).

Wildemania amplissima (Kjellman) Foslie

Synonym: *Porphyra cuneiformis*

Common names: English: Northern pink laver (Bunker et al. 2010), Red cellophane (Gutschmidt 2010).

Description: Is distinguished by its uniformly bright pink to reddish color. Only the margins, where both female and male reproductive structures are intermixed, are somewhat paler. This spring species has a deeply ruffled margin. Despite having two cell layers, it is thinner than *Porphyra perforata* (Purple laver). Ten cm long by 7 cm wide, with marginal ruffles about 3 cm deep and 1 to 1.5 cm apart, but much larger specimens occur, especially in sheltered habitats (the current record is over 6 m). When growing as an epiphyte, it is often longer and narrower than usual (Klinkenberg 2014d).

Geographic distribution: Arctic (Norway); White Sea; NE Atlantic (Scandinavia); Asia (Japan); NE Pacific (Alaska to California).

Uses: Considered edible in the NE Pacific (Gutschmidt 2010, Harrison 2013).

Wildemania miniata (C. Agardh) Foslie (formerly *Porphyra miniata*)

Common names: English: Red nori (Chapman and Chapman 1980); Japanese: Nori (Madlener 1977).

Description: Thallus is light pink, variable in shape, and can reach more than 100 cm in length. It is very thin although it is actually two cell layers thick (distromatic). The margin of the blade often appears filmy because of scattered reproductive cells.

Geographic distribution: N Atlantic; Atlantic Is (Greenland, Iceland); Baltic Sea; NE Pacific (Alaska to California, Chile); Asia (Russia).

Uses: Used as food in NE Pacific (Vancouver, British Columbia, and Puget Sound) (Chapman and Chapman 1980).

References

AAMD. 2004. Asian anti-cancer material resource—*Sargassum pallidum*. Available online at: http://asiancancerherb.info/Hai%20Hao%20Zi.htm

Aaronson, S. 2000. Algae. pp. 231–249. *In*: K.F. Kiple and K.C. Ornelas (eds.). The Cambridge World History of Food. Cambridge University Press, Cambridge.

Abbott, I.A. 1946. The genus *Griffithsia* (Rhodophyceae) in Hawaii. Farlowia—A Journal of Cryptogamic Botany 2(4): 439–453.

Abbott, I.A. 1978. The uses of seaweed as food in Hawaii. Econ. Bot. 32: 409–412.

Abbott, I.A. 1984. Limu: An Ethnobotanical Study of some Hawaiian Seaweeds. Pacific Tropical Botanical Garden, University of California.

Abbott, I.A. 1995. Section III. *Gracilaria*. *In*: I. Abbott (ed.). Taxonomy of Economic Seaweeds, With Reference to some Pacific Species, Volume V. Published by the California Sea Grant College System, University of California, La Jolla, California, 175 pp.

Abbott, I.A. 1996. Ethnobotany of seaweeds: clues to uses of seaweeds. Hydrobiologia 326-327: 15–20.

Abbott, I.A. 1999. Marine Red Algae of the Hawaiian Islands. Bishop Museum Press, Honolulu, Hawai'i.

Abbott, I.A. and G.J. Hollenberg. 1976. Marine Algae of California. Stanford University Press, Redwood City, CA, 844 pp.

Abdallah, M.A.M. 2014. Seaweed uses: human health and agriculture. pp. 71–92. *In*: V.H. Pomin (ed.). Seaweeds—Agriculture Uses, Biological and Antioxidant Agents. Nova Publishers Inc., New York.

Abdel-Kareem, M.S.M. 2009. Phenetic studies and new records of *Sargassum* species (Fucales, Phaeophyceae) from the Arabian Gulf coast of Saudi Arabia. Academic Journal of Plant Sciences 2(3): 173–181.

Abowei, J.F.N. and E.N. Ezekiel. 2013. The potentials and utilization of seaweeds. Scientia Agriculturae 4(2): 58–66.

Abreu, M.H., R. Pereira and J.-F. Sassi. 2014. Marine algae and the global food industry. pp. 300–319. *In*: Leonel Pereira and João M. Neto (eds.). Marine Algae—Biodiversity, Taxonomy, Environmental Assessment and Biotechnology. Science Publishers, an imprint of CRC Press/Taylor & Francis Group, Boca Raton, FL.

Acadian Seaplants. 2014. Hana Tsunomata™. Cultivated Sea Vegetables. Available online at: http://www.acadianseaplants.com/edible-seaweed-nutritional-supplements-ingredients/edible-sea-vegetables

Adami, H.O., L.B. Signorello and D. Trichpoulos. 1998. Towards an understanding of breast cancer etiology. Cancer Biol. Semin. 183: 255–262.

Adharini, R. and H. Kim. 2014. Developmental pattern of crust into upright thalli of *Grateloupia asiatica* (Halymeniaceae, Rhodophyta). Journal of Applied Phycology 26: 1911–1918.

Agngarayngay, Z.M., A.F.C. Llaguno, S.G. Aquino, L.B. Taclan and E.S. Galacgac. 2005. Edible seaweeds of Ilocos Norte: food preparations other local uses and market potentials. Sulu-Celebes Sea Sustainable Fisheries Management Project, South Triangle, Quezon City, 27 pp.

Ahmad, F., M.R. Sulaiman, W. Saimon, C.F. Yee and M. Patricia. 2012. Proximate compositions and total phenolic contents of selected edible seaweed from Semporna, Sabah, Malaysia. Borneo Science, the Journal of Science and Technology 31(1): 85–96.

Airanthi, M.K.W.-A., M. Hosokawa and K. Miyashita. 2011. Comparative antioxidant activity of edible Japanese brown seaweeds. Journal of Food Science 76: 104–111.

Akatsuka, I. 1981. Comparative morphology of the outermost cortical cells in the Gelidiaceae (Rhodophyta) of Japan. Nova Hedwigia 35: 453–463.

Akhtar, P. and V. Sultana. 2002. Biochemical studies of some seaweed species from Karachi coast. Records Zoological Survey of Pakistan 14: 1–4.

Albuquerque, I.R., K.C. Queiroz, L.G. Alves, E.A. Santos, E.L. Leite and H.A. Rocha. 2004. Heterofucans from *Dictyota menstrualis* have anticoagulant activity. Brazilian Journal Of Medical and Biological Research 37: 167.

Albuquerque, I.R.L., S.L. Cordeiro, D.L. Gomes, J.L. Dreyfuss, L.G.A. Filgueira, E.L. Leite, H.B. Nader and H.A.O. Rocha. 2013. Evaluation of anti-nociceptive and anti-inflammatory activities of a heterofucan from *Dictyota menstrualis*. Mar. Drugs 11: 2722–2740.

Aldave-Pajares, A. 1985. Algas andino peruanas como recurso hidrobiologico alimentario. Boletín de Lima 37: 66–72.

AL-Haj, N.A., N.I. Mashan, M.N. Shamsudin, H. Mohamad, C.S. Vairappan and Z. Sekawi. 2009. Antibacterial effect of *Gracilaria changii* and *Euchema denticulatum* on molecular properties of *Staphylococcus aureus* genes mecA, mecR1 and mecI. Research Journal of Biological Sciences 4: 580–584.

Alveal, K. 2006. Seaweed resources of Chile. *In*: A.T. Critchley, M. Ohno and D.B. Largo (eds.). World Seaweed Resources—An Authoritative Reference System. ETI Information Services Ltd. Hybrid Windows and Mac DVD-ROM. ISBN: 90-75000-80-4.

Amimi, A., A. Mouradi, T. Givernaud, N. Chiadmi and M. Lahaye. 2001. Structural analysis of *Gigartina pistillata* carrageenans (Gigartinaceae, Rhodophyta). Carbohydrate Research 333: 271–279.

Amorim, R.C.N., J.A.G. Rodrigues, M.L. Holanda, P.A.S. Mourão and N.M.B. Benevides. 2011. Anticoagulant properties of a crude sulfated polysaccharide from the red marine alga *Halymenia floresia* (Clemente) C. Agardh. Acta Scientiarum—Biological Sciences 33(3): 255–261.

Amornlerdpison, D., Y. Peerapornpisal, T. Taesotikul, T. Noiraksar and D. Kanjanapothi. 2009. Gastroprotective activity of *Padina minor* Yamada. Chiang Mai J. Sci. 36(1): 92–103.

Anderson, R.J., R.H. Simons, N.G. Jarman and G.J. Levitt. 1991. *Gelidium pristoides* in South Africa. Hydrobiologia 221: 55–66.

Anggadiredja, J., R. Andyani, Hayati, Muawanah. 1997. Antioxidant activity of *Sargassum polycystum* (Phaeophyta) and *Laurencia obtusa* (Rhodophyta) from Seribu Islands. Journal of Applied Phycology 9: 477–479.

Applegate, R.D. and P.B. Gray. 1995. Nutritional value of seaweed to ruminants. Rangifer 15: 15–18.

Arasaki, S. and T. Arasaki. 1983. Low Calorie, High Nutrition Vegetables from the Sea. To Help You Look and Feel Better. Japan Publications Inc., Tokyo, 196 pp.

ASSC. 2015. History of salt. America's Sea Salt Company—Salt works webpage. Available online at: http://www.saltworks.us/salt_info/si_HistoryOfSalt.asp

Assreuy, A.M., D.M. Gomes, M.S. da Silva, V.M. Torres, R.C. Siqueira, A.F. Pires, D.N. Criddle, N.M. de Alencar, B.S. Cavada, A.H. Sampaio and W.R. Farias. 2008. Biological effects of a sulfated-polysaccharide isolated from the marine red algae *Champia feldmannii*. Biological and Pharmaceutical Bulletin 31: 691–695.

Astorga-España, M.S. and A. Mansilla. 2014. Sub-antarctic macroalgae: opportunities for gastronomic tourism and local fisheries in the region of magallanes and Chilean Antarctic territory. Journal of Applied Phycology 26: 973–978.

Avila, M., R. Otaíza, R. Norambuena and M. Nuñez. 1996. Biological basis for the management of 'luga negra' (*Sarcothalia crispata* Gigartinales, Rhodophyta) in southern Chile.

pp. 245–252. *In*: S. Lindstrom and D. Chapman (eds.). Fifteenth International Seaweed Symposium, Volume 116, Springer, Netherlands.

Aziz, A. 2009. Coastal and Marine Plant Resources of Bangladesh: Management and Exploitation. Annual Botanical Conference, University of Chittagong, Chittagong, Bangladesh, 17 pp.

Bae, E.H. and I.K. Lee. 2001. *Umbraulva*, a new genus based on *Ulva japonica* (Holmes) Papenfuss (Ulvaceae, Chlorophyta). Algae 16(3): 217–231. Available online at: http://www.e-algae. kr/Upload/files/ALGAE/1Eun%20Hee%20Bae-%EC%9E%AC.pdf

Bagchi, M., M. Mark, W. Casey, B. Jaya, Y. Xumei, S. Sidney and D. Bagachi. 1999. Acute and chronic stress-induced oxidative gastrointestinal injury in rats, and the protective ability of a novel grape seed proanthocyanidin extract. Nutrition Research 19: 1189–1199.

Baik, S.E. and J.W. Kang. 1986. Antimicrobial activity of the volatile and lipid fractions of marine algae. Korean Journal of Phycology 1: 293–310.

Baldock, R.N. 2003. *Padina gymnospora*—Algae Revealed, S Australian State Herbarium. Available online at: http://www.flora.sa.gov.au/efsa/algae_revealed/pdf/Padina_gymnospora.pdf

Baldock, R.N. 2006. *Gracilariopsis lemaneiformis*—Algae Revealed, S Australian State Herbarium. Available online at: http://www.flora.sa.gov.au/efsa/algae_revealed/pdf/Gracilariopsis_lemaneiformis.pdf

Baldock, R.N. 2007. *Chaetomorpha aerea*—Algae Revealed, S Australian State Herbarium. Available online at: http://www.flora.sa.gov.au/efsa/algae_revealed/pdf/Chaetomorpha_aerea.pdf

Baldock, R.N. 2009a. *Ulva flexuosa*—Algae Revealed, S Australian State Herbarium. Available online at: http://www.flora.sa.gov.au/efsa/algae_revealed/pdf/Enteromorpha_flexuosa.pdf

Baldock, R.N. 2009b. *Catenella nipae*—Algae Revealed, S Australian State Herbarium. Available online at: http://www.flora.sa.gov.au/efsa/algae_revealed/pdf/Catenella_nipae.pdf

Baldock, R.N. 2014. *Sarcothalia radula*—Algae Revealed, S Australian State Herbarium. Available online at: http://www.flora.sa.gov.au/efsa/algae_revealed/pdf/Sarcothalia_radula.pdf

Ballesteros, E., D. Martin and M.J. Uriz. 1992. Biological activity of extracts from some mediterranean macrophytes. Botanica Marina 35(6): 481–485.

Bandyopadhyay, S.S., M.H. Navid, T. Ghosh, P. Schnitzler and B. Ray. 2011. Structural features and *in vitro* antiviral activities of sulfated polysaccharides from *Sphacelaria indica*. Phytochemistry 72: 276–283.

Bangmei, X. and I. Abbott. 1987. Edible seaweeds of China and their place in the Chinese diet. Economic Botany 41: 341–353.

Barahona, T., N.P. Chandía, M.V. Encinas, B. Matsuhiro and E.A. Zúñiga. 2011. Antioxidant capacity of sulfated polysaccharides from seaweeds. A kinetic approach. Food Hydrocolloids 25: 529–535.

Barnes, M. 2008. *Chaetomorpha linum*. Spaghetti algae. Marine Life Information Network: Biology and Sensitivity Key Information Sub-programme. Plymouth: Marine Biological Association of the United Kingdom. Available online at: http://www.marlin.ac.uk/speciesinformation.php?speciesID=2944

Barsanti, L. and P. Gualtieri. 2014. Algae: Anatomy, Biochemistry, and Biotechnology, 2nd Edition. CRC Press, Taylor & Francis Group, Boca Raton, FL, 361 pp.

Bast, F. 2014. An illustrated review on cultivation and life history of agronomically important seaplants. pp. 39–70. *In*: V.H. Pomin (ed.). Seaweeds—Agriculture Uses, Biological and Antioxidant Agents. Nova Publishers Inc., New York.

Belleza, D.F.C. and M. Liao. 2007. Taxonomic inventory of the marine green algal genus *Caulerpa* (Chlorophyta, Bryopsidales) at the University of San Carlos (CEBU) herbarium. The Philippine Scientist 44: 71–104.

Benjama, O. and P. Masniyom. 2012. Biochemical composition and physicochemical properties of two red seaweeds (*Gracilaria fisheri* and *G. tenuistipitata*) from the Pattani bay in southern Thailand. Songklanakarin Journal of Science and Technology 34(2): 223–230.

Béress, A., O. Wassermann, T. Bruhn, L. Béress, E.N. Kraiselburd, L.V. Gonzalez, G.E. de Motta and P.I. Chavez. 1993. A new procedure for the isolation of anti-HIV compounds (polysaccharides and polyphenols) from the marine alga *Fucus vesiculosus*. Journal of Natural Products 56: 478–488.

Bersamin, S.V., S.V. Laron, R.R. Gonzales and R.B. Banania. 1973. Some seaweeds consumed fresh in the Philippines. Phil. Jour. of Fish 8(2): 183–189.

Bhosale, R., J. Rout and B. Chaugule. 2012. The Ethnobotanical study of an edible freshwater red alga, *Lemanea fluviatilis* (L.) C. Ag. from Manipur, India. Ethnobotany Journal 10: 69–76.

Bhuvaneswaris, S. and S. Murugesan. 2013. Biochemical composition of seaweeds along south east coast of Tamilnadu, India. Int. Journal of Biology, Pharmacy and Allied Sciences (IJBPAS) 2(7): 1430–1436.

Bianchi, G. 1999. Field Guide to the Living Marine Resources of Namibia. FAO Species Identification Guide for Fishery Purposes. FAO—Food and Agriculture Organization of the United Nations, Rome, 298 pp.

Bixler, H.J. 1996. Recent developments in manufacturing and marketing carrageenan. Hydrobiologia 326/327: 35–57.

Bixler, H.J. and H. Porse. 2011. A decade of change in the seaweed hydrocolloids industry. Journal of Applied Phycology 23(3): 321–335.

Black, W.A.P. 1952. Seaweeds and their value in foodstuffs. Proc. Nutr. Soc. 12: 32–39.

Bleach, J. 2007. *Padina pavonica*. Peacocks tail. Marine Life Information Network: Biology and Sensitivity Key Information Sub-programme. Plymouth: Marine Biological Association of the United Kingdom. Available online at: http://www.marlin.ac.uk/speciesinformation. php?speciesID=4011

Bonotto, S. 1976. Cultivation of plants: multicellular plants. pp. 468–529. *In*: O. Kinne (ed.). Marine Ecology III (I). John Wiley and Sons, London.

Border, K. 2011. Saving the Sargasso. Bermudian Magazine. Available online at: http://www. thebermudian.com/features/412-bermuda-is-leading-the-way-in-ocean-conservation-with-the-formation-of-the-sargasso-sea-alliance

Bottalico, A., G.H. Boo, C. Russo, S.M. Boo and C. Perrone. 2014. *Parviphycus albertanoae* sp. nov. (Gelidiales, Rhodophyta) from the Mediterranean Sea. Phycologia 53(3): 243–251.

Bouhlal, R., C. Haslin, J.C. Chermann, S. Colliec-Jouault, C. Sinquin, G. Simon, S. Cerantola, H. Riadi and N. Bourgougnon. 2011. Antiviral activities of sulfated polysaccharides isolated from *Sphaerococcus coronopifolius* (Rhodophytha, Gigartinales) and *Boergeseniella thuyoides* (Rhodophyta, Ceramiales). Marine Drugs 9: 1187–1209.

Bouyssou, A. 2012. *Grateloupia doryphora*—North Sea Alien Species Database (NORSAS). Available online at: http://www.norsas.eu/species/grateloupia-doryphora

Bowling, B. 2012a. *Gracilariopsis longissima*—Texas Parks and Wildlife Department. Available online at: http://txmarspecies.tamug.edu/vegdetails.cfm?scinameID=Gracilariopsis%20 longissima

Bowling, B. 2012b. *Sargassum fluitans*—Texas Parks and Wildlife Department. Available online at: http://txmarspecies.tamug.edu/vegdetails.cfm?scinameID=Sargassum%20fluitans

Bowling, B. 2012c. *Sargassum natans*—Texas Parks and Wildlife Department. Available online at: http://txmarspecies.tamug.edu/vegdetails.cfm?scinameID=Sargassum%20natans

Braune, W. 2008. Meeresalgen. Ein Farbbildführer zu den verbreiteten benthischen Grün-Braun- und Rotalgen der Weltmeere. A.R.G. Gantner, Verlag, Ruggell, 596 pp.

Braune, W. and M. Guiry. 2011. Seaweeds—A colour guide to common benthic green. *In*: A.R.G. Gatner (ed.). Brown and Red Algae of the World's Oceans. K.G. Verlag, Ruggell, 601 pp.

Brooker, S. and R.C. Cooper. 1961. New Zealand Medicinal Plants. Unity Press, Auckland, New Zealand, 46 pp.

Browman, D.L. 1980. El manejo de la tierra arida del altiplano del Peru y Bolivia. America Indigena 40: 143–59.

Brown, M.T. and W.L. Zemke-White. 2006. New Zealand's seaweed resources. *In*: A.T. Critchley, M. Ohno and D.B. Largo (eds.). World Seaweed Resources—An Authoritative Reference

System. ETI Information Services Ltd. Hybrid Windows and Mac DVD-ROM; ISBN: 90-75000-80-4.

Brownlee, I.A., A.C. Fairclough, A.C. Hall and J.R. Paxman. 2011. The potential health benefits of seaweed and seaweed extract. pp. 119–136. *In*: Vitor H. Pomin (ed.). Seaweed: Ecology, Nutrient Composition and Medicinal Uses. Nova Science Publishers Inc., New York.

Bucholz, C.M., G. Krause and B.H. Buck. 2012. Seaweed and man. pp. 471–492. *In*: C. Wiencke and K. Bischo (eds.). Seaweed Biology: Novel Insights into Ecophysiology, Ecology and Utilization. Springer-Verlag, Berlin Heidelberg.

Bulboa, C., J. Macchiavello, E. Oliveira and K. Véliz. 2008. Growth rate differences between four Chilean populations of edible seaweed *Chondracanthus chamissoi* (Rhodophyta, Gigartinales). Aquaculture Research 39: 1550–1555.

Bunker, F.D., J.A. Brodie, C.A. Maggs and A.R. Bunker. 2010. Guide to Seaweeds of Britain and Ireland. Marine Conservation Society, Seasearch guides, 224p.

Burtin, P. 2003. Nutritional value of seaweeds. Electronic Journal of Environmental, Agricultural and Food Chemistry 2: 498–503.

Buschmann, A.H., M.C. Hernández-González, C. Astudillo, L. De La Fuente, A. Gutierrez and G. Aroca. 2005. Seaweed cultivation, product development and integrated aquaculture studies in Chile. World Aquaculture 36(3): 51–53.

Bushing, B. 2014. The Invasion of the "Devil Weed": *Sargassum horneri* invading California waters. California Diver. Available online at: http://californiadiver.com/the-invasion-of-the-devil-weed-sargassum-horneri-invading-california-waters/

Calado, C.M.B., V.C.C. Pires, A. Pacheco, A.P. Albuquerque, K.M.A. Santos and E.R. Florentino. 2012. Algas comestíveis: comparação nutricional entre espécies de *Gracilaria* (*G. cornea* e *G. domingensis*) de ocorrências no litoral Nordestino. Anais do Encontro Nacional de Educação, Ciência e Tecnologia UEPB 1: 54–62. Available online at: http://www.editorarealize.com.br/revistas/enect/trabalhos/20d170e05f2364bfa4e787ae0e8a5644_54.pdf

Castro-Gonzalez, M.I., F.P.G. Romo, S. Perez-Estrella and S. Carrillo-Dominguez. 1996. Chemical composition of the green alga *Ulva lactuca*. Ciencias Marinas 22: 205–213.

Černá, M. 2011. Seaweed proteins and amino acids as nutraceuticals. pp. 297–312. *In*: K. Se-Kwon (ed.). Marine Medicinal Foods: Implications and Applications, Macro and Microalgae. Advances in Food and Nutrition Research. Academic Press, London.

Cesar, A.O. 2006. Seaweed resources of Peru. *In*: A.T. Critchley, M. Ohno and D.B. Largo (eds.). World Seaweed Resources—An Authoritative Reference System. ETI Information Services Ltd. Hybrid Windows and Mac DVD-ROM; ISBN: 90-75000-80-4.

CEVA. 2014. Edible seaweed and French regulation. Available online at: http://www.ceva.fr/eng/content/download/31974/220188/file/seaweed%20and%20regulation2014.pdf

Chandrasekaran, M., V. Venkatesalu and G.A. Raj. 2014. Antibacterial activity of selected marine macro algae against vancomycin resistant *Enterococcus faecalis*. Journal of Coastal Life Medicine 2(12): 940–946.

Chang, J.-S. 2014. *Codium intricatum*—BiotaTaiwanica, Algae of Taiwan. Available online at: http://algae.biota.biodiv.tw/pages/728

Chang, J.-S. 2015a. *Codium arabicum*—BiotaTaiwanica, Algae of Taiwan. Available online at: http://algae.biota.biodiv.tw/pages/716

Chang, J.-S. 2015b. *Codium cylindricum*—BiotaTaiwanica, Algae of Taiwan. Available online at: http://biota.taibif.tw/pages/20452

Chang, J.-S. 2015c. *Codium papillatum*—BiotaTaiwanica, Algae of Taiwan. Available online at: http://algae.biota.biodiv.tw/pages/736

Chapman, V.J. 1970. Seaweeds and their Uses, 2nd Edition. Methuen, London, 304 pp.

Chapman, V.J. and D.J. Chapman. 1980. Seaweeds and their Uses. Chapman and Hall, New York, pp. 1–334.

Chattopadhyay, K., C.G. Mateu, P. Mandal, C.A. Pujol, E.B. Damonte and B. Ray. 2007. Galactan sulfate of *Grateloupia indica*: isolation, structural features and antiviral activity. Phytochemistry 68: 1428–1435.

Chen, K.-J., C.-K. Tseng, F.-R. Chang, J.-I. Yang, C.-C. Yeh, C.C. Chen, S.F. Wu, H.W. Chang and J.C. Lee. 2013. Aqueous extract of the edible *Gracilaria tenuistipitata* inhibits hepatitis C viral replication via cyclooxygenase-2 suppression and reduces virus-induced inflammation. PLoS ONE 8(2): e57704.

Cheney, D.P. 1986. The genus *Eucheuma* J. Agardh in Florida and the Caribbean. *In*: I.A. Abbott (ed.). Taxonomy of Economic Seaweeds, Vol. II. California Sea Grant College Program, 265 pp. Available online at: http://nsgl.gso.uri.edu/cuimr/cuimrw86003/cuimrw86003_full.pdf

Chennubhotla, V.S.K., N. Kaliaperumal and S. Kalimuthu. 1987. Common seaweed products. CMFRI Bulletin 41: 26–30.

Chevolot, L., B. Mulloy, J. Ratiskol, A. Foucault and S. Colliec-Jouault. 2001. A disaccharide repeat unit is the major structure in fucoidans from two species of brown algae. Carbohydrate Research 330(4): 529–535.

Chiang, Y.-M. 1960. Marine Algae of Northern Taiwan. Taiwania 7: 51–76.

Chiang, Y.-M. 1969. Observations on the reproductive organs of *Gloiopeltis tenax* (Turner) J. Agardh (Cryptonemiales, Endocladiaceae). Phycologia 8: 193–197.

Chiang, Y.-M. 1997. Species of *Hypnea* Lamouroux (Gigartinales, Rhodophyta) from Taiwan. pp. 163–177. *In*: I.A. Abbott (ed.). Taxonomy of Economic Seaweeds, Vol. 6. California Sea Grant College System, La Jolla, California.

Chiao-Wei, C., H. Siew-Ling and W. Ching-Lee. 2011. Antibacterial activity of *Sargassum polycystum* C. Agardh and *Padina australis* Hauck (Phaeophyceae). African Journal of Biotechnology 10(64): 14125–14131.

Choosawad, D., U. Leggat, C. Dechsukhum, A. Phongdara and W. Chotigeat. 2005. Anti-tumour activities of fucoidan from the aquatic plant *Utricularia aurea* Iour. Songklanakarin Journal of Science and Technology 27: 799–807.

Chopin, T. and R. Ugarte. 2006. Seaweed resources of Canada. *In*: A.T. Critchley, M. Ohno and D.B. Largo (eds.). World Seaweed Resources—An Authoritative Reference System. ETI Information Services Ltd. Hybrid Windows and Mac DVD-ROM; ISBN: 90-75000-80-4.

Cian, R.E., M.A. Fajardo, M. Alaiz, J. Vioque, R.J. Gonzalez and S.R. Drago. 2014. Chemical composition, nutritional and antioxidant properties of the red edible seaweed *Porphyra columbina*. International Journal of Food Sciences and Nutrition 65: 299–305.

Conte, E. and C. Payri. 2006. Present day consumption of edible algae in French Polynesia: a study of the survival of pre-European practices. Journal of the Polynesian Society 115: 77–93.

Coppejans, E. and W.F. Prud'Homme Van Reine. 1992. Seaweeds of the snellius-II expedition (E Indonesia): the genus *Caulerpa* (Chlorophyta-Caulerpales). Bulletin des Seances Mededelingen van de Zittingen 37: 667–712.

Coppejans, E., F. Leliaert, O. Dargent, R. Gunasekara and O. De Clerck. 2009. Sri Lankan Seaweeds—Methodologies and field guide to the dominant species. Abc Taxa 6(i-viii): 265 pp.

Cornish, M.L. and D.J. Garbary. 2010. Antioxidants from macroalgae: potential applications in human health and nutrition. Algae 25(4): 1–17.

Cortés, J., J. Samper-Villarreal and A. Bernecker. 2014. Seasonal phenology of *Sargassum liebmannii* J. Agardh (Fucales, Heterokontophyta) in an upwelling area of the Eastern Tropical Pacific. Aquatic Botany 119: 105–110.

Costa, L.S., G.P. Fidelis, S.L. Cordeiro, R.M. Oliveira, D.A. Sabry, R.B.G. Câmara, L.T.D.B. Nobre, M.S.S.P. Costa, J. Almeida-Lima, E.H.C. Farias, E.L. Leite and H.A.O. Rocha. 2010. Biological activities of sulfated polysaccharides from tropical seaweeds. Biomedicine & Pharmacotherapy 64: 21–28.

Costa, M.S.S.P., G.P. Fidelis, H.A.O. Rocha and L.S. Costa. 2014. Antioxidant sulfated polysaccharides from seaweed. pp. 189–208. *In*: V.H. Pomin (ed.). Seaweeds—Agriculture Uses, Biological and Antioxidant Agents. Nova Publishers Inc., New York.

Coura, C.O., I.W.F. De Araújo, E.S.O. Vanderlei, J.A.G. Rodrigues, A.L.G. Quinderé, B.P. Fontes, I.N.L. De Queiroz, D.B. De Menezes, M.M. Bezerra, A.A.R. Silva, H.V. Chaves, R.J.B. Jorge,

J.S.A.M. Evangelista and N.M.B. Benevides. 2012. Antinociceptive and anti-inflammatory activities of sulphated polysaccharides from the red seaweed *Gracilaria cornea*. Basic & Clinical Pharmacology & Toxicology 110: 335–341.

Critchley, A.T. and M. Ohno. 1998. Seaweed Resources of the World. Japan International Cooperation Agency, Yokosuka, Japan, 431 pp.

Critchley, A.T., R.D. Gillespie and K.W.G. Rotmann. 1998. The seaweed resources of South Africa. pp. 413–427. *In*: A.T. Critchley, M. Ohno, D.B. Largo and R.D. Gillespie (eds.). Seaweed Resources of the World. Japan International Cooperation Agency, Yokosuka, Japan.

Davidson, A. 2004. Seafood of Southeast Asia. Ten Speed Press, Berkeley, California, 368 pp.

Dawczynski, C., R. Schubert and G. Jahreis. 2007. Amino acids, fatty acids, and dietary fibre in edible seaweed products. Food Chemistry 103: 891–899.

De Araujo, I.W.F., E.D.O. Vanderlei, J.A.G. Rodrigues, C.O. Coura, A.L.G. Quindere, B.P. Fontes, I.N.L. De Queiroz, R.J.B. Jorge, M.M. Bezerra, A. Silva, H.V. Chavese, H.S.A. Monteiro, R.C.M. De Paula and N.M.B. Benevides. 2011. Effects of a sulfated polysaccharide isolated from the red seaweed *Solieria filiformis* on models of nociception and inflammation. Carbohydrate Polymers 86: 1207–1215.

De Araujo, I.W.F., J.A.G. Rodrigues, E.D.O. Vanderlei, G.A. de Paula, T.D. Lima and N.M.B. Benevides. 2012. Iota-carrageenans from *Solieria filiformis* (Rhodophyta) and their effects in the inflammation and coagulation. Acta Scientiarum Technology 34: 127–135.

De Brito, T.V., R.D.s. Prudêncio, A.B. Sales, F.d.C. Vieira Júnior, S.J.N. Candeira, Á.X. Franco, K.S. Aragão, R.d.A. Ribeiro, M.H.L. Ponte de Souza, L.d.S. Chaves, A.L. Freitas, J.V. Medeiros and A.L. dos Reis Barbosa. 2013. Anti-inflammatory effect of a sulphated polysaccharide fraction extracted from the red algae *Hypnea musciformis* via the suppression of neutrophil migration by the nitric oxide signalling pathway. Journal of Pharmacy and Pharmacology 65: 724–733.

De Clerck, O., H.R. Engledow, J.J. Bolton, J.J. Anderson and E. Coppejans. 2002. Twenty marine benthic algae new to South Africa, with emphasis on the flora of Kwazulu-Natal. Botanica Marina 45: 413–431.

De Kluijver, M., G. Gijswijt, R. de Leon and I. da Cunda. 2014. *Caulerpa cupressoides* (Cactus tree alga). Interactive Guide to Caribbean Diving. Available online at: http://species-identification.org/species.php?species_group=caribbean_diving_guide&menuentry=s oorten&id=491&tab=beschrijving

De Kluijver, M., G. Gijswijt, R. de Leon and I. da Cunda. 2015. *Turbinaria turbinata*—Interactive Guide to Caribbean Diving. Available online at: http://species-identification.org/species. php?species_group=caribbean_diving_guide&id=503

De Oliveira, V.P., D.R.P. Fernandes, N.M. de Figueiredo, Y.Y. Valentin and R.C. Garla. 2009. Four new additions to the marine flora of Fernando de Noronha Archipelago, Tropical western South Atlantic Ocean. Check List 5(2): 210–215.

De Souza, M.C.R., C.T. Marques, C.M.G. Dore, F.R.F. da Silva, H.A.O. Rocha and E.L. Leite. 2007. Antioxidant activities of sulfated polysaccharides from brown and red seaweeds. Journal of Applied Phycology 19: 153–160.

Deane, G. 2014a. *Caulerpa* in Edible Raw, Greens. Available online at: http://www.eattheweeds. com/caulerpa-warm-water-salad-and-pest-2/

Deane, G. 2014b. *Gracilaria*, Graceful Redweed, Edible Raw, Greens. Available online at: http:// www.eattheweeds.com/gracilaria-the-pot-thickens-2/

Deane, G. 2014c. *Sargassum* Sea Vegetable—Eat he Weeds. Available online at: http://www. eattheweeds.com/sargassum-not-just-for-breakfast-any-more-2/

Demirel, Z., Z.D. Yildirim, I. Tuney, K. Kesici and A. Sukuatar. 2012. Biochemical analysis of dome brown seaweeds from the Aegean Sea. Botanica Serbica 36(2): 91–95.

Denis, C., M. Morancais, M. Li, E. Deniaud, P. Gaudin, G. Wielgosz-Collin, G. Barnathan, P. Jaouen and J. Fleurence. 2010. Study of the chemical composition of edible red macroalgae *Grateloupia turuturu* from Brittany (France). Food Chemistry 119: 913–917.

Devi, G.K., G. Thirumaran, K. Manivannan and P. Anantharaman. 2009. Element composition of certain seaweeds from Gulf of Mannar Marine Biosphere Reserve; Southeast Coast of India. World Journal of Dairy and Food Sciences 4(1): 46–55.

Dhargalkar, V.K. and N. Pereira. 2005. Seaweed: promising plant of the millennium. Science & Culture 71(3-4): 60–66.

Dickinson, C.I. 1963. British Seaweeds. Kew Series 3. Eyre and Spottiswoode, London, 232 pp.

Domingo, A.C. and J.A. Corrales. 2001. Inventory and Distribution of Edible Seaweeds in Ilocos Sur. University of Northern Philippines, Tamag, Vigan City, 12 pp.

Dominguez, H. 2013. Functional Ingredients from Algae for Foods and Nutraceuticals. Woodhead Publishing, Sawston, Cambridge, UK, 768 p.

Doshi, Y.A., R.G. Parekh and V.D. Chauhan. 1988. *Sebdenia polydactyla*—a novel source of carrageenan. Phytochemistry 27: 1854–1855.

Drum, R. 2013. Sea vegetables for food and medicine. Well Being Journal, July/August. Available online at: http://www.eidon.com/SeaweedArticle.pdf

Duddington, C.L. 1966. Seaweeds and Other Algae. Faber and Faber, London, 207 pp.

Dumilag, R.V., L.M. Liao and A.O. Lluisma. 2014. Phylogeny of *Betaphycus* (Gigartinales, Rhodophyta) as inferred from COI sequences and morphological observations on *B. philippinensis*. Journal of Applied Phycology 26: 587–595.

Edgar, G. 2008. Australian Marine Life, 2nd Ed. Available online at: http://www.flora.sa.gov. au/efsa/algae_revealed/pdf/Caulerpa_flexilis.pdf

Edibleseaweed. 2014. Harvesting seaweed in the magical waters off of the West Coast of Vancouver Island. Available online at: http://www.edibleseaweed.com/index.html

Edibleseaweed. 2015. *Iridaea cordata* (Rainbow Seaweed). Available online at: http://www. edibleseaweed.com/seaweed-rainbow.html

Edwards, R. 2005. *Chondria coerulescens*. A red seaweed. Marine Life Information Network— Biology and Sensitivity Key Information Sub-programme, Marine Biological Association of the United Kingdom, Plymouth. Available online from: http://www.marlin.ac.uk/ speciesinformation.php?speciesID=2968

Eidlitz, M. 1969. Food and emergency food in the circumpolar area. Studia Ethnographica Upsaliensia 32.

Einav, R. 2014. *Solieria filiformis*—Seaweeds of Eastern Mediterranean coast. Available online at: http://www.blue-ecosystems.com/Solieria_filiformis.html

El-Said, G.F. and A. El-Sikaily. 2013. Chemical composition of some seaweed from Mediterranean Sea coast, Egypt. Environmental Monitoring and Assessment 185: 6089–6099.

El-Sarraf, W. and G. El-Shaarawy. 1994. Chemical composition of some marine macroalgae from the Mediterranean Sea of Alexandria, Egypt. The Bulletin of the High Institute of Public Health 24: 523–534.

Erulan, V., P. Soundarapandian, G. Thirumaran and G. Ananthan. 2009. Studies on the Effect of *Sargassum polycystum* (C. Agardh, 1824) extract on the growth and biochemical composition of *Cajanus cajan* (L.) Mill sp. American-Eurasian Journal of Agricultural & Environmental Sciences 6(4): 392–399.

Etahiri, S., V. Bultel-Poncé, C. Caux and M. Guyot. 2001. New bromoditerpenes from the red alga *Sphaerococcus coronopifolius*. Journal of Natural Products 64: 1024–1027.

Etahiri, S., V. Bultel-Poncé, A.E. Elkouri, O. Assobhei, D. Zaoui and M. Guyot. 2003. Antibacterial activities of marine algae from the Atlantic Coast of Morocco. Marine Life 13(1-2): 3–9.

European Communities. 1998. Multilingual Illustrated Dictionary of Aquatic Animals and Plants. Wiley-Blackwell, Oxford, pp. 608.

Fakoya, K.A., F.G. Owodeinde, S.L. Akintola, M.A. Adewolu, M.A. Abass and P.E. Ndimele. 2011. An exposition on potential seaweed resources for exploitation, culture and utilization in West Africa: a case study of Nigeria. Canadian Journal of Fisheries and Aquatic Sciences 6: 37–47.

FAO. 1990. Training Manual on *Gracilaria* Culture and Seaweed Processing in China. Food and Agriculture Organization, Zhanjiang Fisheries College, People's Republic of China.

FAO. 1996. Taxonomy, Ecology and Processing of Economically Important Red Seaweeds. Food and Agriculture Organization (of the United Nations). Network of Aquaculture Centres in Asia-Pacific and IFREMER, Bangkok, Thailand.

Farasat, M., R.A. Khavari-Nejad, S.M.B. Nabavi and F. Namjooyan. 2013. Antioxidant properties of two edible green seaweeds from northern coasts of the Persian Gulf. Jundishapur Journal of Natural Pharmaceutical Products 8: 47–52.

Farasat, M., R.A. Khavari-Nejad, S.M.B. Nabavi and F. Namjooyan. 2014. Antioxidant activity, total phenolics and flavonoid contents of some edible green seaweeds from northern coasts of the Persian Gulf. Iranian Journal of Pharmaceutical Research 13: 163–170.

Fareeha, A., A. Atika and R. Aliya. 2013. Protein extraction from green seaweeds found at Buleji Coast, Pakistan. International Journal of Phycology and Phycochemistry 9(1): 49–52.

Farid, Y., S. Etahiri and O. Assobhei. 2009. Activité antimicrobienne des algues marines de la lagune d'Oualidia (Maroc): criblage et optimisation de la période de la récolte. Journal of Applied Biosciences 24: 1543–1552.

Farrar, W.V. 1966. Tecuitlatl—a glimpse of Aztec food technology. Nature 211: 341–342.

Fawcett, L. 2015. Sea grape salad from Fiji—a preview of summer! In Seaweed Recipes. Available online at: http://www.seaweedrecipes.co.nz/sea-grape-salad-from-fiji-a-preview-of-summer/

FDA. 1999. GRAS notification with respect to *Phymatolithon calcareum* (*Lithothamnium calcareum*) & *Lithothamnium corallioides*. Available online at: http://www.fda.gov/ucm/groups/fdagov-public/@fdagov-foods-gen/documents/document/ucm260968.pdf

Feldman, S.C., S. Reynaldi, C.A. Stortz, A.S. Cerezo and E.B. Damont. 1999. Antiviral properties of fucoidan fractions from *Leathesia difformis*. Phytomedicine 6: 335–40.

Fidelis, G.P., R.B. Camara, M.F. Queiroz, M.S. Santos, P.C. Santos, H.A. Rocha and L.S. Costa. 2014. Proteolysis, NaOH and ultrasound-enhanced extraction of anticoagulant and antioxidant sulfated polysaccharides from the edible seaweed, *Gracilaria birdiae*. Molecules 19: 18511–18526.

Fikes, R. 2008. *Codium repens*. In Bocas Del Toro, Smithsonian Tropical Research Institute. Available online at: http://biogeodb.stri.si.edu/bocas_database/search/species/6330

Fino, N., A. Horta, S. Pinteus, J. Silva, C. Alves and R. Pedrosa. 2014. Antioxidant activity of *Sphaerococcus coronopifolius* associated bacteria. Front. Mar. Sci. Conference Abstract: IMMR | International Meeting on Marine Research, Peniche, Portugal.

Fitton, J.H. 2003. Brown marine algae—a survey of therapeutic potentials. Alternative and Complementary Therapies February 29–33.

Fitton, J.H. 2011. Therapies from fucoidan; multifunctional marine polymers. Marine Drugs 9: 1731–1760.

Fleurence, J. 1999. Seaweed proteins: biochemical, nutritional aspects and potential uses. Trends in Food Science and Technology 10: 25–28.

Fleurence, J., C. Le Coeur, S. Mabeau, M. Maurice and A. Landrein. 1995. Comparison of different extractive procedures for proteins from the edible seaweeds *Ulva rigida* and *Ulva rotundata*. Journal of Applied Phycology 7: 577–582.

Florent, C. 2014. *Padina boergesenii*—Florent's Guide to the Tropical Reefs. Available online at: http://www.reefguide.org/leafyrolledblade.html

Foon, T.S., L.A. Ai, P. Kuppusamy, M.M. Yusoff and N. Govindan. 2014. Studies on *in vitro* antioxidant activity of marine edible Seaweed from east coastal region, peninsular Malaysia using different extraction methods. Research Journal of Applied Sciences 9: 141–146.

Fortier, J.-F. 2010. *Caulerpa bikinensis*. Aqua Portail. Available online at: http://www.aquaportail.com/fiche-algue-1296-caulerpa-bikinensis.html

Fortner, H.J. 1978. The Limu Eater—A Cookbook of Hawaiian Seaweed. Sea Grant Miscellaneous Report, University of Hawaii, 120 pp.

Fournillier, K. 2015. Seamoss—biodiversity products. Available online at: http://www.biodiversity-products.com/media/documents/products/b9a98ca363151900099600267d22ad58.pdf

França-Pires, V.C., M.T. Tavares, A.P. Albuquerque, C.M.B. Calado, K.M.A. Santos, E.R. Florentino, I.M. Florêncio and M.C. Tejo. 2012. Caracterização físico-química da macroalga *Gracilaria birdiae*. 52° Congresso Brasileiro de Química, 14–18 Octobre 2012, Recife, PE, Brazil. Available online at: http://www.abq.org.br/cbq/2012/trabalhos/10/1154-13804.html

Freile-Pelegrín, Y., J.A. Azamar and D. Robledo. 2011. Preliminary characterization of carrageenan from the red seaweed *Halymenia floresii*. Journal of Aquatic Food Product Technology 20: 73–83.

Fretwell, K. and B. Starzoski. 2015. *Gracilaria pacifica*. Biodiversity of the central coast. Available online at: http://www.centralcoastbiodiversity.org/california-limu-bull-gracilaria-pacifica.html

Fujiwara-Arasaki, T., N. Mino and M. Kuroda. 1984. The protein value in human nutrition of edible marine algae of Japan. Hidrobiolgia 116/117: 513–516.

Funahashi, H., T. Imai, T. Mase, M. Sekiya, K. Yokoi, H. Hayashi, A. Shibata, T. Hayashi, M. Nishikawa, N. Suda, Y. Hibi, Y. Mizuno, K. Tsukamura, A. Hayakawa and S. Tanuma. 2001. Seaweed prevents breast cancer? Japanese Journal of Cancer Research 92(5): 483–487.

Funaki, M., M. Nishizawa, T. Sawaya, S. Inoue and T. Yamagishi. 2001. Mineral composition in the holdfast of three brown algae of the genus *Laminaria*. Fisheries Science 67: 295–300.

Gaia Associates. 2009. Biodiversity species list for county Donegal. County Donegal Heritage Office, Cultural Services, Donegal County Council, Ireland, 90 pp. Available online at: http://www.donegalcoco.ie/media/donegalcountyc/heritage/pdfs/BiodiversitySpeciesListforCountyDonegalMay2009.pdf

Galil, B.S. 2006. *Caulerpa taxifolia*, available online at: http://www.europe-aliens.org/speciesFactsheet.do?speciesId=100166 and http://www.europe-aliens.org/pdf/Caulerpa_taxifolia.pdf

Galutira, E.C. and G.T. Velasquez. 1963. Taxonomy, distribution and seasonal occurrence of edible marine algae in Ilocos Norte, Philippines. Philippine Journal of Science 92: 483–522.

García, I., R. Castroviejo and C. Neira. 1993. Las Algas en Galicia: Alimentación y otros usos. Consellería de Pesca, Marisqueo e Acuicultura—Xunta de Galícia, La Coruña, 231 pp.

Garza, D. 2005. Common Edible Seaweeds in the Gulf of Alaska. Published by the Alaska Sea Grant College Program, University of Alaska, Fairbanks, 57 pp.

GCE. 2010a. *Halimeda discoidea*—Gulf Coast Ecosystems. Available online at: http://marineplantbook.com/marinebookdiscoidea.htm

GCE. 2010b. *Halymenia floresii*—Gulf Coast Ecosystems. Available online at: http://marineplantbook.com/marinebookhalyfloresia.htm

Gerasimenko, N.I., E.A. Martyias, S.V. Logvinov and N.G. Busarova. 2014. Biological activity of lipids and photosynthetic pigments of *Sargassum pallidum* C. Agardh. Prikl Biokhim Mikrobiol 50: 85.

Ghosh, S. and J.P. Keshri. 2009. Observations on the morphology of *Porphyra vietnamensis* Tanaka & P.H. Ho (Bangiales, Rhodophyta) at Visakhapatnam coast, India. Nelumbo 51: 175–178.

Ghosh, T., C.A. Pujol, E.B. Damont, S. Sinha and B. Ray. 2009. Sulfated xylomannans from the red seaweed *Sebdenia polydactyla*: structural features, chemical modification and antiviral activity. Antiviral Chemistry & Chemotherapy 19: 235–242.

Givernaud, T., A. El Gourji, A. Mouradi-Givernaud, Y. Lemoine and N. Chiadmi. 1999. Seasonal variations of growth and agar composition of *Gracilaria multipartita* harvested along the Atlantic coast of Morocco. Proceedings of the International Seaweed Symposium 16: 167–172.

Globinmed. 2011a. *Caloglossa leprieurii*—Global Information Hub on Integrated Medicine. Available at: http://www.globinmed.com/index.php?option=com_content&view=article&id=79506:caloglossa-leprieurii-mont-g-martens&catid=367:c

Globinmed. 2011b. *Gracilaria edulis*—Global Information Hub on Integrated Medicine. Available at: http://www.globinmed.com/index.php?option=com_content&view=article&id=79206:gracilaria-edulis-sg-gmelin-pc-silva&catid=371:g

Goecke, F., M. Escobar and G. Collantes. 2012. Chemical composition of *Padina fernandeziana* (Phaeophyceae, Dictyotales) from Juan Fernandez Archipelago, Chile. Revista Latinoamericana de Biotecnología Ambiental y Algal 3(2): 95–104.

Goldberg, I. 1994. Introduction. pp. 3–16. *In*: I. Goldberg (ed.). Functional Food: Designer Foods, Pharmafoods, Nutraceuticals. Chapman & Hall, London.

Gomez, I.M. and R.C. Westermeier. 1991. Frond regrowth from basal disc in *Iridaea laminarioides* (Rhodophyta, Gigartinales) at Mehuin, southern Chile. Marine Ecology Progress Series 73: 83–91.

Gómez-Gutiérrez, C.M., G. Guerra-Rivas, I.E. Soria-Mercado and N.E. Ayala-Sánchez. 2011. Marine edible algae as disease preventers. pp. 29–40. *In*: Se-Kwon Kim and Steve Taylor (eds.). Marine Medicinal Foods: Implications and Applications, Macro and Microalgae. Academic Press (Elsevier), Waltham, USA.

Gómez-Ordóñez, E. and A.P. Rupérez. 2011. FTIR-ATR spectroscopy as a tool for polysaccharide identification in edible brown and red seaweeds. Food Hydrocolloids 25: 1514–1520.

Gómez-Ordóñez, E., A. Jiménez-Escrig and A.P. Rupérez. 2010. Dietary fibre and physicochemical properties of several edible seaweeds from the northwestern Spanish coast. Food Research International 43: 2289–2294.

Gómez-Ordóñez, E., A. Jiménez-Escrig and P. Rupérez. 2014. Bioactivity of sulfated polysaccharides from the edible red seaweed *Mastocarpus stellatus*. Bioactive Carbohydrates and Dietary Fibre 3: 29–40.

Gonzalez, R., S. Rodriguez, C. Romay, O. Ancheta, A. Gonzalez, J. Armesta, D. Remirez and N. Merino. 1999. Anti-inflammatory activity of phycocyanin extract in acetic acid-induced colitis in rats. Pharmacological Research 39: 55–59.

Goodrich, J., C. Lawson and V.P. Lawson. 1980. Kashaya Pomo Plants. American Indian Monograph Series 2. American Indian Studies Center, Univ. California, Los Angeles, CA.

Gosner, K.L. 1978. A Field Guide to the Atlantic Seashore. The Peterson Field Guide Series. Houghton Mifflin, Boston. 329 pp.

Gray, I. 2015. Edible seaweed and its health benefits. Kill-Tiredness-Now.com. Available online at: http://kill-tiredness-now.com/edible-seaweed.html

Green, S. 2014. Paleo Diet for Beginners: Top 30 Paleo Bread Recipes Revealed! The Blokehead, Create Space Independent Publishing Platform, US, 116 pp.

Greville, R.K. 1830. Algae britannicae, or descriptions of the marine and other inarticulated plants of the British islands, belonging to the order Algae; with plates illustrative of the genera. McLachlan & Stewart, Edinburgh, pp. 218.

Guerra-Rivas, G., C.M. Gómez-Gutiérrez, G. Alarcón-Arteaga, I.E. Soria-Mercado and N.E. Ayala-Sánchez. 2011. Screening for anticoagulant activity in marine algae from the Northwest Mexican Pacific coast. Journal of Applied Phycology 23(3): 495–810.

Guiry, M.D. 1982. *Devaleraea*, a new genus of the Palmariaceae (Rhodophyta) in the North Atlantic and North Pacific. Journal of the Marine Biological Association of the UK 62: 1–13.

Guiry, M.D. 2014a. *Bangia fuscopurpurea*—The Seaweed Site: Information on Marine Algae. Available online at: http://www.seaweed.ie/descriptions/Bangia_fuscopurpurea.php

Guiry, M.D. 2014b. *Eudesme virescens*—The Seaweed Site: Information on Marine Algae. Available online at: http://www.seaweed.ie/descriptions/Eudesme_virescens.php

Guiry, M.D. 2015a. *Rhodymenia pseudopalmata*—The Seaweed Site: Information on Marine Algae. Available online at: http://www.seaweed.ie/descriptions/Rhodymenia_pseudopalmata. php

Guiry, M.D. 2015b. *Vertebrata lanosa*—The Seaweed Site: Information on Marine Algae. Available online at: http://www.seaweed.ie/descriptions/Vertebrata_lanosa.php

Guiry, M.D. and D.J. Garbary. 1991. A geographical and taxonomic guide to European seaweeds of economic importance. pp. 1–19. *In*: M.D. Guiry and G. Blunden (eds.). Seaweed Resources in Europe: Uses and Potential. John Wiley & Sons, Chichester.

Guiry, M.D. and C.C. Hession. 1998. The seaweed resources of Ireland. pp. 210–216. *In*: A.T. Critchley and M. Ohno (eds.). The Seaweed Resources of the World. Japan International Cooperation Agency, Yokosuka, Japan.

Guiry, M.D. and G.M. Guiry. 2014a. *Bangia atropurpurea*—AlgaeBase, World-wide electronic publication, National University of Ireland, Galway. Available online at: http://www.algaebase.org/search/species/detail/?species_id=fbf1682cde1880091

Guiry, M.D. and G.M. Guiry. 2014b. *Boergeseniella thuyoides*—AlgaeBase, World-wide electronic publication, National University of Ireland, Galway. Available online at: http://www.algaebase.org/search/species/detail/?species_id=H6dc86e33efb40743

Guiry, M.D. and G.M. Guiry. 2014c. *Chondracanthus intermedius*—AlgaeBase, World-wide electronic publication, National University of Ireland, Galway. Available online at: http://www.algaebase.org/search/species/detail/?species_id=H200b45e7364e5e5d

Guiry, M.D. and G.M. Guiry. 2014d. *Durvillaea potatorum*—AlgaeBase, World-wide electronic publication, National University of Ireland, Galway. Available online at: http://www.algaebase.org/search/species/detail/?species_id=v25083fd3f6878b28

Guiry, M.D. and G.M. Guiry. 2014e. *Gayralia brasiliensis*—AlgaeBase, World-wide electronic publication, National University of Ireland, Galway. Available online at: http://www.algaebase.org/search/species/detail/?species_id=A541c87aa0f806f57

Guiry, M.D. and W. Guiry. 2014f. *Gelidium pusillum*—AlgaeBase. Available online at: http://www.algaebase.org/search/species/detail/?species_id=k8a2b4322e9527e27

Guiry, M.D. and G.M. Guiry. 2014g. *Gracilaria changii* Uses and Compounds—AlgaeBase, World-wide electronic publication, National University of Ireland, Galway. Available online at: http://www.algaebase.org/search/species/usage/?id=1886

Guiry, M.D. and G.M. Guiry. 2014h. *Gracilaria tenuistipitata* Uses and Compounds—AlgaeBase, World-wide electronic publication, National University of Ireland, Galway. Available online at: http://www.algaebase.org/search/species/usage/?id=18355

Guiry, M.D. and G.M. Guiry. 2014i. *Sargassum polycystum*—AlgaeBase, World-wide electronic publication, National University of Ireland, Galway. Available online at: http://www.algaebase.org/search/species/detail/?species_id=V6d72d0aefbbdc478

Guiry, M.D. and G.M. Guiry. 2014j. *Sphaerococcus coronopifolius*—AlgaeBase, World-wide electronic publication, National University of Ireland, Galway. Available online at: http://www.algaebase.org/search/species/detail/?species_id=123

Guiry, M.D. and G.M. Guiry. 2015a. *Chondracanthus exasperatus*—AlgaeBase, World-wide electronic publication, National University of Ireland, Galway. Available online at: http://www.algaebase.org/search/species/detail/?species_id=g9de5e0a89d563674

Guiry, M.D. and G.M. Guiry. 2015b. *Griffithsia corallinoides*—AlgaeBase. World-wide electronic publication, National University of Ireland, Galway. Available online at: http://www.algaebase.org/search/species/detail/?species_id=b703ebf3b30edf8ac

Gupta, S. and N. Abu-Ghannam. 2011. Bioactive potential and possible health effects of edible brown seaweeds. Trends in Food Science and Technology 22: 315–326.

Gurgel, C.F.D. and S. Fredericq. 2004. Systematics of the Gracilariaceae (Gracilariales, Rhodophyta): a critical assessment based on rbcL sequence analysis. Journal of Phycology 40: 138–159.

Gutschmidt, B. 2010a. *Gracilariopsis andersonii*—Seaweeds of the Pacific Northwest. A Guide to Identifying 25 Common Seaweeds of the Washington Coast. Available online at: http://pendiva.com/seaweed/red-spaghetti/

Gutschmidt, B. 2010b. *Mastocarpus papillatus*—Seaweeds of the Pacific Northwest. A Guide to Identifying 25 Common Seaweeds of the Washington Coast. Available online at: http://pendiva.com/seaweed/turkish-washcloth/

Gutschmidt, B. 2010c. *Mazzaella splendens*—Seaweeds of the Pacific Northwest. A Guide to Identifying 25 Common Seaweeds of the Washington Coast. Available online at: http://pendiva.com/seaweed/rainbow-leaf/

Gutschmidt, B. 2010d. *Nereocystis luetkeana*—Seaweeds of the Pacific Northwest. A Guide to Identifying 25 Common Seaweeds of the Washington Coast. Available online at: http://pendiva.com/seaweed/bull-kelp/

Gutschmidt, B. 2010e. *Saccharina bongardiana* (= *Laminaria bongardiana*)—Seaweeds of the Pacific Northwest. A Guide to Identifying 25 Common Seaweeds of the Washington Coast. Available online at: http://pendiva.com/seaweed/split-kelp/

Gutschmidt, B. 2010f. *Sargassum muticum*—Seaweeds of the Pacific Northwest. A Guide to Identifying 25 Common Seaweeds of the Washington Coast. Available online at: http://pendiva.com/seaweed/sargassum/

Hadad, O.C. and A. Fricke. 2008. *Gracilaria domingensis*—Tropical Field Phycology Field Book, STRI, Bocas de Toro. Available online at: http://biogeodb.stri.si.edu/bocas_database/search/species/6336

Haijin, M., J. Xiaolu and G. Huashi. 2003. A k-carrageenan derived oligosaccharide prepared by enzymatic degradation containing anti-tumor activity. Journal of Applied Phycology 15: 297–303.

Hairon, D. 2014. Edible seaweeds in Jersey. Available online at: http://www.jerseywalkadventures.co.uk/go/press-office/edible-seaweeds-in-jersey/

Hallsson, S.V. 1961. The Uses of Seaweed in Iceland. Fourth International Seaweed Symposium, France.

Han, E. 2015. Red ogo seaweed salad. Recipes from the kitchen. Available online at: http://www.thekitchn.com/la-farmers-market-report-seawe-58761

Hansen, J.E., J.E. Packard and W.T. Doyle. 1981. Mariculture of Red Seaweeds. California Sea Grant College Program, 42 pp.

Harrison, M. 2008. Wild Food School—Edible Seaweeds around the British Isles. Available online at: http://www.countrylovers.co.uk/wfs/wfsseaweed.htm

Harrison, M. 2013. Armageddon Cookbook and Doomsday Kitchen. Wild Food School—Marcus Harrison, Lostwithiel, Cornwall, UK, pp. 127–133.

Hart, G., T. Ticktin, D. Kelman, A. Wright and N. Tabandera. 2014. Contemporary gathering practice and antioxidant benefit of wild seaweeds in Hawai'i. Econ. Bot. 68: 30–43.

Harvey, W.H. 1846–1851. Phycologia Britannica, or, a history of British sea-weeds: containing coloured figures, generic and specific characters, synonymes, and descriptions of all the species of algae inhabiting the shores of the British Islands. Several Volumes. Reeve & Benham, London.

Harvey, W.H. 1858–1863. Phycologia Australica: or, a history of Australian seaweeds; comprising coloured figures and descriptions of the more characteristic marine algae of New South Wales, Victoria, Tasmania, South Australia, and Western Australia, and a synopsis of all known Australian algae. Several Volumes. Lovell Reeve & Co., London.

Hellio, C., H. Thomas-Guyon, G. Culioli, L. Piovetti, N. Bourgougnon and Y. Le Gal. 2001. Marine antifoulants from *Bifurcaria bifurcata* (Phaeophyceae, Cystoseiraceae) and other brown macroalgae. Biofouling 17(3): 189–201.

Heyden, V.V.d. 2013. A coastal foraging expedition seaweed superfoods fresh and free from the sea. The scenic south. Available online at: http://scenicsouth.co.za/2013/11/a-coastal-foraging-expedition-seaweed-superfoods-fresh-and-free-from-the-sea/

Hill, K. 2001a. *Gracilaria tikvahiae*—Smithsonian Marine Station. Available online at: http://www.sms.si.edu/irlspec/Gracil_tikvah.htm

Hill, K. 2001b. *Hypnea cervicornis*—Smithsonian Marine Station. Available online at: http://www.sms.si.edu/irlspec/Hypnea_cervic.htm

Hiscock, K. and P. Pizzolla. 2007. *Ceramium virgatum*. A red seaweed. Marine Life Information Network—Biology and Sensitivity Key Information Sub-programme, Marine Biological Association of the United Kingdom, Plymouth. Available online from: http://www.marlin.ac.uk/speciesinformation.php?speciesID=2922

Hivy, F. and V. Chauhan. 1963. *Caulerpa veravalensis*, a new species from Gujarat, India. Botanica Marina 5(4): 97–100.

Hoek, C. 1995. Algae: An Introduction to Phycology. Cambridge University Press, Cambridge, 626 pp.

Hoek, C. and H.B.S. Womersley. 1984. *Cladophora laetevirens*—Electronic Flora of South Australia Species Fact Sheet. The Marine Benthic Flora of Southern Australia Part I.

Available online at: http://www.flora.sa.gov.au/efsa/Marine_Benthic_Flora_SA/Part_I/Cladophora_laetevirens.shtml

Holmes, E.M. 1895. Marine algae from Japan. Journal of the Linnean Society of London, Botany 31: 248–260.

Holmes, J. 2012. *Costaria costa*—Intertidal organisms EZ-ID Guides. Washington State University Extension, Island County Beach Watchers. Available online at: http://www.beachwatchers.wsu.edu/ezidweb/seaweeds/Costaria.htm

Hommersand, M.H., M.D. Guiry, S. Fredericq and G.L. Leister. 1993. New perspectives in the taxonomy of the Gigartinaceae (Gigartinales, Rhodophyta). Proceedings of the International Seaweed Symposium 14: 105–120.

Hong, D.D. and H.T.M. Hien. 2004. Nutritional analysis of Vietnamese seaweeds for food and medicine. BioFactors 22: 323–325.

Hong, D.D., H.M. Hien and P.N. Son. 2007. Seaweeds from Vietnam used for functional food, medicine and biofertilizer. Journal of Applied Phycology 19: 817–826.

Hoppe, H.A. 1979. Marine algae and their products and constituents in pharmacy. pp. 25–119. *In*: H.A. Hoppe, T. Levring and Y. Tanaka (eds.). Marine Algae in Pharmaceutical Science. Walter de Gruyter, New York.

Hoshino, T., T. Hayashi, K. Hayashi, J. Hamada, J.B. Lee and U. Sankawa. 1998. An antivirally active sulfated polysaccharide from *Sargassum horneri* (Turner) C. Agardh. Biological and Pharmaceutical Bulletin 21(7): 730–734.

Hotchkiss, S. and A. Trius. 2007. Seaweed: the most nutritious form of vegetation on the planet? Food Ingredients—Health and Nutrition January/February 22–33.

Hotta, M., K. Ogata, A. Nita, Hosikawa, M. Yanagi and K. Yamazaki. 1989. Useful Plants of the World. Heibonsha, Tokyo, Japan, 1499 pp.

Hoygaard, A. 1937. Skrifter om Svalhard og Havet. Oslo.

Hu, S.-Y. 2005. Food Plants of China. Chinese University Press, 844 pp.

Huang, S.-F. 2014a. *Acetabularia major*—BiotaTaiwanica, Algae of Taiwan. Available online at: http://algae.biota.biodiv.tw/pages/1418

Huang, S.-F. 2014b. *Caulerpa racemosa* var. *Macrophysa*—BiotaTaiwanica, Algae of Taiwan. Available online at: http://algae.biota.biodiv.tw/pages/530

Huang, S.-F. 2014c. *Grateloupia livida*—BiotaTaiwanica, Algae of Taiwan. Available online at: http://algae.biota.biodiv.tw/pages/1098

Huang, S.-F. 2014d. *Halymenia dilatata*—BiotaTaiwanica, Algae of Taiwan. Available online at: http://algae.biota.biodiv.tw/pages/1103

Huang, S.-F. 2014e. *Sarcodia montagneana*—BiotaTaiwanica, Algae of Taiwan. Available at: http://algae.biota.biodiv.tw/pages/1565

Huang, S.-F. 2014f. *Hypnea japonica*—BiotaTaiwanica, Algae of Taiwan. Available online at: http://algae.biota.biodiv.tw/pages/1188

Huang, S.-F. 2014g. *Polyopes lancifolius*—BiotaTaiwanica, Algae of Taiwan. Available online at: http://algae.biota.biodiv.tw/pages/1112

Huang, S.-F. 2014h. *Sebdenia flabellata*—BiotaTaiwanica, Algae of Taiwan. Available online at: http://algae.biota.biodiv.tw/pages/1693

Huang, S.-F. 2015a. *Boodlea composita*—BiotaTaiwanica, Algae of Taiwan. Available online at: http://algae.biota.biodiv.tw/pages/449

Huang, S.-F. 2015b. *Portieria hornemannii*—BiotaTaiwanica, Algae of Taiwan. Available online at: http://algae.biota.biodiv.tw/pages/1460

Huang, S.-F. 2015c. *Valonia aegagropila*—BiotaTaiwanica, Algae of Taiwan. Available online at: http://algae.biota.biodiv.tw/pages/1848

Huisman, J. and C. Parker. 2011a. *Codium dwarkense*—FloraBase, The Western Australian Flora. Available online at: https://florabase.dpaw.wa.gov.au/browse/profile/35857

Huisman, J. and C. Parker. 2011b. *Padina boryana*—FloraBase, The Western Australian Flora. Available online at: http://florabase.dpaw.wa.gov.au/browse/profile/27115

Huisman, J. and C. Parker. 2011c. *Trichogloea requienii*—FloraBase, The Western Australian Flora. Available online at: http://florabase.dpaw.wa.gov.au/browse/profile/27339

Huynh, Q.N. and H.D. Nguyen. 1998. The seaweed resources of Vietnam. pp. 62–69. *In*: A.T. Critchley, M. Ohno, D.B. Largo and R.D. Gillespie (eds.). Seaweed Resources of the World. Japan International Cooperation Agency, Yokosuka, Japan.

Hwang, P.-A., C.-H. Wu, S.-Y. Gau, S.-Y. Chien and D.-F. Hwang. 2010. Antioxidant and immune-stimulating activities of hot-water extract from seaweed *Sargassum hemiphyllum*. Journal of Marine Science and Technology 18(1): 41–46.

Hwang, P.-A., Y.-L. Hung and S.-Y. Chien. 2014. Inhibitory activity of *Sargassum hemiphyllum* sulfated polysaccharide in arachidonic acid-induced animal models of inflammation. Journal of Food and Drug Analysis 1–8. http://dx.doi.org/10.1016/j.jfda.2014.05.004

Ibrahim, A.M.M., M.H. Mostafa, M.H. El-Masry and M.M.A. El-Naggar. 2005. Active biological materials inhibiting tumor initiation extracted from marine algae. Egyptian Journal of Aquatic Research 31(1): 146–155.

IIMHO. 2015. *Sarcodiotheca gaudichaudii* (Red String Seaweed)—Indian Island Marine Health Observatory Identification Guides (Seaweeds and Sea Grasses). Available online at: https://sites.google.com/site/indianislandproject/identification-guides/seaweeds

IMR. 2011. Value Adding and Supply Chain Development for Fisheries and Aquaculture Products in Fiji, Samoa and Tonga: Scoping Study for *Caulerpa* (Sea Grapes). Institute of Marine Resources School of Marine Studies Faculty of Science, Technology and Environment University of the South Pacific, Australia.

Indergaard, M. and J. Minsaas. 1991. Animal and human nutrition. pp. 21–64. *In*: M.D. Guiry and G. Blunden (eds.). Seaweed Resources in Europe: Uses and Potential. John Wiley & Sons, Chichester.

INII. 1966. Aproveitamento de algas na costa portuguesa. Instituto Nacional de Investigação Industrial, Lisbon, 261 pp.

Ioannou, E. and V. Roussis. 2009. Natural products from seaweeds. pp. 51–81. *In*: A.E. Osbourn and V. Lanzotti (eds.). Plant-Derived Natural Products. Springer Science + Business Media, LLC, London.

Irianto, E.H. and Syamdidi. 2015. Production, handling and processing of seaweeds. pp. 359–375. *In*: Se-Kwon Kim (ed.). Seafood Science: Advances in Chemistry, Technology and Applications. CRC Press, Taylor & Francis Group, Boca Raton, FL.

Irving, F. 1957. Wild and emergency foods of Australian and Tasmanian aborigines. Oceania 28: 113–142.

Islam, A.K.M.N. 1998. The seaweed resources of Bangladesh. pp. 106–109. *In*: A.T. Critchley, M. Ohno, D.B. Largo and R.D. Gillespie (eds.). Seaweed Resources of the World. Japan International Cooperation Agency, Yokosuka, Japan.

Islam, M.N., S.H. Choi, H.E. Moon, J.J. Park, H.A. Jung, M.H. Woo, H.C. Woo and J.S. Choi. 2014. The inhibitory activities of the edible green alga *Capsosiphon fulvescens* on rat lens aldose reductase and advanced glycation end products formation. European Journal of Nutrition 53(1): 233–42.

Istini, S., A. Zatnika and W. Sujatmiko. 1998. The seaweed resources of Indonesia. pp. 92–98. *In*: A.T. Critchley, M. Ohno, D.B. Largo and R.D. Gillespie (eds.). Seaweed Resources of the World. Japan International Cooperation Agency, Yokosuka, Japan.

Istini, S., A. Zatnika and W. Sujatmiko. 2006. The seaweed resources of Indonesia. *In*: A.T. Critchley, M. Ohno and D.B. Largo (eds.). World Seaweed Resources—An Authoritative Reference System. ETI Information Services Ltd. Hybrid Windows and Mac DVD-ROM; ISBN: 90-75000-80-4.

Itono, H. and R.T. Tsuda. 1981. *Titanophora marianensis* sp. nov. (Nemastomataceae, Rhodophyta) from Guam. Pacific Science 34(1): 21–24.

Ivanova, V., M. Stancheva and D. Petrova. 2013. Fatty acid composition of Black Sea *Ulva rigida* and *Cystoseira crinita*. Bulgarian Journal of Agricultural Science 19(1): 42–47.

Iyer, R., O. De Clerck, J.J. Bolton and V.E. Coyne. 2004. Morphological and taxonomic studies of *Gracilaria* and *Gracilariopsis* species (Gracilariales, Rhodophyta) from South Africa. South African Journal of Botany 70(4): 521–539.

Jeon, H.-J., M.-J. Seo, H.-S. Choi, O.-H. Lee and B.-Y. Lee. 2014. *Gelidium elegans*, an edible red seaweed, and hesperidin inhibit lipid accumulation and production of reactive oxygen species and reactive nitrogen species in 3T3-L1 and RAW 264.7 cells. Phytotherapy Research 28(11): 1701–1709.

Jha, B., C.R.K. Reddy, M.C. Thakur and M.U. Rao. 2009. The Diversity and Distribution of Seaweeds of Gujarat Coast Series: Developments in Applied Phycology, Vol. 3. Springer (India) Private Limited, Bangalore, Mumbai, New Delhi, 215 p.

Jiao, G., G. Yu, J. Zhang and H.S. Ewart. 2011. Chemical structures and bioactivities of sulfated polysaccharides from marine algae. Marine Drugs 9(2): 196–223.

Jimenez-Escrig, A. and I.G. Cambrodon. 1999. Nutritional evaluation and physiological effects of edible marine macroalgae. Archivos Latinoamericanos de Nutricion 49: 114–120.

Jimenez-Escrig, A. and F.J. Sanchez-Muniz. 2000. Dietary fibre from edible seaweeds: chemical structure, physicochemical properties and effects on cholesterol metabolism. Nutrition Research 20: 585–598.

Jimenez-Escrig, A., E. Gomez-Ordonez and P. Ruperez. 2012. Brown and red seaweeds as potential sources of antioxidant nutraceuticals. Journal of Applied Phycology 24: 1123–1132.

JIRCAS—Japan International Research Center for Agricultural Science. 2012. *Chondrophycus cartilagineus*—Common Underwater Plants in Coastal Areas of Thailand. Available at: http://www.jircas.affrc.go.jp/project/aquacult_Thailand/data/chondrophycus_cartilagineus.html

JIRCAS—Japan International Research Center for Agricultural Science. 2012a. *Dictyota friabilis*—Common Underwater Plants in Coastal Areas of Thailand. Available online at: http://www.jircas.affrc.go.jp/project/aquacult_Thailand/data/dictyota_friabilis.html

JIRCAS—Japan International Research Center for Agricultural Science. 2012b. *Dictyopteris repens*—Common Underwater Plants in Coastal Areas of Thailand. Available online at: http://www.jircas.affrc.go.jp/project/aquacult_Thailand/data/dictyopteris_repens.html

JIRCAS—Japan International Research Center for Agricultural Sciences. 2012c. *Gracilaria blodgettii*—Common Underwater Plants in Coastal Areas of Thailand. Available online at: http://www.jircas.affrc.go.jp/project/aquacult_Thailand/data/gracilaria_blodgettii.html

JIRCAS—Japan International Research Center for Agricultural Sciences. 2012d. *Halimeda opuntia*—Common Underwater Plants in Coastal Areas of Thailand. Available online at: http://www.jircas.affrc.go.jp/project/aquacult_Thailand/data/halimeda_opuntia.html

JIRCAS—Japan International Research Center for Agricultural Sciences. 2012e. *Halymenia durvillei*—Common Underwater Plants in Coastal Areas of Thailand. Available online at: http://www.jircas.affrc.go.jp/project/aquacult_Thailand/data/halymenia_durvillei.html

JIRCAS—Japan International Research Center for Agricultural Sciences. 2012f. *Hormophysa cuneiformis*—Common Underwater Plants in Coastal Areas of Thailand. Available online at: http://www.jircas.affrc.go.jp/project/aquacult_Thailand/data/hormophysa_cuneiformis.html

JIRCAS—Japan International Research Center for Agricultural Sciences. 2012g. *Sargassum swartzii*—Common Underwater Plants in Coastal Areas of Thailand. Available online at: http://www.jircas.affrc.go.jp/project/aquacult_Thailand/data/sargassum_swartzii.html

JIRCAS—Japan International Research Center for Agricultural Sciences. 2012h. *Turbinaria conoides*—Common Underwater Plants in Coastal Areas of Thailand. Available online at: http://www.jircas.affrc.go.jp/project/aquacult_Thailand/data/turbinaria_conoides.html

JIRCAS—Japan International Research Center for Agricultural Sciences. 2012i. *Turbinaria decurrens*—Common Underwater Plants in Coastal Areas of Thailand. Available online

at: http://www.jircas.affrc.go.jp/project/aquacult_Thailand/data/turbinaria_decurrens.
html

John, D.M. 2014. *Macrocystis pyrifera* (Giant Kelp)—The Trustees of the Natural History
Museum, London. Available online at: http://www.nhm.ac.uk/nature-online/species-
of-the-day/biodiversity/climate-change/macrocystis-pyrifera/

Johnston, H.W. 1966. The biological and economic importance of algae, 2. Tuatara 14(1): 30–62.

Johnston, H.W. 1970. The biological and economic importance of algae, 3. Edible algae of fresh
and brackish waters. Tuatara 18: 19–35.

Joly, A.B. 1965. Flora marinha do litoral norte do Estado de São Paulo e regiões circunvizinhas.
Boletim 294 da Faculdade de Ciências e Letras da USP—Botânica 21.

Jung, H.A., S.K. Hyun, H.R. Kim and J.S. Choi. 2006. Angiotensin-converting enzyme I
inhibitory activity of phlorotannins from *Ecklonia stolonifera*. Fisheries Science 72:
1292–1299.

Kaas, R. 2005. Mapping and biomass estimation for a harvested population of *Gelidium
sesquipedale* (Rhodophyta, Gelidiales) along the Atlantic coast of Morocco. Phycologia
44: 66–71.

Kai Ho. 2014. *Lessonai corrugata*—Ocean Treasure. Tasmanian Sea Vegetables. Available online
at: http://oceantreasure.com.au/?product=tasmanian-kombu-dried

Kai Ho. 2015a. *Chaetomorpha coliformis*—Ocean Treasure. Tasmanian Sea Vegetables. Available
online at: http://oceantreasure.com.au/?product=mermaids-necklace

Kai Ho. 2015b. *Grateloupia turuturu* (Red Lettuce)—Ocean Treasure. Tasmanian Sea Vegetables.
Available online at: http://oceantreasure.com.au/?product=red-lettuce-grateloupia-
turuturu

Kalesh, N.S. 2003. Phycochemical Distinctiveness of Selected Marine Macrophytes of Kerala
Coast. Ph.D. Thesis, Department of Chemical Oceanography, School of Marine Sciences,
Faculty of Marine Sciences, Cochin University of Science and Technology, Kochi, India,
368 pp.

Kaliaperumal, N., S. Kalimuthu and J.R. Ramalingam. 1995. Economically important seaweeds.
pp. 1–35. *In*: M. Devaraj (ed.). Special Publication 62. Central Marine Fisheries Research
Institute, Indian Council of Agricultural Research, Cochin, India.

Kanagasabhapathy, M., H. Sasaki and S. Nagata. 2008. Phylogenetic identification of epibiotic
bacteria possessing antimicrobial activities isolated from red algal species of Japan. World
J. Microbiol. Biotechnol. 24: 2315–2321.

Kang, J.W. 1968. Illustrated encyclopedia of fauna and flora of Korea, Vol. 8—Marine algae.
Ministry of Education, Seoul, 465 pp.

Kanke, Y., Y. Itoi and M. Iwasaki. 1996. Effects of human diets of two different Japanese
populations on cancer incidence in rat hepatic drug metabolizing and antioxidant enzyme
systems. Nutrition and Cancer 26: 63–71.

Karen, B. 2014. *Sargassum* spp. Seaweeds. The Bermudian Magazine. Available online at:
http://www.conservation.bm/sargassum-seaweed/http://www.conservation.bm/
sargassum-seaweed/

Karthick, N. 2014. Biopotentials of Organic Fertilizers Produced from Three Different Seaweeds
on Soil Biota. Ph.D. Thesis, Manonmaniam Sundaranar University, Tirunelveli, India,
pp. 1–23.

Karthikeyan, R., S.T. Somasundaram, T. Manivasagam, T. Balasubramanian and P.
Anantharaman. 2010. Hepatoprotective activity of brown alga *Padina boergesenii* against
CCl4 induced oxidative damage in Wistar rats. Asian Pacific Journal of Tropical Medicine
696–701.

Kawaguchi, S., S. Shimada, H.W. Wang and M. Masuda. 2004. The new genus *Yonagunia*
Kawaguchi and Masuda (Halymeniaceae, Rhodophyta), based on *Y. tenuifolia* Kawaguchi
and Masuda sp. nov. from southern Japan and including *Y. formosana* (Okamura)
Kawaguchi and Masuda comb. nov. from southeast Asia. Journal of Phycology 40:
180–192.

Kazutosi, N. 2002. Seaweeds Kaiso: bountiful harvest from the seas. Sustenance for health and well being. Japan Seaweed Association, 106 pp.

Kellogg, J. and M.A. Lila. 2013. Chemical and *in vitro* assessment of Alaskan coastal vegetation antioxidant capacity. J. Agric. Food Chem. 61: 11025–11032.

Kelly, J., D.W. Freshwater, A.P. Orellana and C. Fernandez. 2014. *Lobophora variegata*. In Bocas Del Toro, Smithsonian Tropical Research Institute. Available online at: http://biogeodb. stri.si.edu/bocas_database/search/species/3406

Kendall, A., K. Kusic, E. Maloney and M. Williams. 2002. List of species to be discussed at the 2002 MMS Taxonomic Workshop. Available online at: http://www.marine.gov/ research/coresurveys/postelsia.htm

Kent, H.W. 1986. Treasury of Hawaiian Words in One Hundred and One Categories. University of Hawaii Press, Honolulu, Hawaii, 475 pp.

Khan, M.N.A., J.S. Choi, M.C. Lee, E. Kim, T.J. Nam, H. Fujii and Y.K. Hong. 2008. Anti-inflammatory activities of methanol extracts from various seaweed species. Journal of Environmental Biology 29(4): 465–469.

Khanavi, M., P.B. Toulabi, M.R. Abai, N. Sadati, F. Hadjiakhoondi, A. Hadjiakhoondi and H. Vatandoost. 2011. Larvicidal activity of marine algae, *Sargassum swartzii* and *Chondria dasyphylla*, against malaria vector *Anopheles stephensi*. J. Vector Borne Dis. 48(4): 241–244.

Kılınç, B., S. Cirik, G. Turan, H. Tekogul and E. Koru. 2013. Seaweeds for food and industrial applications. *In*: I. Muzzalupo (ed.). Food Industry. Available online at: http://www. intechopen.com/books/food-industry/seaweeds-for-food-and-industrial-applications

Kim, C.S. and H.J. Humm. 1965. The red alga, *Gracilaria foliifera*, with special reference to the cell wall polysaccharides. Bulletin of Marine Science 15: 1036–1050.

Kim, D.-W., K. Sapkota, J.-H. Choi, Y.-S. Kim, S. Kim and S.-J. Kim. 2013. Direct acting anti-thrombotic serine protease from brown seaweed *Costaria costata*. Process Biochemistry 48(2): 340–350.

Kim, I.H. and J.H. Lee. 2008. Antimicrobial activities against methicillin resistant *Staphylococcus aureus* from macroalgae. Journal of Industrial Engineering Chemistry 14: 568–572.

Kim, M.M., N. Rajapakse and S.K. Kim. 2009. Anti-inflammatory effect of *Ishige okamurae* ethanolic extract via inhibition of NF-kappaB transcription factor in RAW 264.7 cells. Phytother. Res. 23(5): 628–34.

Kim, S.-K. (ed.). 2011. Handbook of Marine Macroalgae: Biotechnology and Applied Phycology. Wiley, Weinheim, 592 pp.

Kim, S.Y., T.H. Seo, J.K. Park and G.H. Boo. 2012. *Cryptonemia rotunda* (Halymeniales) and *Schizymenia apoda* (Nemastomatales), two new records of red algae from Korea. Algae 27(1): 1–8.

Kim, Y., D. Kim, T. Kim, M.-K. Shin, Y.J. Kim, J.-J. Yoon and I.S. Chang. 2013. Use of red algae, Ceylon moss (*Gelidium amansii*), hydrolyzate for clostridial fermentation. Biomass and Bioenergy 56: 38–42.

Kim, Y.A., C.S. Kong, Y.R. Um, J.I. Lee, T.J. Nam and Y. Seo. 2008. Antioxidant efficacy of extracts from a variety of seaweeds in a cellular system. Ocean Science Journal 43: 31–37.

Kiple, K.F. and K.C. Ornelas (eds.). 2000. The Cambridge World History of Food, Vol. 2. Cambridge University Press, Cambridge, UK, 1131 pp.

Kipp, R.M., M. McCarthy and A. Fusaro. 2014. *Bangia atropurpurea*. USGS Nonindigenous Aquatic Species Database, Gainesville, FL. Available online at: http://nas.er.usgs.gov/ queries/GreatLakes/SpeciesInfo.asp?NoCache=2%2F28%2F2012+12%3A45%3A08+A M&SpeciesID=1700&State=&HUCNumber=4060200

Kirby, R.H. 1953. Seaweeds in Commerce. Her Majesty's Stationery Office, London, 157 pp.

Klinkenberg, B. 2014a. *Gracilariopsis andersonii*, E-Flora BC: Electronic Atlas of the Plants of British Columbia. Lab for Advanced Spatial Analysis, Department of Geography, University of British Columbia, Vancouver. Available at: http://linnet.geog.ubc.ca/ Atlas/Atlas.aspx?sciname=Gracilariopsis%20andersonii

Klinkenberg, B. 2014b. *Porphyra torta*, E-Flora BC - Electronic Atlas of the Plants of British Columbia. Lab for Advanced Spatial Analysis, Department of Geography, University of British Columbia, Vancouver. Available online at: http://linnet.geog.ubc.ca/Atlas/Atlas.aspx?sciname=Porphyra%20torta

Klinkenberg, B. 2014c. *Stephanocystis geminata*, E-Flora BC - Electronic Atlas of the Plants of British Columbia. Lab for Advanced Spatial Analysis, Department of Geography, University of British Columbia, Vancouver. Available online at: http://linnet.geog.ubc.ca/Atlas/Atlas.aspx?sciname=Stephanocystis%20geminata

Klinkenberg, B. 2014d. *Wildemania amplissima* (= *Porphyra cuneiformis*), E-Flora BC - Electronic Atlas of the Plants of British Columbia. Lab for Advanced Spatial Analysis, Department of Geography, University of British Columbia, Vancouver. Available online at: http://linnet.geog.ubc.ca/Atlas/Atlas.aspx?sciname=Porphyra%20cuneiformis&redblue=Both&lifeform=13

Klinkenberg, B. 2015. *Laminaria setchellii*, E-Flora BC - Electronic Atlas of the Plants of British Columbia. Lab for Advanced Spatial Analysis, Department of Geography, University of British Columbia, Vancouver. Available online at: http://linnet.geog.ubc.ca/Atlas/Atlas.aspx?sciname=Laminaria%20setchellii&redblue=Both&lifeform=13

Klinkenberg, B. 2016. *Pterosiphonia bipinnata*, E-Flora BC - Electronic Atlas of the Plants of British Columbia. Lab for Advanced Spatial Analysis, Department of Geography, University of British Columbia, Vancouver. Available online at:http://linnet.geog.ubc.ca/Atlas/Atlas.aspx?sciname=Pterosiphonia%20bipinnata

Kluijver, M.D., G. Gijswijt, R. de Leon and I. da Cunda. 2014. Sea Grapes (*Caulerpas racemosa*)—Interactive Guide to Caribbean Diving. Marine Species Identification Portal. Available online at: http://species-identification.org/species.php?species_group=caribbean_diving_guide&menuentry=soorten&id=489&tab=beschrijving

Koh, C.-H., Y. Kim and S.-G. Kang. 1993. Size distribution, growth and production of *Sargassum thunbergii* in an intertidal zone of Padori, west coast of Korea. Hydrobiologia 260-261: 207–214.

Kolb, N., L. Vallorani and V. Stocchi. 1999. Chemical composition and evaluation of protein quality by amino acid score method of edible brown marine algae Arame (*Eisenia bicyclis*) and Hijiki (*Hijikia fusiforme*). Acta Alimentaria 28: 213–222.

Kolb, N., L. Vallorani, N. Milanovi and V. Stocchi. 2004. Evaluation of marine algae Wakame (*Undaria pinnatifida*) and Kombu (*Laminaria digitata*, *L. japonica*) as food supplements. Food Technology and Biotechnology 42: 57–61.

Kraft, G.T. 1972. Preliminary studies of Philippine *Eucheuma* species (Rhodophyta) Part 1, taxonomy and ecology of *Eucheuma arnoldii* Weber-van Bosse. Pacific Science 26: 319–334.

Kraft, G.T. 2009. Algae of Australia: Marine Benthic Algae of Lord Howe Island and the Southern Great Barrier Reef, 2 - Brown Algae. CSIRO Publishing/Australian Biological Resources, Australia, 370 pp.

Kremb, S., M. Helfer, B. Kraus, H. Wolff, C. Wild et al. 2014. Aqueous extracts of the marine brown alga *Lobophora variegata* inhibit HIV-1 infection at the level of virus entry into Cell. PLoS ONE 9(8): e103895.

Kuda, T. and T. Ikemori. 2009. Minerals, polysaccharides and antioxidant properties of aqueous solutions obtained from macroalgal beach-casts in the Noto Peninsula, Ishikawa, Japan. Food Chemistry 112: 575–581.

Kuda, T., M. Tsunekawa, H. Goto and Y. Araki. 2005. Antioxidant properties of four edible algae harvested in the Noto Peninsula, Japan. Journal of Food Composition and Analysis 18: 625–633.

Kuhnlein, H.V. and N.J. Turner. 1996. Traditional Plant Foods of Canadian Indigenous Peoples—Nutrition, Botany and Uses. Gordon and Breach Publishers, Amsterdam, The Netherlands, 635 pp.

Kumar, M., V. Gupta, P. Kumari, C.R.K. Reddy and B. Jha. 2011a. Assessment of nutrient composition and antioxidant potential of Caulerpaceae seaweeds. Journal of Food Composition and Analysis 24: 270–278.

Kumar, M., P. Kumari, N. Trivedi, M.K. Shukla, V. Gupta, C.R.K. Reddy and B. Jha. 2011b. Minerals, PUFAs and antioxidant properties of some tropical seaweeds from Saurashtra coast of India. Journal of Applied Phycology 23(5): 797–810.

Kumar, S.A., A.K. Das, B. Nilima, A.K. Siddhanta, K.H. Mody, B.K. Ramavat, V.D. Chauhan, J.R. Vedasiromoni and D.K. Ganguly. 1994. A new sulfated polysaccharide with potent blood anti-coagulant activity from the red seaweed *Grateloupia Indica*. International Journal of Biological Macromolecules 16(5): 279–280.

Kumar, V.V. and P. Kaladharan. 2007. Source of Protein for Animal Feed. Journal of the Marine Biological Association of India 49(1): 35–40.

Kumari, P., M. Kumar, V. Gupta, C.R.K. Reddy and B. Jha. 2010. Tropical marine macroalgae as potential sources of nutritionally important PUFAs. Food Chemistry 120: 749–757.

Kylin, H. 1913. Biochemistry of sea algae. Zeitschrift für Physikalische Chemie 83: 171–197.

Lahaye, M. 1991. Marine algae as sources of fibers: determination of soluble and insoluble dietary fiber contents in some, sea vegetables. Journal of the Science of Food and Agriculture 54: 587–594.

Lahaye, M. 2001. Chemistry and physico-chemistry of phycocolloids. Cahiers de Biologie Marine 42(1-2): 137–157.

Lahaye, M. and J.F. Thibault. 1990. Chemical and physio-chemical properties of fibers from algal extraction by-products. pp. 68–72. *In*: D.A.T. Southgate, K. Waldron, I.T. Johnson and G.R. Fenwick (eds.). Dietary Fibre: Chemical and Biological Aspects. Royal Society of Chemistry, Cambridge.

Lahaye, M. and A. Robic. 2007. Structure and functional properties of Ulvan, a polysaccharide from green seaweeds. Biomacromolecules 8: 1765–1774.

Larsen, B. 1978. Fucoidan. pp. 152–156. *In*: J.A. Hellebust and J.S. Craigie (eds.). Handbook of Phycological Methods: Physiological and Biological Methods. Cambridge University Press, Cambridge.

Larsen, B., D.M.S.A. Salem, M.A.E. Sallam, M.M. Mishrikey and A.I. Beltagy. 2003. Characterization of the alginates from algae harvested at the Egyptian red sea coast. Carbohydrate Research 338: 2325–2336.

Lawson, J.S., D. Tran and W.D. Rawlinson. 2001. From bittner to barr: a viral, diet and hormone breast cancer aetiology hypothesis. Breast Cancer Research 3: 81–85.

Leal, D., B. Matsuhiro, M. Rossi and F. Caruso. 2008. FT-IR Spectra of alginic acid block fractions in three species of brown seaweeds. Carbohydrate Research 343: 308–316.

Lee, B. 2008. Seaweed—Potential as a Marine Vegetable and other Opportunities. Rural Industries Research and Development Corporation. Barton, Australia.

Lee, E.Y., I.K. Lee and H.G. Choi. 2003. Morphology and nuclear small-subunit rDNA sequences of *Ishige* (Ishigeaceae, Phaeophyceae) and its phylogenetic relationship among selected brown algal orders. Botanica Marina 46: 193–201.

Lee, J.W., B.-G. Oh and H.-B. Lee. 1996. Morphological variations of *Gloiopeltis furcata* (Postels et Ruprecht) J. Agardh (Rhodophyta) in the east coast of Korea. Algae (The Korean Journal of Phycology) 11: 91–94.

Lee, K.-Y. 1965. Some studies on the marine algae of Hong Kong. II. Rhodophyta. New Asia College Academic Annual 7: 63–110.

Lee, M.S., T. Shin, T. Utsuki, J.S. Choi, D.S. Byun and H.R. Kim. 2012. Isolation and identification of phlorotannins from *Ecklonia stolonifera* with antioxidant and hepatoprotective properties in tacrine-treated HepG2 cells. Journal of Agricultural and Food Chemistry 60(21): 5340–5349.

Lee, Y.-P. and J.-T. Yoon. 1998. Taxonomy and Morphology of *Undaria* (Alariaceae, Phaeophyta) in Korea. Algae 13: 427–446.

Lembi, C.A. and J.R. Waaland. 1988. Algae and Human Affairs. Cambridge University Press, Cambridge, 590 pp.

Leon-Alvarez, D., E. Serviere-Zaragoza and J. Gonzalez-Gonzalez. 1997. Crustose and erect phases of *Ahnjeltiopsis gigartinoides* from Mexico. Botanica Marina 40: 397–404.

Levring, T., H.A. Hoppe and O.J. Schmid. 1969. Marine algae. A survey of research and utilization. Cram, De Gruyter & Co., Hamburg 1–421.

Lewmanomont, K. 1978. Some edible algae of Thailand. The Kasetsart Journal 12(2): 119–129.

Lewmanomont, K. 1998. The seaweed resources of Thailand. pp. 62–69. *In*: A.T. Critchley, M. Ohno, D.B. Largo and R.D. Gillespie (eds.). Seaweed Resources of the World. Japan International Cooperation Agency, Yokosuka, Japan.

Leyte, P. 2015. Our edible seaweeds. Available online at: http://www.palompon-leyte.gov. ph/index.php?option=com_content&view=article&id=326&Itemid=346

Li, B., F. Lu, X. Wei and R. Zhao. 2008. Fucoidan: structure and bioactivity. Molecules 13: 1671–1695.

Li, J.-Y., Y. Agatsuma, T. Nagai, Y. Sato and K. Taniguchi. 2009. Differences in resource storage pattern between *Laminaria longissima* and *Laminaria diabolica* (Laminariaceae; Phaeophyta) reflecting their morphological characteristics. Journal of Applied Phycology 21: 215–224.

Li, L., R. Ni, Y. Shao and S. Mao. 2014. Carrageenan, and its applications in drug delivery. Carbohydrate Polymers 103: 1–11.

Lideman, L., G.N. Nishihara, T. Noro and R. Terada. 2011. *In vitro* growth and photosynthesis of three edible seaweeds, *Betaphycus gelatinus*, *Eucheuma serra* and *Meristotheca papulosa* (Solieriaceae, Rhodophyta). Aquaculture Science 59(4): 563–571.

Lim, P.E. and S.M. Phang. 2004. *Gracilaria* species (Gracilariales, Rhodophyta) of Malaysia, including two new records. Malaysian Journal of Science 23(2): 71–80.

Lin, S.-M. 2014a. *Gracilaria canaliculata*—BiotaTaiwanica, Algae of Taiwan. Available online at: http://algae.biota.biodiv.tw/pages/1066

Lin, S.-M. 2014b. *Hydropuntia eucheumatoides*—BiotaTaiwanica, Algae of Taiwan. Available online at: http://algae.biota.biodiv.tw/pages/1086

Lin, S.-M. 2015a. *Centroceras clavulatum*—BiotaTaiwanica, Algae of Taiwan. Available online at: http://algae.biota.biodiv.tw/pages/548

Lin, S.-M. 2015b. *Gracilaria gigas*—BiotaTaiwanica, Algae of Taiwan. Available online at: http://algae.biota.biodiv.tw/pages/1068

Lindeberg, M. and S.C. Lindstrom. 2014a. Seaweeds of Alaska—*Analipus japonicus*. Available online at: http://www.seaweedsofalaska.com/species.asp?SeaweedID=21

Lindeberg, M. and S.C. Lindstrom. 2014b. Seaweeds of Alaska—*Neorhodomela larix*. Available online at: http://www.seaweedsofalaska.com/species.asp?SeaweedID=95

Lindeberg, M. and S.C. Lindstrom. 2014c. Seaweeds of Alaska—*Odonthalia floccosa*. Available online at: http://www.seaweedsofalaska.com/species.asp?SeaweedID=97

Lindeberg, M. and S.C. Lindstrom. 2014d. Seaweeds of Alaska—*Pterosiphonia bipinnata*. Available online at: http://www.seaweedsofalaska.com/species.asp?SeaweedID=113

Lindeberg, M. and S.C. Lindstrom. 2014e. Seaweeds of Alaska—*Saundersella simplex*. Available online at: http://www.seaweedsofalaska.com/species.asp?SeaweedID=52

Lindeberg, M. and S.C. Lindstrom. 2014f. Seaweeds of Alaska—*Turnerella mertensiana*. Available online at: http://www.seaweedsofalaska.com/species.asp?SeaweedID=121

Lindeberg, M. and S.C. Lindstrom. 2015a. Seaweeds of Alaska—*Cymathaere triplicata*. Available online at: http://www.seaweedsofalaska.com/species.asp?SeaweedID=29

Lindeberg, M. and S.C. Lindstrom. 2015b. Seaweeds of Alaska—*Laminaria yezoensis*. Available online at: http://www.seaweedsofalaska.com/species.asp?SeaweedID=42

Lindeberg, M. and S.C. Lindstrom. 2015c. Seaweeds of Alaska—*Pleurophycus gardneri*. Available online at: http://www.seaweedsofalaska.com/species.asp?SeaweedID=238

Lindeberg, M. and S.C. Lindstrom. 2015d. Seaweeds of Alaska—*Polyneura latissima*. Available online at: http://www.seaweedsofalaska.com/species.asp?SeaweedID=260

Lindeberg, M. and S.C. Lindstrom. 2015e. Seaweeds of Alaska—*Porphyra nereocystis*. Available online at: http://www.seaweedsofalaska.com/species.asp?SeaweedID=177

Lindstrom, S.C. 1998. The seaweed resources of British Columbia, Canada. pp. 266–272. *In*: A.T. Critchley, M. Ohno, D.B. Largo and R.D. Gillespie (eds.). Seaweed Resources of the World. Kanagawa International Fisheries Training Centre. Japan International Cooperation Agency, Japan.

Lindstrom, S.C. 2006. Seaweed resources of British Columbia. *In*: A.T. Critchley, M. Ohno and D.B. Largo (eds.). World Seaweed Resources—An Authoritative Reference System. ETI Information Services Ltd. Hybrid Windows and Mac DVD-ROM; ISBN: 90-75000-80-4.

Lins, K.O., D.P. Bezerra, A.P. Alves, N.M. Alencar, M.W. Lima, V.M. Torres, W.R. Farias, C. Pessoa, M.O. de Moraes and L.V. Costa-Lotufo. 2009. Antitumor properties of a sulfated polysaccharide from the red seaweed *Champia feldmannii* (Diaz-Pifferer). Journal of Applied Toxicology 29: 20.

Lipkin, Y. and M. Friedlander. 2006. The seaweed resources of Israel and other eastern Mediterranean coun. *In*: A.T. Critchley, M. Ohno and D.B. Largo (eds.). World Seaweed Resources—An Authoritative Reference System. ETI Information Services Ltd. Hybrid Windows and Mac DVD-ROM; ISBN: 90-75000-80-4.

Littler, D.S. and M.M. Littler. 2015a. *Dictyota stolonifera*—Marine Plants of Pacific Panama. Available online at: http://biogeodb.stri.si.edu/pacificalgae/specie/46

Littler, D.S. and M.M. Littler. 2015b. *Padina crispata*—Marine Plants of Pacific Panama. Available online at: http://biogeodb.stri.si.edu/pacificalgae/specie/52

Lobban, C.S. and D.J. Chapman. 1988. Experimental Phycology: A Laboratory Manual. Phycological Society of America. Cambridge University Press, Cambridge, UK.

Lodeiro, P., M. Lopez-Garcia, L. Herrero, J.L. Barriada, R. Herrero, J. Cremades, I. Barbara and M.E. Sastre de Vicente. 2012. A physicochemical study of Al(+3) interactions with edible seaweed biomass in acidic waters. Journal of Food Science 77: 1750–3841.

López-López, I., S. Bastida, C. Ruiz-Capillas, L. Bravo, M.T. Larrea, F. Sánchez-Muniz, S. Cofrades and F. Jiménez-Colmenero. 2009. Composition and antioxidant capacity of low-salt meat emulsion model systems containing edible seaweeds. Meat Science 83: 492–498.

Lohrmann, N.L., B.A. Logan and A.S. Johnson. 2004. Seasonal acclimatization of antioxidants and photosynthesis in *Chondrus crispus* and *Mastocarpus stellatus*, two co-occurring red algae with differing stress tolerances. The Biological Bulletin 207: 225–232.

LTS. 2005. Health and Food Technology. Home Economics, Learning and Teaching Scotland, Scotland.

Lubke, R. and I. Moor. 1998. Field Guide to Eastern and Southern Cape Coasts. University of Cape Town Press, SA, 520 pp.

MacArtain, P., C.I.R. Gill, M. Brooks, R. Campbell and I.R. Rowland. 2007. Nutritional value of edible seaweeds. Nutrition Reviews 65(12): 535–543.

MacCaughey, V. 1918. Algae of the Hawaiian archipelago, II. The Botanical Gazette 65(2): 121–149.

Madhusudan, C., S. Manoj, K. Rhaul and C. Rishi. 2011. Seaweeds: a diet with nutritional, medicinal and industrial value. Research Journal of Medicinal Plant 5: 153–157.

Madlener, J.C. 1977. The Sea Vegetable Book: Foraging and Cooking Seaweeds. Clarkson N. Potter, Inc. Publishers, New York, 288 pp.

Mæhre, H.K., M.K. Malde, K.-E. Eilertsen and E.O. Elvevoll. 2014. Characterization of protein, lipid and mineral contents in common Norwegian seaweeds and evaluation of their potential as food and feed. Journal of the Science of Food and Agriculture 94: 3281–3290.

Mamatha, B.S., K.K. Namitha, A. Senthil, J. Smitha and G.A. Ravishankar. 2007. Studies on use of *Enteromorpha* in snack food. Food Chemistry 101: 1707–1713.

Magalhaes, K.D., L.S. Costa, G.P. Fidelis, R.M. Oliveira, L.T. Nobre, N. Dantas-Santos et al. 2011. Anticoagulant, antioxidant and antitumor activities of heterofucans from the seaweed *Dictyopteris delicatula*. International Journal of Molecular Sciences 12: 33–52.

Mandal, P., C.A. Pujol, M.J. Carlucci, K. Chattopadhyay, E.B. Damonte and B. Ray. 2008. Anti-herpetic activity of a sulfated xylomannan from *Scinaia hatei*. Phytochemistry 69: 2193–2199.

Maneveldt, G.M. and R. Frans. 2015. Common red seaweeds of the Cape Peninsula. Available online at: http://www.bcb.uwc.ac.za/staff/gavin/Pop_articles/Popart12.htm

Manivannan, K., G.K. Devi, G. Thirumaran and P. Anantharaman. 2008. Mineral composition of marine macroalgae from Mandapam coastal regions; southeast coast of India. American-Eurasian Journal of Botany 1(2): 58–67.

Manivannan, K., G.K. Devi, P. Anantharaman and T. Balasubramanian. 2011. Antimicrobial potential of selected brown seaweeds from Vedalai coastal waters, Gulf of Mannar. Asian Pacific Journal of Tropical Biomedicine 1(2): 114–120.

Marques, C.T., T.C.G. Azevedo, M.S. Nascimento, V.P. Medeiros, L.G. Alves, N.M.B. Benevides et al. 2012. Sulfated fucans extracted from algae *Padina gymnospora* have anti-inflammatory effect. Revista Brasileira de Farmacognosia 22: 115.

Martinez, B., F. Arenas, L. Duarte, C. Fernandez, M. Garcia-Tasende, J.M. Gorostiaga, C. Porteiro and M. Snson. 2014. Declive de los bosques de peninsulares y canarias. Quercus 345: 32–39.

Maruyama, H., K. Watanabe and I. Yamamoto. 1991. Effect of dietary kelp on lipid peroxidation and glutathione peroxidase activity in livers of rats given breast carcinogen DMBA. Nutrition and Cancer 15: 221–228.

Masterson, J. 2007. *Caulerpa brachypus*—Smithsonian Marine Station, available online at: http://www.sms.si.edu/irlspec/Caulerpa_brachypus.htm

Masuda, S. 1981a. Cochayuyo. Macha camaron y higos chargueados. Estudios Etnográficos del Perú Meridional, 173–192, Tokyo.

Masuda, M. 1981b. Taxonomic notes on *Odonthalia lyallii* (Harvey) J. Aagardh and related species (Rhodophyta). Journal of the Faculty of Science, Hokkaido University, Botany 12(3): 147–158.

Masuda, M. and K. Horiguchi. 1988. Additional notes on the life history of *Nemalion vermiculare* Suringar (Nemaliales, Rhodophyta). Japanese Journal of Phycology 36: 231–236.

Matanjun, P., S. Mohamed, N. Mustapha and K. Muhammad. 2009. Nutrient content of tropical edible seaweeds, *Eucheuma cottonii*, *Caulerpa lentillifera* and *Sargassum polycystum*. Journal of Applied Phycology 21: 75–80.

Mathew, S.S. 1991. Some Observations on the Ecology and Biochemical Aspects of the Seaweeds of Kerala Coast. Ph.D. Thesis, Centre of Advanced Studies in Mariculture, Central Marine Fisheries Research Institute, Kochi, India, 237 pp.

Matos, S. 2013. Designing food cultures: propagating the consumption of seaweed in the Azores Islands through recipes. Iridescent: Icograda Journal of Design Research 2(3). Available online at: http://iridescent.icograda.org/database/rte/files/Iridescent2_3_Matos.pdf

Matsuhiro, B., A.F. Conte, E.B. Damonte, A.A. Kolender, M.C. Matulewicz, E.G. Mejias, C.A. Pujol and E.A. Zuniga. 2005. Structural analysis and antiviral activity of a sulfated galactan from the red seaweed *Schizymenia binderi* (Gigartinales, Rhodophyta). Carbohydrate Research 340: 2392–2402.

Mayhew, E. 2002. *Gracilaria multipartita*. A red seaweed. Marine Life Information Network. Biology and Sensitivity Key Information Sub-programme. Plymouth, Marine Biological Association of the United Kingdom. Available online at: http://www.marlin.ac.uk/speciesinformation.php?speciesID=3417

McConnaughey, E. 2012. Sea Vegetables, Harvesting Guide. Naturegraph Publishers, California, 235 pp.

McDermid, K.J. and B. Stuercke. 2003. Nutritional composition of edible Hawaiian seaweeds. Journal of Applied Phycology 15: 513–524.

McDermid, K.J., B. Stuercke and O.J. Haleakala. 2005. Total dietary fiber content in Hawaiian marine algae. Botanica Marina 48(5-6): 437–440.

McHugh, D.J. 2003. A guide to the seaweed industry. FAO Fisheries Technical Paper 441: 1–105.

Mendes, G.d.S., I.C. Bravin, Y. Yoneshigue-Valentin, N.S. Yokoya and M.T.V. Romanos. 2012. Anti-HSV activity of *Hypnea musciformis* cultured with different phytohormones. Revista Brasileira de Farmacognosia 22: 789–794.

Merriwether. 2014. *Sargassum* Seaweed—Merriwether's Guide to Edible Wild Plants of Texas and the Southwest. Available online at: http://www.foragingtexas.com/2011/11/sargassum.html

Michanek, G. 1975. Seaweed Resources of the Ocean. FAO, Fisheries Technical Paper n. 138.

Miguel, T.B.A.R., E.C. Schmidt, Z.L. Bouzon, F.E.P. Nascimento, M. Da Cunha, S.F. Pireda, K.S. Nascimento, C.S. Nagano, S. Saker-Sampaio, B.S. Cavada, E.C. Miguel and A.H. Sampaio. 2014. Morphology, ultrastructure and immunocytochemistry of *Hypnea cervicornis* and

Hypnea musciformis-(Hypneaceae, Rhodophyta) from the coastal waters of Ceará, Brazil. Journal of Microscopy and Ultrastructure 2: 104–116.

Milchakova, N. 2011. Marine Plants of the Black Sea. An illustrated Field Guide. Digit Print Press, Sevastopol, Ukraine, 144 pp.

Milstein, D., A.S. Medeiros, E.C. Oliveira and M.C. Oliveira. 2015. Native or introduced? A re-evaluation of *Pyropia* species (Bangiales, Rhodophyta) from Brazil based on molecular analyses. European Journal of Phycology 50(1): 37–45.

Min, K.-H., H.-J. Kim, Y.-J. Jeon and J.-S. Han. 2011. *Ishige okamurae* ameliorates hyperglycemia and insulin resistance in C57BL/KsJ-db/db mice. Diabetes Research and Clinical Practice 93(1): 70–6.

Minghou, J. 1990. Processing and Extraction of Phycocolloids. Regional Workshop on the Culture and Utilization of Seaweeds, Volume II. FAO Technical Resource Papers. Cebu City, Philippines.

Mišurcová, L., S. Kráčmar, B. Klejdus and J. Vacek. 2010. Nitrogen content, dietary fiber, and digestibility in algal food products. Czech Journal of Food Sciences 28: 27–35.

Mišurcová, L., J. Ambrožová and D. Samek. 2011. Seaweed lipids as nutraceuticals. pp. 339–355. *In*: K. Se-Kwon (ed.). Marine Medicinal Foods Implications and Applications, Macro and Microalgae. Advances in Food and Nutrition Research. Academic Press, London.

Mitchell, K. 2000. Keith Michell's Practically Macrobiotic Cookbook. Healing Arts Press, Rochester, Vermont, 248 pp.

Modelo, R.B., I. Umezaki and L.M. Liao. 1998. Morphology of *Sargassum corderoi* (Phaeophyta, Fucales), a new seaweed species from central Philippines. Algae 13(1): 79–83.

Mohammadi, M., H. Tajik and P. Hajeb. 2013. Nutritional composition of seaweeds from the Northern Persian Gulf. Iranian Journal of Fisheries Sciences 12(1): 232–240.

Mole, M. and A. Sabale. 2013. Antioxidant Potential of Seaweeds from Kunakeshwar along the West Coast Maharashtra. Asian Journal of Biomedical and Pharmaceutical Sciences, North America. Available at: http://www.jbiopharm.com/index.php/ajbps/article/view/322

Molloy, F.J. 1990. Utilized and potentially utilizable seaweeds on the Namibian coast: biogeography and accessibility. Hydrobiologia 204/205: 293–299.

Monotilla, W. and M. Notoya. 2010. Growth and development of *Porphyra marcosii* (Bangiales, Rhodophyta) under different temperatures and photoperiod. Philippine Journal of Science 139(2): 197–206.

Mori, K., T. Ooi, M. Hiraoka, N. Oka, H. Hamada, M. Tamura and T. Kusumi. 2004. Fucoxanthin and its metabolites in edible brown algae cultivated in deep seawater. Marine Drugs 2(2): 63–72.

Morris, C., S. Bala, G.R. South, J. Lako, M. Lober and T. Simos. 2014. Supply chain and marketing of sea grapes, *Caulerpa racemosa* (Forsskål) J. Agardh (Chlorophyta: Caulerpaceae) in Fiji, Samoa and Tonga. Journal of Applied Phycology 26: 783–789.

Morrissey, J., S. Kraan and M.D. Guiry. 2001. A guide to commercially important seaweeds on the Irish coast. Bord Iascaigh Mhara/Irish Fisheries Board, Dublin, pp. 5–40.

Morton, B. and J. Morton. 1983. The Sea Shore Ecology of Hong Kong. Hong Kong University Press, Hong Kong, 366 pp.

Motshakeri, M., M. Ebrahimi, Y.M. Goh, P. Matanjun and S. Mohamed. 2013. *Sargassum polycystum* reduces hyperglycaemia, dyslipidaemia and oxidative stress via increasing insulin sensitivity in a rat model of type 2 diabetes. J. Sci. Food Agric. 93: 1772–1778.

Mouritsen, O.G. 2013. Seaweeds: Edible, Available, and Sustainable. University of Chicago Press, Chicago, USA, 304 pp.

Mouritsen, O.G., C. Dawczynski, L. Duelund, G. Jahreis, W. Vetter and M. Schröder. 2013. On the human consumption of the red seaweed dulse (*Palmaria palmata* (L.) Weber & Mohr. Journal of Applied Phycology 25: 1777–1791.

Mshigeni, K.E. and R. Jahn. 2014. *Eucheuma* (Rhodophyta, Gigartinales) in the W Indian Ocean region—notes on collections kept in Berlin-Dahlem and Hamburg. Available online at: http://www.bgbm.org/sites/default/files/documents/Vol%2B25%2Bp%2B399-407.pdf

Munda, I.M. 2006. The seaweed resources of the Adriatic Sea. *In*: A.T. Critchley, M. Ohno and D.B. Largo (eds.). World Seaweed Resources—An Authoritative Reference System. ETI Information Services Ltd. Hybrid Windows and Mac DVD-ROM; ISBN: 90-75000-80-4.

Munier, M., J. Dumay, M. Morançais, P. Jaouen and J. Fleurence. 2013. Variation in the biochemical composition of the edible seaweed *Grateloupia turuturu* Yamada harvested from two sampling sites on the Brittany coast (France): the influence of storage method on the extraction of the seaweed pigment r-phycoerythrin. Journal of Chemistry, Volume 2013, Article ID 568548, 8 pp. Available online at: http://www.hindawi.com/journals/jchem/2013/568548/

Murata, M. and J. Nakazoe. 2001. Production and use of marine algae in Japan. Japan Agricultural Research Quartery 35: 281–290.

N'Yeurt, A.D.R. 1999. A preliminary illustrated field guide to the common marine algae of the Cook Islands (Rarotonga and Aitutaki). World Wide Fund for Nature, 74 pp.

N'Yeurt, A.D.R. 2002. A Revision of *Amansia glomerata* C. Agardh, *Amansia rhodantha* (Harvey) J. Agardh and *Melanamansia glomerata* (C. Agardh) R.E. Norris (Rhodophyta: Rhodomelaceae). Botanica Marina 45: 231–242.

Nagayama, K., Y. Iwamura, T. Shibata, I. Hirayama and T. Nakamura. 2002. Bactericidal activity of phlorotannins from the brown alga *Ecklonia kurome*. Journal of Antimicrobial Chemotherapy 50: 889–89.

Nagayama, K., T. Shibata, K. Fujimoto, T. Honjo and T. Nakamura. 2003. Algicidal effect of phlorotannins from the brown alga *Ecklonia kurome* on red tide microalgae. Aquaculture 218: 601–611.

Naidu, K.A., A. Tewari, H.V. Joshi, S. Viswanath, H.P. Ramesh and S.V. Rao. 1993. Evaluation of nutritional quality and food safety of seaweeds of India. J. Food Saf. 13(2): 77–90.

Nakahara, H. and Y. Nakamura. 1973. Parthenogenesis, apogamy and apospory in *Alaria crassifolia* (Laminariales). Marine Biology 18: 327–332.

Nakamura, T., K. Nagayama, K. Uchida and R. Tanaka. 1996. Antioxidant activity of phlorotannins isolated from the brown alga *Eisenia bicyclis*. Fisheries Science 62: 923–926.

Nam, B., H. Jin, S. Kim and Y. Hong. 1998. Quantitative viability of seaweed tissues assessed with 2,3,5triphenyltetrazolium chloride. Journal of Applied Phycology 10: 31–36.

Namikawa, S. 1906. Fresh water algae as an article of human food. Bulletin of the College of Agricultural. Tokyo Imperial University 7: 123–124.

Nang, H.Q. 2006. The seaweed resources of Vietnam. *In*: A.T. Critchley, M. Ohno and D.B. Largo (eds.). World Seaweed Resources—An Authoritative Reference System. ETI Information Services Ltd. Hybrid Windows and Mac DVD-ROM; ISBN: 90-75000-80-4.

Nang, H.Q. and N.H. Dinh. 1998. The seaweed resources of Vietnam. pp. 62–69. *In*: A.T. Critchley, M. Ohno, D.B. Largo and R.D. Gillespie (eds.). Seaweed Resources of the World. Japan International Cooperation Agency, Yokosuka, Japan.

Naoko, K.N., Y. Egashira and H. Sanada. 2007. Effects of various kinds of edible seaweeds in diets on the development of D-Galactosamine-induced hepatopathy in rats. Journal of Nutritional Science and Vitaminology 53(4): 315–323.

Narasimman, S. and K. Murugaiyan. 2013. Proximate and elemental composition of *Amphiroa fragilissima* (Linnaeus) Lamouroux from Tharuvaikulam coast, Gulf of Mannar region, south-east coast of Tamil Nadu. International Journal of Research in Botany 3(3): 44–47.

Naylor, M. 2006. Factsheet on *Gracilaria vermiculophylla*. Available online at: http://www.frammandearter.se/0/2english/pdf/Gracilaria_vermiculophylla.pdf

Neeta, V.M.J. and S. Srisudha. 2012. Biochemical characterization, haemagglutinating activity and cytotoxic activity of *Padina gymnospora* (Kutzing) Sonder. International Journal of Biological and Pharmaceutical Research 3(8): 956–961.

NETALGAE. 2012. Seaweed industry in Europe. Available online at: http://www.netalgae.eu/uploadedfiles/Filieres_12p_UK.pdf

Neto, A.I., I. Titley and P.M. Raposeiro. 2005. Rocky Shore Marine Flora of the Azores. Secrataria Regional do Ambiente e do Mar, Azores, Portugal, 159 pp.

Nishino, T., C. Nishioka, H. Ura and T. Nagumo. 1994. Isolation and partial characterization of a novel amino sugar-containing fucan sulfate from commercial *Fucus vesiculosus* fucoidan. Carbohydr. Res. 255: 213–224.

Nisizawa, I., H. Noda, R. Kikuchi and T. Watanabe. 1987. The main seaweed foods in Japan. Hydrobiologia 151(1): 5–29.

Nisizawa, K. 2006. Seaweeds kaiso—bountiful harvest from the seas. *In*: A.T. Critchley, M. Ohno and D. Largo (eds.). World Seaweed Resources—An Authoritative Reference System: ETI Information Services Ltd. Hybrid Windows and Mac DVD-ROM; ISBN: 90-75000-80-4.

Nizamuddin, M. 1963. Studies on the green alga, *Udotea indica*. Pacific Science XVII: 243–245.

Noda, H. 1993. Health benefits and nutritional properties of nori. Journal of Applied Phycology 5: 255–258.

Norziah, M.H. and C.Y. Ching. 2000. Nutritional composition of edible seaweed *Gracilaria changgi*. Food Chemistry 68: 69–76.

Novaczek, I. 2001. A Guide to the Common and Edible and Medicinal Sea Plants of the Pacific Islands. University of the South Pacific, 40 pp.

Novaczek, I. and A. Athy. 2001. Sea Vegetable Recipes for the Pacific Islands. USP Marine Studies Programme/SPC Coastal Fisheries Programme—Training Materials for Pacific Community Fisheries.

Nussinovitch, A. 1997. Hydrocolloid Applications: Gum Technology in the Food and other Industries. Chapman and Hall, London, 354 pp.

N'Yeurt, A.D.R. 1995. *Meristotheca procumbens* P. Gabrielson et kraft (Gigartinales, Solieriaceae): an edible seaweed from Rotuma Island. The South Pacific Journal of Natural Sciences 14: 243–250.

Oceanharvest. 2015. Seaweed: a rich source of vitamins and bioactive compounds. Ocean Harvest Technology. Available online at: https://oceanharvesttechnology.wordpress.com/2013/02/25/seaweed-a-rich-source-of-vitamins-and-bioactive-compounds/

O'Clair, R. and S. Lindstrom. 2000. North Pacific Seaweeds. Alaska Sea Grant College Program, 162 pp.

Oh, Y.S., I.K. Lee and S.M. Boo. 1990. An annotated account of Korean economic seaweeds for food, medical and industrial uses. The Korean Journal of Phycology 5(1): 57–71.

Ohni, H. 1968. Edible seaweeds in Chile. Japanese Society of Physiology Bulletin 16: 52–54.

Ohno, M. and D.B. Largo. 1998. The seaweeds resources of Japan. pp. 1–14. *In*: A.T. Critchley, M. Ohno, D.B. Largo and R.D. Gillespie (eds.). Seaweed Resources of the World. Japan International Cooperation Agency, Yokosuka, Japan.

Ohno, M. and D.B. Largo. 2006. The seaweed resources of Japan. *In*: A.T. Critchley, M. Ohno and D.B. Largo (eds.). World Seaweed Resources—An Authoritative Reference System. ETI Information Services Ltd. Hybrid Windows and Mac DVD-ROM; ISBN: 90-75000-80-4.

OHSVC. 2015. *Laminaria dentigera*—Ocean Harvest Sea Vegetable Company. Available online at: http://www.ohsv.net/selection.html

Okaih, Y., K. Higashi-Okaia, Y. Yanob and S. Otanib. 1996. Identification of antimutagenic substances in an extract of edible red alga, *Porphyra tenera* (Asadusa-nori). Cancer Letters 100: 235–240.

Okamura, K. 1915. *Undaria* and its species. Botanical Magazine, Tokyo 29: 266–278.

Okazaki, A. 1971. Seaweeds and their uses in Japan. Tokai Univ. Press, 165 pp.

Okuzumi, J., T. Takahashi, T. Yamane, Y. Kitao, M. Inagake, K. Ohya, H. Nishino and Y. Tanaka. 1993. Inhibitory effects of fucoxanthin, a natural carotenoid, on N-ethyl-N'-nitro-N nitrosoguanidine induced mouse duodenal carcinogenesis. Cancer Letters 68: 159–168.

Oliveira, E.C. 2006. Seaweed resources of Brazil. *In*: A.T. Critchley, M. Ohno and D.B. Largo (eds.). World Seaweed Resources—An Authoritative Reference System. ETI Information Services Ltd. Hybrid Windows and Mac DVD-ROM; ISBN: 90-75000-80-4.

Oliveira, E.C., C.J. Bird and J. McLachlan. 1983. The genus *Gracilaria* (Rhodophyta, Gigartinales) in the western Atlantic. *Gracilaria domingensis*, *G. cervicornis*, and *G. ferox*. Canadian Journal of Botany 61(12): 2999–3008.

Omar, H.H., H.M. Shiekh, N.M. Gumgumjee, M.M. El-Kazan and A.M. El-Gendy. 2012a. Antibacterial activity of extracts of marine algae from the red sea of Jeddah, Saudi Arabia. African Journal of Biotechnology 11(71): 13576–13585.

Omar, H.H., N.M. Gumgumji, H.M. Shiek, M.M. El-Kazan and A.M. El-Gendy. 2012b. Inhibition of the development of pathogenic fungi by extracts of some marine algae from the red sea of Jeddah, Saudi Arabia. African Journal of Biotechnology 11(72): 13697–13704.

Oneill, A.N. 1955a. 3,6-anhydro-D-galactose as a constituent of κ-carrageenin. Journal of the American Chemical Society 77(10): 2837–2839.

Oneill, A.N. 1955b. Derivatives of 4-O-beta-D-galactopyranosyl-3,6-anhydro-D-galactose from κ-carrageenin. Journal of the American Chemical Society 77(23): 6324–6326.

Onsoyen, E. 1997. Alginates. pp. 22–44. In: A. Imeson (ed.). Thickening and Gelling Agents for Food. Blackie Academic and Professional, London, UK.

Ortega, M.M., J.L. Godinez and G.G. Solórzano. 2001. Catálogo de algas bénticas de las costas Mexicanas del Golfo de México y Mar Caribe. Cuadernos 34. Instituo de Biologia, Universidad Nacional Autonoma de México, Tlalnepantla, Mexico, 600 pp.

Ortiz, J., E. Uquiche, P. Robert, N. Romero, V. Quitral and C. Llantén. 2009. Functional and nutritional value of the Chilean seaweeds Codium fragile, Gracilaria chilensis and Macrocystis pyrifera. European Journal of Lipid Science and Technology 111: 320–327.

Osman, M.E.H., A.M. Abushady and M.E. Elshobary. 2010. In vitro screening of antimicrobial activity of extracts of some macroalgae collected from Abu-Qir bay Alexandria. Egypt, Afr. J. Biotechnol. 9(42): 7203–7208.

Ostermann, H. 1938. Knud Rasmussen's posthumous notes on the life and doings of cast Greenlanders in olden times. Meddelelser Om Gronland, 109.

Ostraff, M. 2003. Contemporary uses of Limu (marine algae) in the Vava'u Island Group, Kingdom of Tonga: An Ethnobotanical Study. Ph.D. Thesis, University of Victoria, Canada, 261 pp.

Özvarol, Y., O.O. Ertan and I.I. Turna. 2010. Determination macrobentic flora of Northeastern Mediterranean. E-Journal of New World Sciences Academy 5(2): 5A0035. Available online at: http://www.newwsa.com/download/gecici_makale_dosyalari/NWSA-2998-1-6.pdf

Padula, M. and S. Boiteux. 1999. Photodynamic DNA damage induces by phycocyanin and its repair in Saccharomyces cerevisiae. Brazilian Journal of Medical and Biological Research 32: 1063–1071.

Panayotova, V. and M. Stancheva. 2013. Mineral composition of marine macroalgae from the bulgarian black sea coast. Scripta Scientifica Medica 45(6): 42–45.

Papenfuss, G.F. 1947. Generic names of algae proposed for conservation. I. Madroño 9: 8–17.

Parente, M.I., A.I. Neto and R.L. Fletcher. 2003. Morphology and life history studies of Endarachne binghamiae (Scytosiphonaceae, Phaeophyta) from the Azores. Aquatic Botany 76: 109–116.

Parish, E., R. Kennison and K. Goodman. 2014. Dictyota menstrualis. In Bocas Del Toro, Smithsonian Tropical Research Institute. Available online at: http://biogeodb.stri.si.edu/bocas_database/search/species/6646

Parker, B. and M.D. Guiry. 2014a. Arthrothamnus—Seaweed Africa website and database. Available online at: http://www.seaweedafrica.org/search/genus/detail/?genus_id=41248

Parker, B. and M.D. Guiry. 2014b. Callophyllis—Seaweed Africa website and database. Available online at: http://www.seaweedafrica.org/search/genus/detail/?genus_id=115

Parker, B. and M.D. Guiry. 2014c. Papenfussiella—Seaweed Africa website and database. Available online at: http://www.seaweedafrica.org/search/genus/detail/?genus_id=33397

Patarra, R.F., L. Paiva, A.I. Neto, E. Lima and J. Baptista. 2011. Nutritional value of selected macroalgae. Journal of Applied Phycology 23: 205–208.

Pattama, R.A. and A. Chirapart. 2006. Nutritional evaluation of tropical green seaweeds Caulerpa lentillifera and Ulva reticulata. Kasetsart Journal - Natural Science 40: 75–83.

Paula, D. 2014a. *Gracilaria arcuata*—Seaweeds of the central west coast of India. National Institute of Oceanography, Goa. Available online at: http://www.niobioinformatics.in/seaweed/system_Gracilaria%20arcuata.htm

Paula, D. 2014b. *Grateloupia indica*—Seaweeds of the central west coast of India. National Institute of Oceanography, Goa. Available online at: http://www.niobioinformatics.in/seaweed/system_Grateloupia%20indica.htm

Paula, D. 2014c. *Hypnea pannosa*—Seaweeds of the central west coast of India. National Institute of Oceanography, Goa. Available online at: http://www.niobioinformatics.in/seaweed/system_Hypnea%20pannosa.htm

Paula, D. 2014d. *Padina tetrastromatica*—Seaweeds of the central west coast of India. National Institute of Oceanography, Goa. Available online at: http://www.niobioinformatics.in/seaweed/system_Padina%20tetrastromatica.htm

Paula, D. 2014e. *Sargassum polycystum*—Seaweeds of the central west coast of India. National Institute of Oceanography, Goa. Available online at: http://www.niobioinformatics.in/seaweed/system_Sargassum%20polysystem.htm

Paula, D. 2015. *Gracilaria foliifera*—Seaweeds of the central west coast of India. National Institute of Oceanography, Goa. Available online at: http://www.niobioinformatics.in/seaweed/system_Gracilaria%20foliifera.htm

Payri, C., A.D.R. N'Yeurt and J. Orempuller. 2000. Algae of French Polynesia. Algues de Polynésie Française. Au Vent des Iles Editions, Tahiti, 320 pp.

Pedroche, F.F. and A.A. Ortiz. 1996. Aspectos morfológicos vegetativos y reproductivos de *Dermonema* (Rhodophyceae: Liagoraceae) en México. Acta Botánica Mexicana 34: 63–80.

Peng, J., J.P. Yuan, C.F. Wu and J.H. Wang. 2011. Fucoxanthin, a marine carotenoid present in brown seaweeds and diatoms: metabolism and bioactivities relevant to human health. Mar. Drugs 9(10): 1806–1828.

Peng, Y., E. Xie, K. Zheng, M. Fredimoses, X. Yang, X. Zhou, Y. Wang, B. Yang, X. Lin, J. Liu and Y. Liu. 2013. Nutritional and chemical composition and antiviral activity of cultivated seaweed *Sargassum naozhouense* Tseng et Lu. Marine Drugs 11: 20–32.

Pengzhan, Y., P.Z. Quanbin, L. Ning, X. Zuhong, W. Yanmei and L. Zhi'en. 2003. Polysaccharides from *Ulva pertusa* (Chlorophyta) and preliminary studies on their antihyperlipidemia activity. Journal of Applied Phycology 15: 21–27.

Pereira, H., L. Barreira, F. Figueiredo, L. Custodio, C. Vizetto-Duarte, C. Polo, E. Resek, A. Engelen and J. Varela. 2012. Polyunsaturated fatty acids of marine macroalgae: potential for nutritional and pharmaceutical applications. Marine Drugs 10(9): 1920–1935.

Pereira, H.S., L.R. Leão-Ferreira, N. Moussatché, V.L. Teixeira, D.N. Cavalcanti, L.J. Costa, R. Diaz and I.C. Frugulhetti. 2004. Antiviral activity of diterpenes isolated from the Brazilian marine alga *Dictyota menstrualis* against human immunodeficiency virus type 1 (HIV-1). Antiviral Research 64: 69–76.

Pereira, L. 2004. Estudos em macroalgas carragenófitas (Gigartinales, Rhodophyceae) da costa portuguesa—aspectos ecológicos, bioquímicos e citológicos. Ph.D. Thesis, Coimbra, Universidade de Coimbra, Portugal, pp. 293.

Pereira, L. 2008. As algas marinhas e respectivas utilidades. Monografias. Available online at: http://br.monografias.com/trabalhos913/algas-marinhas-utilidades/algas-marinhas-utilidades.pdf

Pereira, L. 2009. Guia ilustrado das macroalgas—conhecer e reconhecer algumas espécies da flora Portuguesa. Imprensa da Universidade de Coimbra, Coimbra, Portugal, 90 pp.

Pereira, L. 2010a. Littoral of Viana do Castelo—ALGAE (Bilingual). Câmara Municipal de Viana do Castelo, Portugal, 68 pp.

Pereira, L. 2010b. Littoral of Viana do Castelo—ALGAE (Bilingual). Uses in Agriculture, Gastronomy and Food Industry (Bilingual). Câmara Municipal de Viana do Castelo, VC, Portugal, 68 pp.

Pereira, L. 2010c. Algas à mesa. Brochura do Workshop sobre algas comestíveis, Litoral Norte, Polis Litoral, Viana do castelo, Portugal, 16 pp. Available online at: http://macoi.ci.uc.pt/include/downloadContentDoc.php?id=24

Pereira, L. 2011. Chapter 2—A review of the nutrient composition of selected edible seaweeds. pp. 15–47. *In*: Vitor H. Pomin (ed.). Seaweed: Ecology, Nutrient Composition and Medicinal Uses. Nova Science Publishers Inc., New York.

Pereira, L. 2012. Algas. Um alimento equilibrado. pp. 72–73. *In*: P. Ramos Nogueira (ed.). Conversas com Ciência—Tertúlias FNACiências, Escola de Ciências da Universidade do Minho, Braga, Portugal.

Pereira, L. 2013. Population studies and carrageenan properties in eight Gigartinales (Rhodophyta) from western coast of Portugal. The Scientific World Journal 2013: 11 pp. Available online at: http://www.hindawi.com/journals/tswj/2013/939830/

Pereira, L. 2014a. *Amphiroa cryptarthrodia*—Portuguese Seaweeds Website, available online at: http://macoi.ci.uc.pt/spec_list_detail.php?spec_id=502

Pereira, L. 2014b. *Asparagopsis taxiformis*—Portuguese Seaweeds Website, available online at: http://macoi.ci.uc.pt/spec_list_detail.php?spec_id=434

Pereira, L. 2014c. *Canistrocarpus cervicornis*—Portuguese Seaweeds Website, available online at: http://macoi.ci.uc.pt/spec_list_detail.php?spec_id=300

Pereira, L. 2014d. *Corallina officinalis*—Portuguese Seaweeds Website, available online at: http://macoi.ci.uc.pt/spec_list_detail.php?spec_id=51

Pereira, L. 2014e. *Dictyota dichotoma* var. *intricata*—Portuguese Seaweeds Website, available online at: http://macoi.ci.uc.pt/spec_list_detail.php?spec_id=258

Pereira, L. 2014f. *Halopteris scoparia*—Portuguese Seaweeds Website. Available online at: http://macoi.ci.uc.pt/spec_list_detail.php?spec_id=164

Pereira, L. 2014g. *Laurencia obtusa*—Portuguese Seaweeds Website, available online at: http://macoi.ci.uc.pt/spec_list_detail.php?spec_id=110

Pereira, L. 2014h. *Ulva linza*—Portuguese Seaweeds Website, available online at: http://macoi.ci.uc.pt/spec_list_detail.php?spec_id=204

Pereira, L. 2015a. Seaweed flora of the European north Atlantic and Mediterranean. pp. 65–178. *In*: Se-Kwon Kim (ed.). Handbook of Marine Biotechnology, Part A. Springer, London.

Pereira, L. 2015b. *Fucus vesiculosus*—Portuguese Seaweeds Website, available online at: http://macoi.ci.uc.pt/spec_list_detail.php?spec_id=78

Pereira, L. 2015c. *Codium adhaerens*—Portuguese Seaweeds Website, available online at: http://macoi.ci.uc.pt/spec_list_detail.php?spec_id=44

Pereira, L. and F. van de Velde. 2011. Portuguese carrageenophytes: carrageenan composition and geographic distribution of eight species (Gigartinales, Rhodophyta). Carbohydrate Polymers 84: 614–623.

Pereira, L. and P.J. Ribeiro-Claro. 2014. Analysis by vibrational spectroscopy of seaweed with potential use in food, pharmaceutical and cosmetic industries. pp. 225–247. *In*: Leonel Pereira and João M. Neto (eds.). Marine Algae—Biodiversity, Taxonomy, Environmental Assessment and Biotechnology. Science Publishers, an Imprint of CRC Press/Taylor & Francis Group, Boca Raton, FL.

Pereira, L. and F. Correia. 2015. Algas marinhas da costa Portuguesa—ecologia, biodiversidade e utilizações. Nota de Rodapé Edições, Paris, 340 pp.

Pereira, L., A. Sousa, H. Coelho, A.M. Amado and P.J.A. Ribeiro-Claro. 2003. Use of FTIR, FT-Raman and ¹³C-NMR spectroscopy for identification of some seaweed phycocolloids. Biomolecular Engineering 20: 223–228.

Pereira, L., A.M. Amado, A.T. Critchley, F. van de Velde and P.J.A. Ribeiro-Claro. 2009a. Identification of selected seaweed polysaccharides (phycocolloids) by vibrational spectroscopy (FTIR-ATR and FT-Raman). Food Hydrocolloids 23: 1903–1909.

Pereira, L., A.T. Critchley, A.M. Amado and P.J.A. Ribeiro-Claro. 2009b. A comparative analysis of phycocolloids produced by underutilized *versus* industrially utilized carrageenophytes (Gigartinales, Rhodophyta). Journal of Applied Phycology 21: 599–605.

Pereira, L., F.S. Gheda and P.J.A. Ribeiro-Claro. 2013. Analysis by vibrational spectroscopy of seaweed polysaccharides with potential use in food, pharmaceutical and cosmetic industries. International Journal of Carbohydrate Chemistry Volume 2013, Article ID 537202, 7 pp. Available online at: http://www.hindawi.com/journals/ijcc/2013/537202/

Pereira, M.G., N.M.B. Benevides, M.R.S. Melo, A.P. Valente, F.R. Melo and P.A.S. Mourão. 2005. Structure and anticoagulant activity of a sulfated galactan from the red alga, *Gelidium crinale*. Is there a specific structural requirement for the anticoagulant action? Carbohydrate Research 340: 2015–2023.

Pereira, S.M.B. and Y. Ugadim. 1979. *Champia feldmannii* Diaz-Piferrer and *Spermothamnion gymnocarpum* Howe, Duas Novas Referências para o Litoral Brasileiro. Boletim de Botânica da Universidade de São Paulo 7: 39–42.

Pereira-Pacheco, F., D. Robledo, L. Rodriguez-Carvajal and Y. Freile-Pelegrin. 2007. Optimization of native agar extraction from *Hydropuntia cornea* from Yucatan, Mexico. Bioresource Technology 98(6): 1278–1284.

Piazza, V., V. Roussis, F. Garaventa, G. Greco, V. Smyrniotopoulos, C. Vagias and M. Faimali. 2011. Terpenes from the red alga *Sphaerococcus coronopifolius* inhibit the settlement of barnacles. Marine Biotechnology 13: 764–772.

Pizzolla, P. 2003. *Catenella caespitosa*. A Red Seaweed. Marine Life Information Network: Biology and Sensitivity Key Information Sub-programme. Plymouth, Marine Biological Association of the United Kingdom. Available online at: http://www.marlin.ac.uk/speciesinformation.php?speciesID=2904

Polo, J.A. 1977. Nombres vulgares y usos de las algas en el Perú. Serie de Divulgación, Universidad Nacional Mayor de San Marcos, Museo de Historia Natural Javier Prado, Departamento de Botánica, Lima, 7.

Prabhasankar, P., P. Ganesan and N. Bhaskar. 2009. Influence of Indian brown seaweed (*Sargassum marginatum*) as an ingredient on quality, biofunctional, and microstructure characteristics of pasta. Food Science and Technology International 15: 471–479.

Preskitt, L. 2002. Edible Limu... Gifts from the Sea. Botany Department, University of Hawaii. Available online at: http://www.hawaii.edu/reefalgae/publications/ediblelimu/

Pujol, C.A., S. Ray, B. Ray and E.B. Damonte. 2012. Antiviral activity against dengue virus of diverse classes of algal sulfated polysaccharides. Int. J. Biol. Macromol. 51: 412–416.

Pushpamali, W.A., C. Nikapitiya, M.D. Zoysa, I. Whang, S.J. Kim and J. Lee. 2008. Isolation and purification of an anticoagulant from fermented red seaweed *Lomentaria catenata*. Carbohydrate Polymers 73: 274–279.

Quirós, A.R.B., C. Ron, J. López-Hernández and M.A. Lage-Yusty. 2004. Determination of folates in seaweeds by high-performance liquid chromatography. Journal of Chromatography A 1032: 135–139.

Radulovich, R., S. Umanzor and R. Cabrera. 2013. Algas marinas: cultivo y uso como alimento humano. University of Costa Rica, San Jose. Available online at: http://www.maricultura.net/wp-content/uploads/file/ALGAS%20TROPICALES.pdf

Radulovich, R., S. Umanzor, R. Cabrera and R. Mata. 2015. Tropical seaweeds for human food, their cultivation and its effect on biodiversity enrichment. Aquaculture 436: 40–46.

Rahila, N., S.P. Ahmed and I. Azhar. 2010. Pharmacological Activities of *Hypnea musciformis*. African Journal of Biomedical Research 13(1): 69–74.

Rahman, S. 2002. Molluscan Fauna of Intertidal Rocky Ledges of Karachi—A Comparative Ecological Study. Ph.D. Thesis, Department of Zoology, University of Karachi, Pakistan, 636 pp. Available online at: http://prr.hec.gov.pk/Chapters/10-8.pdf

Rajamani, K., V.C. Renju, S. Sethupathy and S.S. Thirugnanasambandan. 2014. Ameliorative effect of polyphenols from *Padina boergesenii* against ferric nitrilotriacetate induced renal oxidative damage: with inhibition of oxidative hemolysis and *in vitro* free radicals. Environmental Toxicology 24: 1–12.

Rao, G.M.N., G.S. Rangaiah and S.V.V.S.N. Dora. 2008. Spore shedding in *Catenella impudica* from the Godavari estuary at Bhiravapalem, India. Algae 23(1): 71–74.

Rao, K.R. 1977. Species of *Gracilaria* and *Hypnea* as potential sources of agar. Seaweed Res. Util. 2: 95–101.

Rao, M.U. 1970. The economic seaweeds of India. Bulletin of the Central Marine Fisheries Research Institute 20: 93 pp.

Rao, P.V.S., V.A. Mantri, K. Ganesan and K.S. Kumar. 2007. Seaweeds as a human diet: an emerging trend in the new millennium. pp. 85–96. *In*: R.K. Gupta and V.D. Pandey (eds.). Advances in Applied Phycology. Daya Publishing House, New Delhi.

Rao, P.V.S., K. Ganesan and K.S. Kumar. 2010. Application of seaweeds as food: a scenario. pp. 77–92. *In*: K.M. Das (ed.). Algal Biotechnology. New Vistas, Daya Publishing House, New Delhi.

Rath, J. and S.P. Adhikary. 2006. Marine Macro-algae of Orissa, East Coast of India. Algae 21(1): 49–59.

Ravikumar, S., L. Anburajan, G. Ramanathan and N. Kaliaperumal. 2002. Screening of seaweed extracts against antibiotic resistant post operative infectious pathogens. Seaweed Research and Utilisation 24(1): 95–99.

Rayment, W. 2004. *Ahnfeltia plicata*. A red seaweed. Marine Life Information Network: Biology and Sensitivity Key Information Sub-programme, Marine Biological Association of the United Kingdom, available online at: http://www.marlin.ac.uk/speciesfullreview.php?speciesID=2417

Redmond, S., L. Green, C. Yarish, J. Kim and C. Neefus. 2014. New England seaweed culture handbook—Nursery systems. Connecticut Sea Grant CTSG-14-01, 92 pp. Available online at: http://seagrant.uconn.edu/publications/aquaculture/handbook.pdf

Reed, M. 1906. The Economic Seaweeds of Hawaii and their Food Value. Hawaii Agricultural Experiment Station, Honolulu, Hawaii, 88 pp.

Remirez, D., A. Gonzalez, N. Merino, R. Gonzalez, O. Ancheta, C. Romay and S. Rodriguez. 1999. Effect of phycocyanin in Zymosan-induced arthritis in mice-phycocyanin as an antiarthritic compound. Drug Development Research 48: 70–75.

Rhimou, B., H. Camille, C. Jean-Claude, C.-J. Sylvia, S. Corinne, S. Gaelle, C. Stephane, R. Hassane and B. Nathalie. 2011. Antiviral activities of sulfated polysaccharides isolated from *Sphaerococcus coronopifolius* (Rhodophytha, Gigartinales) and *Boergeseniella thuyoides* (Rhodophyta, Ceramiales). Marine Drugs 9(7): 1187–1209.

Rhimou, B., R. Hassane and B. Nathalie. 2013. Antioxidant activity of Rhodophyceae extracts from Atlantic and Mediterranean coasts of Morocco. African Journal of Plant Science 7(3): 110–117.

Rhoads, S.A. and P. Zunic. 1978. Cooking with Sea Vegetables. Autumn Press Ltd., Massachusetts, 136 pp.

Ribier, J. and J.C. Godineau. 1984. Les algues. La Maison Rustique, Flammarion, France, pp. 15–26.

Riley, K. 2005. *Rhodothamniella floridula*. A red seaweed. Marine Life Information Network: Biology and Sensitivity Key Information Sub-programme. Plymouth, Marine Biological Association of the United Kingdom. Available from: http://www.marlin.ac.uk/speciesfullreview.php?speciesID=4246

Riouall, R. 1985. Sur la presence dans l'etang de Thau (Herault-France) de *Sphaerotrichia divaricata* (C. Ag.) Kylin et *Chorda filum* (L.) Stackhouse. Botanica Marina 28: 83–86.

Risso, S., C. Escudero, S. Estevao-Belchior, M.L. de Portelaand and M.A. Fajardo. 2003. Chemical composition and seasonal fluctuations of the edible green seaweed, *Monostroma undulatum* Wittrock, from the Southern Argentina coast. Arch. Latinoam. Nutr. 53(3): 306–311.

Rizvi, M.A. and M. Shameel. 2005. Pharmaceutical biology of seaweeds from the Karachi Coast of Pakistan. Pharmaceutical Biology 43(2): 97–107.

Robledo, D. and Y.F. Pelegrin. 1997. Chemical and mineral composition of six potentially edible seaweed species of Yucatan. Botanica Marina 40: 301–306.

Robledo, D. and Y.F. Pelegrín. 2009. Chemical and mineral composition of six potentially edible seaweed species of Yucatán. Botanica Marina 40(1-6): 1–577.

Rodrigues, D., S. Sousa, A. Silva, M. Amorim, L. Pereira, T.A. Rocha-Santos, A.M. Gomes, A.C. Duarte and A.C. Freitas. 2015a. Impact of enzyme and ultrasound assisted extraction methods on biological properties of red, brown and green seaweeds from the Central West Coast of Portugal. Journal of Agricultural and Food Chemistry 10: 10.

Rodrigues, D., A.C. Freitas, L. Pereira, T.A.P. Rocha-Santos, M.W. Vasconcelos, M. Roriz, L.M. Rodríguez-Alcalá, A.M.P. Gomes and A.C. Duarte. 2015b. Chemical composition of red, brown and green macroalgae from Buarcos bay in Central West Coast of Portugal. Food Chemistry 183: 197–207.

Rodrigues, J.A.G., I.W.F.d. Araújo, G.A.d. Paula, É.F. Bessa, T.d.B. Lima and N.M.B. Benevides. 2010. Isolamento, fracionamento e atividade anticoagulante de iota-carragenanas da *Solieria filiformis*. Ciência Rural 40: 2310–2316.

Roo, P.V.S., V.A. Mantri, K. Ganesan and K.S. Kumar. 2007. Seaweeds as a human diet: an emerging trend in the new millennium. pp. 85–96. *In*: R.K. Gupta and V.D. Pandey (eds.). Advances in Applied Phycology. Daya Publishing House, New Delhi.

Roos-Collins, M. 1990. The Flavors of Home. Heyday Books, Berkeley, CA.

Rossano, R., N. Ungaro, A. D'Ambrosio, G.M. Liuzzi and P. Riccio. 2003. Extracting and purifying R-phycoerythrin from Mediterranean red algae *Corallina elongata* Ellis Solander. Journal of Biotechnology 101(3): 289–293.

Rumphius, G.E. 1750. Her Amboinsch Kruidboek. Herbarium Amboinense. Amsterdam 6: 1–256.

Ruperez, P. 2002. Mineral content of edible marine seaweeds. Food Chemistry 79: 23–26.

Ruperez, P. and F. Saura-Calixto. 2001. Dietary fiber and physicochemical properties of edible Spanish seaweeds. European Food Research and Technology 212: 349–354.

Russell, D.J. and G.H. Balazs. 2000. Identification manual for dietary vegetation of the Hawaiian green turtle, *Chelonia mydas*. NOAA TM-NMFS-SWFSC-294. 49 pp.

Sáa, C.F. 2002. Atlantic Sea Vegetables—Nutrition and Health: Properties, Recipes, Description. Algamar, Redondela-Pontevedra, Spain, 300 pp.

Sabina, H., S. Tasneem, S.Y. Kausar, M.I. Choudhary and R. Aliya. 2005. Antileishmanial activity in the crude extract of various seaweed from the coast of Karachi, Pakistan. Pakistan Journal of Botany 37: 163–168.

Sahoo, D. 2010. Common Seaweeds of India. I.K. International Pvt. Ltd., New Delhi, India, 196 pp.

Saito, H., C. Xue, R. Yamashiro, Moromizato and Y. Itabashi. 2010. High polyunsaturated fatty acid levels in two subtropical macroalgae, *Cladosiphon okamuranus* and *Caulerpa lentillifera*. Journal of Phycology 46(4): 665–673.

Saito, P. 1969. The algal genus *Laurencia* from the Hawaiian Islands, the Philippine Islands and adjacent Aareas. Pacific Science 23: 148–160.

Saito, Y. 1967. Studies on Japanese Species of *Laurencia*, with special reference to their comparative morphology. Mem. Fac. Fish., Hokkaido Univ. 15(1): 1–81.

Salas, R.A. 2008. *Codium taylorii*—Tropical field phycology field book, STRI, Bocas de Toro. Available online at: http://biogeodb.stri.si.edu/bocas_database/search/species/3588

Sanderson, C. 2000. Assessment of the distribution of the alga *Cystoseira trinodis* (Forsskal) C. Agardh in Blackman Bay, unpublished report to the Threatened Species Unit, Department of Primary Industries, Water & Environment, Hobart, Australia.

Sanderson, J.C. and R.D. Benedetto. 1988. Tasmanian Seaweeds for the Edible Market. Marine Laboratories. Department of Sea Fisheries, Tasmania, Australia, 47 pp.

Sangeetha, R.K., N. Bhaskar and V. Baskaran. 2009. Comparative effects of β-carotene and fucoxanthin on retinol deficiency induced oxidative stress in rats. Molecular and Cellular Biochemistry 331: 59–67.

Santelices, B. 1977. A taxonomic review of Hawaiian Gelidiales (Rhodophyta). Pacific Science 31(1): 61–84.

Santelices, B. and J.G. Stewart. 1985. Pacific species of *Gelidium* Lamouroux and other Gelidiales (Rhodophyta), with keys and descriptions to the common or economically important species. pp. 17–31. *In*: I.A. Abbott and J.N. Norris (eds.). Taxonomy of Economic Seaweeds with Reference to some Pacific and Caribbean Species. California Sea Grant College Program, La Jolla, California.

Santelices, B. and M. Miyata. 1995. Observations on *Gelidium pacificum* Okamura. pp. 55–79. *In*: I.A. Abbott (ed.). Taxonomy of Economic Seaweeds. With Reference to some Pacific

Species, Volume IV, California Sea Grant College. University of California, La Jolla, California.

Santhanam, R. 2015. Nutritional Marine Life. CRC Press, Taylor & Francis Group, Boca Raton, FL, 290 pp.

Santos, G.A. 1990. A manual for the processing of agar from *Gracilaria*. ASEAN FAO Manual n. 5, Regional Small-Scale Coastal Fisheries Development Project Manila, Philippines, 37 pp.

Santoso, J., Y. Yoshie-Stark and T. Suzuki. 2006. Comparative contents of minerals and dietary fibres in several tropical seaweeds. Bulletin Teknologi Hasil Perikanan 9: 1–11.

Sasidharan, S., I. Darah and K. Jain. 2011. *In vitro* and *in situ* antiyeast activity of *Gracilaria changii* methanol extract against *Candida albicans*. Eur. Rev. Med. Pharmacol. Sci. 15(9): 1020–1026.

Sathya, R., N. Kanaga, P. Sankar and S. Jeeva. 2013. Antioxidant properties of phlorotannins from brown seaweed *Cystoseira trinodis* (Forsskål) C. Agardh. Arabian Journal of Chemistry 2013: 1–7.

Schmitt, S. 2008. Tropical Field Phycology Field Book, STRI, Bocas Del Toro, Smithsonian Tropical Research Institute. Available online at: http://biogeodb.stri.si.edu/bocas_database/search/species/6340

Schmitt, S. and N. Mamoozadeh. 2014. *Dictyopteris delicatula*. Tropical Field Phycology Field Book, STRI, Bocas Del Toro, Smithsonian Tropical Research Institute. Available online at: http://biogeodb.stri.si.edu/bocas_database/search/species/3527

Schneider, C.W. and M.J. Wynne. 2007. A synoptic review of the classification of red algal genera a half a century after Kylin's "Die Gattungen der Rhodophyceen". Botanica Marina 50: 197–249.

Schwarzfluss, M. 2014. Mako's food diary in Japan. Available online at: http://arigatoformyfood.blogspot.pt/2013/05/my-mother-and-my-fathers-mother-are.html

Scönfeld-Leber, B. 1979. Marine algae as human food in Hawaii, with note on other Polynesian islands. Ecology of Food and Nutrition 8: 47–59.

Selvin, J. and A.P. Lipton. 2004. Biopotentials of *Ulva fasciata* and *Hypnea musciformis* collected from the Peninsular Coast of India. Journal of Marine Science and Technology 12(1): 1–6.

Senate, N.Y. 1869. Documents of the Senate of the State of New York. Albany 6: 73.

Senthilkumar, P., S. Sudha and S. Prakash. 2014. Antidiabetic activity of aqueous extract of *Padina boergesenii* in streptozotocin-induced diabetic rats. International Journal of Pharmacy and Pharmaceutical Sciences 6(5): 419–422.

Seo, M.-J., H.-S. Choi, O.-H. Lee and B.-Y. Lee. 2013. *Grateloupia lanceolata* (Okamura) Kawaguchi, the edible red seaweed, inhibits lipid accumulation and reactive oxygen species production during differentiation in 3T3-L1 cells. Phytother. Res. 27: 655–663.

Serra, D.R. 2013. *Gracilariopsis tenuifrons* (Gracilariales—Rhodophyta) Responde to Irradiance *In vitro*. Ph.D. Thesis, University of S. Paulo, Brazil, 109 pp.

Serrão, E.A., L.A. Alice and S.H. Brawley. 1999. Evolution of the Fucaceae (Phaeophyceae) Inferred from nrDNA-ITS. Journal of Phycology 35(2): 382–94.

Setchell, W.A. 1924. American Samoa: Part I. Vegetation of Tutuila Island. Part II. Ethnobotany of the Samoans. Part III. Vegetation of Rose Atoll. Publications of the Carnegie Institution of Washington 341: 275.

Sethi, P. 2012. Biochemical composition of marine algae, *Padina tetrastromatica* Hauck. International Journal of Current Pharmaceutical Research 4(1): 117–118.

Shanmugam, A. and C. Palpandi. 2008. Biochemical composition and fatty acid profile of the green alga *Ulva reticulata*. Asian Journal of Biochemistry 3: 26–31.

Shelar, P.S., V.K. Reddy, S.G.S. Shelar, M. Kavitha, G.P. Kumar and G.V.S. Reddy. 2012. Medicinal value of seaweeds and its applications—a review. Cont. J. Pharm. Toxicol. Res. 5(2): 1–22.

Shibata, T., K. shimaru, S. Kawaguchi, H. Yoshikawa and Y. Hama. 2008. Antioxidant activities of phlorotannins isolated from Japanese Laminariaceae. Journal of Applied Phycology 20: 705–711.

Shimabukuro, H., R. Terada, T. Noro and T. Yoshida. 2008. Taxonomic study of two *Sargassum* species (Fucales, Phaeophyceae) from the Ryukyu Islands, southern Japan: *Sargassum ryukyuense* sp. nov. and *Sargassum pinnatifidum* Harvey. Botanica Marina 51: 26–33.

Shimamura, N. 2010. Agar. *In*: Rediscovering the Treasures of Food. The Tokyo Foundation. Available online at: http://www.tokyofoundation.org/en/topics/japanese-traditional-foods/vol.-4-agar

Showe-Mei, L. 2014. *Gracilaria firma*—BiotaTaiwanica, Algae of Taiwan. Available online at: http://mras.taibif.tw/pages/395

SIA. 2014. *Fucus spiralis*—Seaweed Industry Association, available online at: https://seaweedindustry.com/seaweed/type/fucus-spiralis

SIA. 2015. *Sargassum muticum*—Seaweed Industry Association, available online at: https://seaweedindustry.com/seaweed/type/sargassum-muticum

Siddique, M.A.M., M.S.K. Khan and M.K.A. Bhuiyan. 2013. Nutritional composition and amino acid profile of a sub-tropical red seaweed *Gelidium pusillum* collected from St. Martin's Island, Bangladesh. International Food Research Journal 20(5): 2287–2292.

Sidik, B.J., Z.M. Harah and S. Kawaguchi. 2012. Historical review of seaweed research in Malaysia before 2001. Coastal Marine Science 35(1): 169–177.

Silva, C.L., F.G. Pereira, S. Telles, N. Dantas-Santos, R.B.C. Camara, S.L. Cordeiro, M.S.S.P. Costa, J. Almeida-Lima, R.F. Melo-Silveira, R.M. Oliveira, I.R.L. Albuquerque, G.P.V. Andrade and H.A.O. Rocha. 2011. Antioxidant and antiproliferative activities of heterofucans from the seaweed *Sargassum filipendula*. Marine Drugs 9(6): 952–966.

Simié, S. 2007. Morphological and ecological characteristics of rare and endangered species *Lemanea fluviatilis* (L.) C. Ag. (Lemaneaceae, Rhodophyta) on new localities in Serbia. Kragujevac J. Sci. 29: 97–106.

Simoons, F.J. 1990. Food in China: A Cultural and Historical Inquiry. CRC Press, New York, 600 pp.

Sivasankari, S., V. Venkatesalu, M. Anantharaj and M. Chandrasekaran. 2006. Effect of seaweed extracts on the growth and biochemical constituents of *Vigna sinensis*. Bioresource Technology 97: 1745–1751.

Skelton, P., G.R. South and A.J.K. Millar. 2004. *Gracilaria ephemera* sp. nov. (Gracilariales, Rhodophyceae), a flattened species from Samoa, South Pacific. pp. 231–242. *In*: I.A. Abbott and K.J. McDermid (eds.). Taxonomy of Economic Seaweeds with Reference to the Pacific and other Locations Volume IX. University of Hawaii Sea Grant College Program, Hawaii.

Skelton, P.A. 2003. Marine Plants of American Samoa. Department of Marine & Wildlife Resources Government of American Samoa, 103 pp.

Skewes, M. 2008. *Bifurcaria bifurcata*. A brown seaweed. Marine Life Information Network. Biology and Sensitivity Key Information Sub-programme. Marine Biological Association of the United Kingdom, Plymouth. Available online at: http://www.marlin.ac.uk/speciesinformation.php?speciesID=2760

Škrovánková, S. 2011. Seaweed vitamins as nutraceuticals. pp. 357–369. *In*: K. Se-Kwon (ed.). Marine Medicinal Foods Implications and Applications, Macro and Microalgae. Advances in Food and Nutrition Research. Academic Press, London.

Skvortzov, V.B. 1919. The use of *Nostoc* as food in N China. Royal Society of Great Britain and Ireland 13: 67.

Smit, A.J. 2004. Medicinal and pharmaceutical uses of seaweed natural products: a review. Journal of Applied Phycology 16: 245–262.

Smith, D.B., A.N. Oneill and A.S. Perlin. 1955. Studies on the heterogeneity of carrageenin. Canadian Journal of Chemistry 33(8): 1352–1360.

Smith, H.M. 1904. The seaweed industries of Japan. Bulletin of the Bureau of Fisheries (USA) 24: 133–81.

Smith, J.L., G. Summers and R. Wong. 2010. Nutrient and heavy metal content of edible seaweeds in New Zealand. New Zealand Journal of Crop and Horticultural Science 38(1): 19–28.

Smyrniotopoulos, V., C. Vagias, C. Bruyere, D. Lamoral-Theys, R. Kiss and V. Roussis. 2010a. Structure and *in vitro* antitumor activity evaluation of brominated diterpenes from the red alga *Sphaerococcus coronopifolius*. Bioorganic & Medicinal Chemistry 18: 1321–1330.

Smyrniotopoulos, V., C. Vagias, M.M. Rahman, S. Gibbons and V. Roussis. 2010b. Structure and antibacterial activity of brominated diterpenes from the red alga *Sphaerococcus coronopifolius*. Chemistry & Biodiversity 7: 186–195.

Snakenberg, R.L. 1987. Limu: Learning about Hawaii's Edible Seaweeds. Department of Education, State of Hawaii, 126 pp.

Soegiarto, A. and Sulustijo. 1990. Utilization and farming of seaweeds in Indonesia. pp. 9–19. *In*: I.J. Dogma Jr., G.C. Trono Jr. and R.A. Tabbada (eds.). Culture and Use of Algae in Southeast Asia. Southeast Asian Fisheries Development Center, Tigbauan, Iloilo, Philippines.

Soetan, K.O., C.O. Olaiya and O.E. Oyewole. 2010. The importance of mineral elements for humans, domestic animals and plants: a review. African Journal of Food Science 4(5): 200–222.

Sohn, C.H. 1998. The seaweed resources of Korea. pp. 62–69. *In*: A.T. Critchley, M. Ohno, D.B. Largo and R.D. Gillespie (eds.). Seaweed Resources of the World. Japan International Cooperation Agency, Yokosuka, Japan.

Sohn, C.H. 2006. The seaweed resources of Korea. *In*: A.T. Critchley, M. Ohno and D.B. Largo (eds.). World Seaweed Resources—An Authoritative Reference System. ETI Information Services Ltd. Hybrid Windows and Mac DVD-ROM; ISBN: 90-75000-80-4.

Song, M.-J., H. Kim, H. Brian, K.H. Choi and B.-Y. Lee. 2013. Traditional knowledge of edible plants on Jeju Island, Korea. Indian Journal of Traditional Knowledge 12(2): 177–194.

Sosa, P.A., J.L.G. Pinchetti and J.A. Juanes. 2006. The seaweed resources of Spain. *In*: A.T. Critchley, M. Ohno and D.B. Largo (eds.). World Seaweed Resources—An Authoritative Reference System. ETI Information Services Ltd. Hybrid Windows and Mac DVD-ROM; ISBN: 90-75000-80-4.

Sousa, A.A.S. 2010. Galactana sulfatada da alga marinha vermelha *Gelidium crinale* (Turner) Lamouroux: respostas na inflamação e nocicepção. Universidade Estadual do Ceará (UECE), Forteleza, Ceará, Brazil.

South, G.R. 1993a. Edible seaweeds of Fiji: an ethnobotanical study. Bot. Mar. 36: 335–349.

South, G.R. 1993b. Edible seaweeds—an important source of food and income to indigenous Fijians. NAGA, the Iclarm Quarterly 16(2-3): 4–6.

South, G.R. 1993c. Seaweed. pp. 683–710. *In*: A. Wright and L. Hill (eds.). Nearshore Marine Resources of the South Pacific: Information for Fisheries Development and Management. University of the South Pacific, Institute of Pacific Studies, Suva, Fiji.

South, G.R. 1995. Edible seaweeds: an important source of food and income to indigenous Fijians. pp. 43–48. *In*: E. Matthews (ed.). Fishing for Answers: Women and Fisheries in the Pacific Islands. Women and Fisheries Network, Suva, Fiji.

South, G.R. and T. Pickering. 2006. The seaweed resources of the Pacific Islands. *In*: A.T. Critchley, M. Ohno and D.B. Largo (eds.). World Seaweed Resources—An Authoritative Reference System. ETI Information Services Ltd. Hybrid Windows and Mac DVD-ROM; ISBN: 90-75000-80-4.

South, G.R., C. Morris, S. Bala and M. Lober. 2012. Value adding and supply chain development for fisheries and aquaculture products in Fiji, Samoa and Tonga: Scoping study for *Caulerpa* (Sea grapes). Suva, Fiji: Institute of Marine Resources, School of Marine Studies, FSTE, USP. IMR Technical Report 03/2012, 19 pp.

Souza, B.W.S., M.A. Cerqueira, J.T. Martins, M.A.C. Quintas, A.C.S. Ferreira, J.A. Teixeira et al. 2011. Antioxidant potential of two red seaweeds from the Brazilian coasts. Journal of Agricultural and Food Chemistry 59: 5589.

Spalding, H. 2009. Got Limu? Uses for algae in Hawaii and beyond. Material developed for the University of Hawaii-Manoa GK-12 program (NSF grant #05385500). Available online at: http://www.hawaii.edu/gk-12/evolution/pdfs/algae.uses.highschool.pdf

Spieler, R. 2002. Seaweed compound's anti-HIV efficacy will be tested in southern Africa. Lancet 359: 16–75.

Stanley, N. 1987. Production, properties and uses of carrageenan. FAO Fisheries Technical Paper, pp. 116–146.

Stefanov, K., I. Tsvetkova, S. Dimitrova-Konaklieva and S. Popov. 2009. Antibacterial, antiviral, and cytotoxic activities of some red and brown seaweeds from the Black Sea. Botanica Marina 52(1): 80–86.

Stender, K. and Y. Stender. 2014a. Marinelife Photography—*Ahnfeltiopsis flabelliformis*. Available online at: http://www.marinelifephotography.com/marine/seaweeds/ahnfeltiopsis-flabelliformis.htm

Stender, K. and Y. Stender. 2014b. Marinelife Photography—*Hypnea musciformis*. Available online at: http://www.marinelifephotography.com/marine/seaweeds/hypnea-musciformis.htm

Stender, K. and Y. Stender. 2014c. Marinelife Photography—*Liagora albicans*. Available online at: http://www.marinelifephotography.com/marine/seaweeds/liagora-albicans.htm

Stender, K. and Y. Stender. 2014d. Marinelife Photography—*Sargassum obtusifolium*. Available online at: http://www.marinelifephotography.com/marine/seaweeds/sargassum-obtusifolium.htm

Subba Rao, G.N. 1965. Uses of seaweed directly as human food. Indo-Pacific Fisheries Council Regional Studies 2: 1–32.

Subramaniam, D., T. Menon, H.L. Elizabeth and S. Swaminathan. 2014. Anti-HIV-1 activity of *Sargassum swartzii* a marine brown alga. BMC Infectious Diseases 14(S3): E43. Available online at: http://www.biomedcentral.com/1471-2334/14/S3/E43

Subsea, 2015. Edible bubble algae—*Valonia macrophysa*. Reef Central, available online at: http://www.reefcentral.com/forums/showthread.php?t=2315138

Sugawa-Katayama, Y. and M. Katayama. 2009. Release of minerals from dried Hijiki, *Sargassum fusiforme* (Harvey) Setchell, during water-soaking. Trace Nutrients Research 24: 106–109.

Sugawara, I., W. Itoh, S. Kimura, S. Mori and K. Shimada. 1989. Further characterization of sulfated homopolysaccharides as anti-HIV agents. Cellular and Molecular Life Sciences 45: 996–998.

Sugawara, T., K. Matsubara, R. Akagi, M. Mori and T. Hirata. 2006. Antiangiogenic activity of brown algae fucoxanthin and its deacetylated product, fucoxanthinol. Journal of Agricultural and Food Chemistry 54: 9805–9810.

Sujuan, W. and Z. Jingrong. 1980. On *Porphyra monosporangia*, a New Species from China. Oceanologia et Limnologia Sinica 11(2): 141–149.

Surialink. 2014. *Betaphycus*. Seaplants handbook. Available online at: http://surialink.seaplant.net/HANDBOOK/Genera/reds/Betaphycus/Betaphycus.htm

Surialink. 2015. *Catenella caespitosa*. Seaplants handbook. Available online at: http://surialink.seaplant.net/HANDBOOK/Genera/reds/Catenella/Catenella.htm

Suzuki, H., T. Higuchi, K. Sawa, S. Ohtaki and J. Tolli. 1965. Endemic coast goitre in Hokkaido, Japan. Acta Endocrinologica 50: 161–176.

Suzuki, M. 2011. *Saccharina japonica* var. *religiosa*. Natural History Japan. Available online at: http://natural-history.main.jp/Seaweeds_list_English/Brown_photo/Saccharina_japonica_religiosa/Saccharina_japonica_religiosa.html

Syad, A.N., K.P. Shunmugiah and P.D. Kasi. 2013. Antioxidant and anti-cholinesterase activity of *Sargassum wightii*. Pharm. Biol. 51: 1401–1410.

Taboada, C., R. Millán and I. Míguez. 2010. Composition, nutritional aspects and effect on serum parameters of marine algae *Ulva rigida*. Journal of the Science of Food and Agriculture 90: 445–449.

Takagi, M. 1975. Seaweeds as medicine. pp. 321–325. *In*: J. Tokida and H. Hirose (eds.). Advance of Phycology in Japan. Veb Gustav Fischer Verlag Jena.

Takaichi, S. 2011. Carotenoids in algae: distributions, biosyntheses and functions. Marine Drugs 9(6): 1101–1118.

Takano, R., Y. Nose, K. Hayashi, S. Hara and S. Hirase. 1994. Agarose-carrageenan hybrid polysaccharides from *Lomentaria catenata*. Phytochemistry 37: 1615–1619.

Tan, R. 2008a. Wildfactsheets wildsingapore homepage—*Caulerpa lentillifera*, available online at: http://www.wildsingapore.com/wildfacts/plants/seaweed/chlorophyta/lentillifera.htm

Tan, R. 2008b. Wildfactsheets wildsingapore homepage—*Caulerpa serrulata*, available online at: http://www.wildsingapore.com/wildfacts/plants/seaweed/chlorophyta/serrulata.htm

Tan, R. 2008c. Wildfactsheets wildsingapore homepage—*Halymenia maculata*, available online at: http://www.wildsingapore.com/wildfacts/plants/seaweed/rhodophyta/maculata. htm

Tan, R. 2013. Wildfactsheets wildsingapore homepage—*Caulerpa taxifolia*, available online at: http://www.wildsingapore.com/wildfacts/plants/seaweed/chlorophyta/taxifolia.htm

Taskin, E., M. Ozturk, E. Taskin and O. Kurt. 2007. Antibacterial activities of some marine algae from the Aegean Sea (Turkey). African Journal of Biotechnology 6(24): 2746–2751.

Taylor, W.R. 1960. Marine Algae of the Eastern Tropical and Subtropical Coasts of the Americas. The University of Michigan Press, Ann Arbor, MI.

Teas, J. 1983. The dietary intake of *Laminaria*, a brown seaweed, and breast cancer prevention. Nutrition and Cancer 4(3): 217–222.

Teo, L.W. and Y.C. Wee. 1983. Seaweeds of Singapore. Singapore University Press, 123 pp.

Terada, R. and M. Ohno. 2000. Notes on *Gracilaria* (Gracilariales, Rhodophyta) from Tosa Bay and adjacent waters I: *Gracilaria chorda*, *Gracilaria gigas* and *Gracilaria incurvata*. Bulletin of Marine Sciences and Fisheries, Kochi University 20: 81–88.

Terasaki, M., B. Narayan, H. Kamogawa, M. Nomura, N.M. Stephen, C. Kawagoe, M. Hosokawa and K. Miyashita. 2012. Carotenoid profile of edible Japanese seaweeds: an improved HPLC method for separation of major carotenoids. Journal of Aquatic Food Product Technology 21: 468–479.

TFJ. 1893. The fisheries of Japan. Bulletin of the United States Fish Commission 13: 419–435.

Therkelsen, G.H. 1993. Carrageenan. pp. 145–180. *In*: R.L. Whistler and J.N. Bemiller (eds.). Industrial Gums: Polysaccharides and their Derivatives. Academic Press, San Diego, CA.

Thiamdao, S., G.H. Boo, S.M. Boo and Y. Peerapornpisal. 2012. Diversity of edible *Cladophora* (Cladophorales, Chlorophyta) in northern and northeastern Thailand based on morphology and nuclear ribosomal DNA sequences. Chiang Mai J. Sci. 39(2): 300–310.

Thivy, F. and V.D. Chauhan. 1963. *Caulerpa veravalensis*, a new species from Gujarat, India. Botanica Marina 5(4): 97–100.

Tito, O.D. and L.M. Liao. 2000. Ethnobotany of *Solieria robusta* (Gigartinales, Rhodophyta) in Zamboanga, Philippines. Science Diliman 12(2): 75–77.

Tittley, I. and A.I. Neto. 2005. The marine algal (seaweed) flora of the Azores: additions and amendments. Botanica Marina 48(3): 248–255.

Tokida, J. 1954. The marine algae of south Saghalien. Mem. Fac. Fish., Hokkaido Univ. 2(1): 1–264.

Tokida, J. and H. Hirose. 1975. *Porphyra* cultivation in Japan. pp. 282–283. *In*: Advance of Phycology in Japan. VEB Gustav Fischer Verlag Jena, Germany.

Tokuda, H., M. Ohno and H. Ogawa. 1986. The resources and cultivation of seaweeds. Midorishobou, Tokyo, 354 pp.

Tolstoy, A. and K. Österlund. 2003. Alger vid Sveriges östersjökust—en fotoflora, pp. 1–282. ArtDatabanken, Uppsala.

Toma, T. 1994. Seasonal, topographical and geographical distribution of the red alga *Gloiopeltis complanata* (Harvey) Yamada in the Islands of Okinawa. Aquaculture Science 42: 553–561.

Toma, T. 1997. Cultivation of the brown alga, *Cladosiphon okamuranus* "Okinawa-mozuku". pp. 51–56. *In*: M. Ohno and A.T. Critchely (eds.). Seaweed Cultivation and Marine Ranching. JICA, Yokosuka.

Trono, G.C., Jr. 1998. The seaweed resources of the Philippines. pp. 47–61. *In*: A.T. Critchley, M. Ohno, D.B. Largo and R.D. Gillespie (eds.). Seaweed Resources of the World. Japan International Cooperation Agency, Yokosuka, Japan.

Trono, G.C., Jr. 2001. Seaweeds. pp. 19–99. *In*: K.E. Carpenter and V.H. Niem (eds.). The Living Marine Resources of the Western Central Pacific, Vol. 1. FAO Species Identification Guide for Fishery Purposes. FAO, Rome.

Trono, G.C. and T. Toma. 1997. Cultivation of the green alga *Caulerpa lentilifera*. pp. 17–23. *In*: M. Ohno and A.T. Cricthley (eds.). Seaweed Cultivation and Marine Ranching. JICA, Yokosuka, Japan.

Trono, G.C., Jr. and N.E. Montano. 2006. The seaweed resources of the Philippines. *In*: A.T. Critchley, M. Ohno and D.B. Largo (eds.). World Seaweed Resources—An Authoritative Reference System. ETI Information Services Ltd. Hybrid Windows and Mac DVD-ROM; ISBN: 90-75000-80-4.

Troyano, J. and T.S. Talley. 2014. Red ogo seaweed: *Gracilaria pacifica*—Biology Fact Sheet. California Sea Grant. Univ. of San Diego under a Grant Awarded from Collaborative Research Fisheries Research West. Available online at: http://ca-sgep.ucsd.edu/sites/ca-sgep.ucsd.edu/files/advisors/tstalley/files/fact%20sheet-red-ogo.pdf

Truus, K., M. Vaher and I. Taure. 2001. Algal biomass from *Fucus vesiculosus* (Phaeophyta): investigation of the mineral and alginate components. Proceedings of the Estonian Academy of Sciences—Chemistry 50: 95–103.

Tsuchiya, Y. 1950. Physiological studies on the vitamin C content of marine algae. Tohoku Journal of Agricultural Research 1: 97–102.

Tseng, C.-K. 1933. *Gloiopeltis* and other economic seaweeds of Amoy, China. Lingnan Science Journal 12: 43–63.

Tseng, C.-K. 1935. Economics seaweeds of Kwangtung Province, S China. Lingnan Science Journal 14: 93–104.

Tseng, C.-K. 1945. The terminology of seaweed colloids. Science 101(2633): 597–602.

Tseng, C.K. 1983. Common seaweeds of China. Science Press, 316 pp.

Tseng, C.K. and T.J. Chang. 1958. On *Phorphyra marginata* sp. nov. and its systematic position. Acta Botanica Sinica 7(1): 15–25.

Tseng, C.K. and J.F. Zhang. 1984. Chinese seaweeds in herbal medicine. Proceedings of the International Seaweed Symposium 11: 152–154.

Tsiamis, K. and M. Verlaque. 2011. A new contribution to the alien red macroalgal flora of Greece (Eastern Mediterranean) with emphasis on *Hypnea* species. Cryptogamie 32(4): 393–410.

Tsuda, R.T. 1985. *Gracilaria* from Micronesia: key, list and distribution of the species. pp. 91–92. *In*: I.A. Abbott and J.N. Norris (eds.). Taxonomy of Economic Seaweeds with Reference to some Pacific and Caribbean Species, Vol. I. California Sea Grant College Program, California.

Tsutsui, I., S. Arai, T. Terawaki and M. Ohno. 1996. A morphometric comparison of *Ecklonia kurome* (Laminariales, Phaeophyta) from Japan. Phycological Research 44: 215–222.

Turner, N.J. 1974. Plant taxonomic systems and ethnobotany of three contemporary Indian groups of the Pacific North-west (Haida, Bella Costa, and Lillooet). Syesis 7: 1–104.

Tyler-Walters, H. 2008. *Alaria esculenta*. Dabberlocks. Marine Life Information Network: Biology and Sensitivity Key Information Sub-programme. Plymouth: Marine Biological Association of the United Kingdom. Available online at: http://www.marlin.ac.uk/speciesinformation.php?speciesID=2431

UHDB. 2014. ReefWatcher's Field Guide to Alien and Native Hawaiian Marine Algae—Marine Algae of Hawai`i. Available online at: http://www.hawaii.edu/reefalgae/natives/sgfieldguide.htm

Umezaki, I. 1990. Life history of *Acrothrix pacifica* and *Sphaerotrichia divaricata* in laboratory cultures. pp. 31–38. *In*: I.J. Dogma Jr., G.C. Trono Jr. and R.A. Tabbada (eds.). Culture and use of algae in Southeast Asia, Proceedings of the Symposium on Culture and Utilization of Algae in Southeast Asia. Aquaculture Dept., Southeast Asian Fisheries, Development Center, Tigbauan, Iloilo, Philippines.

Urta Islandica. 2015. Knotted kelp and ginger—herbal tea. Available online at: http://www.urta.is/index.php?main_page=product_info&cPath=1&products_id=34

Usov, A.I. and M.Y. Elashvili. 1991. Polysaccharides of algae. 44. Investigation of sulfated galactan from *Laurencia nipponica* Yamada (Rhodophyta, Rhodomelaceae) using partial reductive hydrolysis. Botanica Marina 34: 553–560.

Usov, A.I. and N.D. Zelinsky. 2013. Chemical structures of algal polysaccharides. pp. 23–86. *In*: H. Domínguez (ed.). Functional Ingredients from Algae for Foods and Nutraceuticals. Woodhead Publishing Limited, Cambridge, UK.

Uzair, B., Z. Mahmood and S. Tabassum. 2011. Antiviral activity of natural products extracted from marine organisms. BioImpacts BI 1: 203–211.

van de Velde, F. and G.A. de Ruiter. 2002. Carrageenan. pp. 245–274. *In*: E.J. Vandamme, S.D. Baets and A. Steinbèuchel (eds.). Biopolymers Vol. 6: Polysaccharides II, polysaccharides from eukaryotes. Wiley-VCH, Weinheim, Germany.

van de Velde, F., L. Pereira and H.S. Rollema. 2004. The revised NMR chemical shift data of carrageenans. Carbohydrate Research 339: 2309–2313.

Venugopal, V. 2009. Marine Products for Healthcare: Functional and Bioactive Nutraceutical Compounds from the Ocean. CRC Press, Taylor and Francis Group, New York, 528 pp.

Wang, B., H. Huang, H.P. Xiong, E.Y. Xie and Z.M. Li. 2010. Analysis on nutrition constituents of *Sargassum naozhouense* sp. nov. Food Res. Dev. 31: 195–197.

Wang, H., L.C.M. Chiu, V.E.C. Ooi and P.O. Ang Jr. 2010. A potent antitumor polysaccharide from the edible brown seaweed *Hydroclathrus clathratus*. Botanica Marina 53: 265–274.

Wang, W.-L. 2014. *Padina australis*—BiotaTaiwanica, Algae of Taiwan. Available online at: http://algae.biota.biodiv.tw/pages/922

Wang, W.-L. 2015. *Padina minor*. BiotaTaiwanica. Available online at: http://algae.biota.biodiv.tw/pages/924

Wassilieff, M. 2012a. Seaweed—Traditional use of seaweeds, Te Ara—the Encyclopedia of New Zealand. Available online at: http://www.TeAra.govt.nz/en/seaweed/

Wassilieff, M. 2012b. Seaweed—Types of seaweed, Te Ara—the Encyclopedia of New Zealand. Available online at: http://www.TeAra.govt.nz/en/seaweed/page-2

Watanabe, F., S. Takenaka, H. Katsura, E. Miyamoto, K. Abe, Y. Tamura, T. Nakatsuka and Y. Nakano. 2000. Characterization of a vitamin B_{12} compound in the edible purple laver, *Porphyra yezoensis*. Bioscience, Biotechnology, and Biochemistry 64: 2712–2715.

Wegeberg, S. 2010. Potential for farming edible seaweeds in Denmark, Faroe Is and Greenland. National Environmental Research Institute, Aarhus University. Available online at: http://www.seaweedbook.net/wp-content/uploads/wegeberg-21.pdf

Wei, T.L. and W.Y. Chin. 1983. Seaweeds of Singapore. Singapore University Press Pte Ltd., Kent Ridge, Singapore, 125 pp.

Wei, Y., C. Xu, Q. Liu, H. Xiao, A. Zhao, Y. Hu, L. Liu and L. Zhao. 2012. Effect of phlorotannins from *Sargassum thunbergii* on blood lipids regulation in mice with high-fat diet. Natural Product Research 26(22): 2117–2120.

White, N. 2006. *Chorda filum*. Sea lace or Dead man's rope. Marine Life Information Network, Biology and Sensitivity Key Information Sub-programme. Plymouth, Marine Biological Association of the United Kingdom. Available online at: http://www.marlin.ac.uk/speciesinformation.php?speciesID=2974

White, N. 2007. *Fucus ceranoides*. Horned wrack. Marine Life Information Network, Biology and Sensitivity Key Information Sub-programme. Plymouth, Marine Biological Association of the United Kingdom. Available online from: http://www.marlin.ac.uk/speciesinformation.php?speciesID=3344

White, S. and M. Keleshian. 1994. A Field Guide to Economically Important Seaweeds of Northern New England. University of Maine and University of New Hampshire Sea Grant Marine Advisory Program. MSG-E-93-16. Available online at: http://www.noamkelp.com/technical/handbook.html

Wijesekara, I., N.Y. Yoon and S. Kim. 2010. Phlorotannins from *Ecklonia cava* (Phaeophyceae): Biological activities and potential health benefits. BioFactors 36: 408–414.

Witvrouw, M. and E. De Clercq. 1997. Sulfated polysaccharides extracted from sea algae as potential antiviral drugs. General Pharmacology 29: 497–511.

Womersley, H.B.S. 1984. The Marine Benthic Flora of Southern Australia—Part I. Board of the Botanic Gardens and State Herbarium, Government of South Australia.

Womersley, H.B.S. 1987. The Marine Benthic Flora of Southern Australia. Part II. South Australian Government Printing Division, Adelaide, Australia, 481 pp.

Womersley, H.B.S. 1994. The Marine Benthic Flora of Southern Australia, Rhodophyta—Part IIIA. Australian Biological Resources Study (ABRS), Australia, pp. 133.

Womersley, H.B.S. 2003. Marine benthic flora of southern Australia. Available online at: http://www.flora.sa.gov.au/algae_flora/The_Marine_Benthic_Flora_of_SA_static_index.shtml

Womersley, H.B.S. and A. Bailey. 1970. Marine algae of the Solomon Islands. Philosophical Transactions of the royal society of London, B. Biological Sciences 259: 257–352.

Womersley, H.B.S. and J.A. Lewis. 1994. *Halymenia floresia* subsp. *Harveyana*—Electronic Flora of South Australia Species Fact Sheet. Available online at: http://www.flora.sa.gov.au/efsa/Marine_Benthic_Flora_SA/Part_IIIA/Halymenia_floresia_harveyana.shtml

Worthington, V. 2014. Edible Cape Cod. Available online at: http://ediblecapecod.com/online-magazine/seaweed-pudding/

Wynne, M.J. and O. De Clerck. 1999. First reports of *Padina antillarum* and *P. glabra* (Phaeophyta—Dictyotaceae) from Florida, with a key to the Western Atlantic species of the Genus. Caribbean Journal of Science 35(3-4): 286–295.

Wynne, M.J. and C.W. Schneider. 2010. Addendum to the synoptic review of red algal genera. Botanica Marina 53: 291–299.

Wynne, M.J.W., D. Serio, M. Cormaci and G. Furnari. 2005. *Cholldrophycus* and *Laurencia* from Dhofar, the Sultanate of Oman. Phycologia 44(5): 497–509.

Xie, E., D. Liu, C. Jia, X. Chen and B. Yang. 2013. Artificial seed production and cultivation of the edible brown alga *Sargassum naozhouense* Tseng et Lu. Journal of Applied Phycology 25: 513–522.

Yabuta, Y., H. Fujimura, C.S. Kwak, T. Enomoto and F. Watanabe. 2010. Antioxidant activity of the phycoerythrobilin compound formed from a dried Korean purple laver (*Porphyra* sp.) during *in vitro* digestion. Food Science and Technology Research 16: 347–351.

Yamada, Y. 1932. Notes on Some Japanese Algae III. Journal of the Faculty of Science, Hokkaido Imperial University. Ser. 5, Botany 1(3): 109–123.

Yamada, Y. 1938. Observations on *Arthrothamnus bifidus* J. Ag. Papers of the Institute of Algological Research, Faculty of Science, Hokkaido Imperial University 2(1): 113–118.

Yamada, Y., T. Miyoshi, S. Tanada and M. Imaki. 1991. Digestibility and energy availability of Wakame (*Undaria pinnatifida*) seaweed (in Japanese). Nippon Eiseigaku Zasshi 46: 788–94.

Yamaguchi, M. and T. Matsumoto. 2012. Marine algae *Sargassum horneri* bioactive factor stimulates osteoblastogenesis and suppresses osteoclastogenesis *in vitro*. Medical Biotechnology 1(1): 1–7.

Yamori, Y., A. Miura and K. Taira. 2001. Implications from and for food cultures for cardiovascular diseases: Japanese food, particularly Okinawan diets. Asia Pacific Journal of Clinical Nutrition 10(2): 144–145.

Yan, X., Y. Chuda, M. Suzuki and T. Nagata. 1999. Fucoxanthin as the major antioxidant in *Hijikia fusiformis*, a common edible seaweed. On the Bioscience, Biotechnology and Biochemistry 63: 605–607.

Yan, X.H., L. Li, J.H. Chen and Y.S. Aruga. 2007. Parthenogenesis and isolation of genetic pure strains in *Porphyra haitanensis* (Bangiales, Rhodophyta). High Technology Letters 17(2): 205–210.

Yang, E.J., J.Y. Moon, M.J. Kim, D.S. Kim, W.J. Lee, N.H. Lee and C.G. Hyun. 2010. Anti-inflammatory effect of *Petalonia binghamiae* in LPS-induced macrophages is mediated by suppression of iNOS and COX-2. International Journal of Agricultural and Biological Engineering 12: 754–758.

Yang, H.-N. 2014a. *Sargassum hemiphyllum*—BiotaTaiwanica, Algae of Taiwan. Available at: http://algae.biota.biodiv.tw/pages/1598

Yang, H.-N. 2014b. *Sargassum ilicifolium*—BiotaTaiwanica, Algae of Taiwan. Available at: http://algae.biota.biodiv.tw/pages/1608

Yang, H.-N. 2014c. *Sargassum mcclurei*—BiotaTaiwanica, Algae of Taiwan. Available at: http://algae.biota.biodiv.tw/pages/1612

Yang, H.-N. 2014d. *Sargassum oligocystum*—BiotaTaiwanica, Algae of Taiwan. Available at: http://algae.biota.biodiv.tw/pages/1614

Yang, H.-N. 2014e. *Sargassum siliquosum*—BiotaTaiwanica, Algae of Taiwan. Available at: http://algae.biota.biodiv.tw/pages/1624

Yang, H.-N. 2015. *Sargassum henslowianum*—BiotaTaiwanica, Algae of Taiwan. Available at: http://algae.biota.biodiv.tw/pages/1602

Yang, J.-I., C.-C. Yeh, J.-C. Lee, S.-C. Yi, H.-W. Huang, C.-N. Tseng and H.-W. Chang. 2012. Aqueous extracts of the edible *Gracilaria tenuistipitata* are protective against H_2O_2-Induced DNA damage, growth inhibition, and cell cycle arrest. Molecules 17: 7241–7254.

Yanovsky, E. 1936. Food Plants of the North American Indians. US Department of Agriculture Miscellaneous Publication 237, Washington, D.C.

Ye, H., C. Zhou, Y. Sun, X. Zhang, J. Liu, Q. Hu and X. Zeng. 2009. Antioxidant activities *in vitro* of ethanol extract from brown seaweed *Sargassum pallidum*. European Food Research and Technology 230: 101.

Yendo, K. 1902. Uses of marine alga on Japan. New Phytologist 2: 115–116.

Yendo, K. 1907. The Fucaceae of Japan. Coll. Journal of the College of Science, Imperial University of Tokyo 21: 1–174.

Yong-Xin, L., Y. Li, Z.-J. Qian, M.-M. Kim and S.-K. Kim. 2009. *In vitro* antioxidant activity of 5-HMF isolated from marine red alga *Laurencia undulata* in free radical mediated oxidative systems. Journal of Microbiology and Biotechnology 19(11): 1319–1327.

Yoon, N.Y., S.-H. Lee, L. Yong and S.-K. Kim. 2009. Phlorotannins from *Ishige okamurae* and their acetyl- and butyrylcholinesterase inhibitory effects. Journal of Functional Foods 1(4): 331–335.

Yoshida, T. 1992. Typification of *Eucheuma amakusaense* Okamura. pp. 235–236. *In*: I.A. Abbott (ed.). Taxonomy of Economic Seaweeds, Vol. 3. California Sea Grant College, La Jolla, California.

Yoshie-Stark, Y., Y.-P. Hsieh and T. Suzuki. 2003. Distribution of flavonoids and related compounds from seaweeds in Japan. Journal of Tokyo University of Fisheries 89: 1–3.

Yu, Q., J. Yan, S. Wang, L. Ji, K. Ding, C. Vella, Z. Wang and Z. Hu. 2012. Antiangiogenic effects of GFP08, an agaran-type polysaccharide isolated from *Grateloupia filicina*. Glycobiology 22: 1343–1352.

Yuan, H. and J. Song. 2005. Preparation, structural characterization and *in vitro* antitumor activity of kappa-carrageenan oligosaccharide fraction from *Kappaphycus striatum*. Journal of Applied Phycology 17: 7–13.

Yuan, H., J. Song, X. Li, N. Li and S. Liu. 2011. Enhanced immunostimulatory and antitumor activity of different derivatives of κ-carrageenan oligosaccharides from *Kappaphycus striatum*. Journal of Applied Phycology 23: 59–65.

Yuan, Y.V. and N.A. Walsh. 2006. Antioxidant and antiproliferative activities of extracts from a variety of edible seaweeds. Food and Chemical Toxicology 44: 1144–1150.

Yuan, Y.V., D.E. Bone and M.F. Carrington. 2005. Antioxidant activity of dulse (*Palmaria palmata*) extract evaluated *in vitro*. Food Chemistry, 2005, 91: 485–494.

Zaixsoa, A.B., M. Ciancia and A.S. Cerezo. 2006. Seaweed resources of Argentina. *In*: A.T. Critchley, M. Ohno and D.B. Largo (eds.). World Seaweed Resources—An Authoritative Reference System. ETI Information Services Ltd. Hybrid Windows and Mac DVD-ROM; ISBN: 90-75000-80-4.

Zaneveld, J.S. 1950. A review of three centuries on phycological work and collectors in Indonesia (1650–1950). Organ. Sci. Res. Indon. Public. 21: 1–16.

Zaneveld, J.S. 1951. Economic marine algae of Malaysia and their applications, II. The Phaeophyta. Proceedings of the Indo-Pacific Fisheries Council 129–133.

Zaneveld, J.S. 1955. Economic Marine Algae of Tropical South and East Asia and their Utilization. Indo-pacific Publications, 3. Bangkok.

Zaneveld, J.S. 1959. The utilization of marine algae in tropical South and East Asia. Econ. Bot. 12: 89–131.

Zbakh, H., I. Chiheb, V. Motilva and H. Riadi. 2014. Antibacterial, cytotoxic and antioxidant potentials of *Cladophora prolifera* (Roth) Kutzing collected from the Mediterranean coast of Morocco. American Journal of Phytomedicine and Clinical Therapeutics 2(10): 1187–1199. Available online at: http://www.ajpct.org/index.php/AJPCT/article/download/201/201

Zemke-White, W.L. and M. Ohno. 1999. World seaweed utilisation: an end-of-century summary. Journal of Applied Phycology 11: 369–376.

Zhang, R.L., W.D. Luo, T.N. Bi and S.K. Zhou. 2012. Evaluation of antioxidant and immunity-enhancing activities of *Sargassum pallidum* aqueous extract in gastric cancer rats. Molecules 17: 8419.

Zhao, X., G. Jiao, J. Wu, J. Zhang and G. Yu. 2015. *Laminaria japonica* Aresch. and *Ecklonia kurome* Okam. (Kunbu, Kelp). pp. 767–779. *In*: Y. Liu, Z. Wang and J. Zhang (eds.). Dietary Chinese Herbs—Chemistry, Pharmacology and Clinical Evidence. Springer, London.

Zheng, B.P. 1981. *Porphyra oligospermatangia*, a new species of *Porphyra*. Oceanologia Et limnologia Sinica 12(5): 447–451.

Zhou, G., Y. Sun, H. Xin, Y. Zhang, Z. Li and Z. Xu. 2004. *In vivo* antitumor and immunomodulation activities of different molecular weight lambda-carrageenans from *Chondrus ocellatus*. Pharmacological Research 50: 47–53.

Zhou, G., W. Ma and P. Yuan. 2014. Chemical Characterization and antioxidant activities of different sulfate content of λ-carrageenan fractions from edible red seaweed *Chondrus ocellatus*. Cellular and Molecular Biology 60: 107.

Zhou, Y., H. Yang, H. Hu, Y. Liu, Y. Mao, H. Zhou, X. Xu and F. Zhang. 2006. Bioremediation potential of the macroalga *Gracilaria lemaneiformis* (Rhodophyta) integrated into fed fish culture in coastal waters of north China. Aquaculture 252: 264–276.

Zhuang, C., H. Itoh, T. Mizuno and H. Ito. 1995. Antitumor active fucoidan from the brown seaweed, umitoranoo (*Sargassum thunbergii*). Biosci. Biotechnol. Biochem. 59: 563–567.

Zimmerman, E. 2012. *Aegagropila linnaei*—The plight of the spheroid seaweed. Quest on Able Evolution. Available online at: http://questionableevolution.com/2012/04/25/the-plight-of-the-wish-granting-spherical-alga/

Zou, Y., Z.-J. Qian, Y. Li, M.-M. Kim, S.-H. Lee and S.-K. Kim. 2008. Antioxidant effects of phlorotannins isolated from *Ishige okamurae* in free radical mediated oxidative systems. Journal of Agricultural and Food Chemistry 56(16): 7001–7009.

Zuccarello, G.C., W.F. Prud'homme van Reine and H. Stegenga. 2004. Recognition of *Spyridia griffithsiana* comb. nov. (Ceramiales, Rhodophyta): a taxon previously misidentified as *Spyridia filamentosa* from Europe. Botanica Marina 47: 481–489.

Glossary

Abaxially—in an abaxial manner

Aculeate—sharp pointed tip

Acuminate—provided with sharp points

Acute—sharp at the end; ending in a point

A.D.—after death (see also B.C.)

Aerocyst—gas-filled bladder

Agar—phycocolloid extracted from various genera and species of red seaweed (Gigartinales and Gelidiales) which consists of a heterogeneous mixture of two polysaccharides, agarose and agaropectin

Agarose—one of two separable components of agar found in some red algae; has a remarkable gelling property

Air vesicle/air-bladder/a small bladder containing a variety of gases (nitrogen, oxygen and carbon dioxide) and aiding in the buoyancy of certain seaweeds (e.g., brown algae)

Akinetes—special cells of certain Cyanobacteria

Alginate—phycocolloid extract from brown algae from the class of Phaeophyceae; are linear macromolecules made up of two types of monomers, the manuronic acid (M) and guluronic acid (G)

Alternate branching—the branches, leaves, etc. are placed singly at different heights on the axis on opposite sides, or at definite angular distances from one another

Amorphous—having no definite shape or form

Anastomose—joined together irregularly to form a network

Anastomosing—to be connected (e.g., the lateral connecting of cells or filaments to form a network)

Annulate—marked with rings; surrounded by rings or bands

Anthelmintic—acting to expel or destroy parasitic intestinal worms

Antheridial—relative to Antheridium

Antheridium (pl., Antheridia)—male sex organ containing male motile gametes

Anticlinial—perpendicular to the circumference

Antioxidant—a molecule capable of inhibiting the oxidation of other molecules; may be of natural origin and also laboratory-synthesized

Apex—the tip or summit

Aphthae (ulcers)—aphthous ulcers are ulcers that form on the mucous membranes; they are also called aphthosis, aphthous stomatitis and canker sores

Apical—located at the tip or highest point

Apiculate—terminating in a short point

Applanate—having a horizontally flattened form

Aplanosporangia—sporangia producing non-motile spores

Aplanospore—a non-motile spore

Apressed—pressed closely against, but not joined to, a surface

Arcuate—having the form of a bow; curved

Arcuated—curved like a bow

Apophysis—region between the stipe and the expanding leafy area

Assimilatory filaments—pigmented or photosynthetic filaments

Attenuate—tapering gradually to a narrow extremity

Asexual reproduction—the reproduction without the involvement of originated fertilization, usually by diploid spores, propagation by division, stolons, etc.

Axial cell—the central cell of an axis, sometimes being distinguishable among medullary cells in transverse section

Axil—the angle between the upper side of a branchlet (or stem or leaf) and the supporting stem or branch

Axilla (pl., Axillae)—an axil

Axis (pl., axes)—a stem-like stalk on which parts or structures are arranged, a line or point forming the center of an object

Basal disk—the structure flat, disk-shaped, which ensures the fixation of algae to the substrate

B.C.—before Christ (see also A.D.)

Bifurcate—divided or forked into two branches

Bistratose—in two parallel layers, one placed on top of the other

Blade—a broad, thin flat part of the thallus

Branchlet—a small secondary or higher-order branch, usually the ultimate branch in a system of branching (ramulus)

Bullate—puckered or blistered in appearance

Bulbous—bulging, enlarged

Caespitose—forming dense tufts or clumps

Calcified—made calcareous or hardened by the deposition of calcium salts

Carpogonium—the one-celled female sex organ of some red algae that gives rise to the carpospores when fertilized

Carposporangium (pl., carposporangia)—in red algae, the cell in a carposporophyte that produces carpospores

Carpospore—a non-motile, usually diploid spore of red algae produced by the carposporophyte that in most red algae develops into a free-living tetrasporophyte

Carposporophyte—formed by the union of haploid gametes in red algae; diploid ($2n$) stage of the life cycle that is carried by the haploid (n) female gametophyte in the cystocarp; produces carpospores

Carrageenan—carrageenans are a special type of polysaccharide extracted from red algae (Rhodophyta, Gigartinales), consisting of galactose units and variable substitution of sulfate groups

Cartilaginous—seaweed with cartilage-like consistency; firm

Cenocyte—a multinucleate thallus with contiguous cytoplasm without division by septa

Cervicorn—resembling a deer's horn

Clavate—club-shaped

Claviform—club-shaped; clavate

Coenocytic (Cenocytic)—thallus formed of a cenocyte

Complanete—structures arranged in one plane, uniform flattened

Compressed—dorsi-ventrally flattened

Connivent—converging and touching but not fused

Constriction—strangulation

Cordate—heart-shaped; generally refers to a leaf base

Coriaceous—leathery in texture

Cortex—the outer tissue of a thallus (with or without an enclosing epidermis) that contains photosynthetic pigments

Corticate—having a cortex or a similar specialized outer layer

Corticated—having cortex

Corymbosely—a usually flat-topped flower cluster in which the individual flower stalks grow upward from various points of the main stem to approximately the same height

Costate—having veins or ridges

Cruciate—shaped like, or resembling, a cross

Crustose—having a crustlike appearance

Cryptostomata—sterile cavities in the surface of the thallus, containing hairs, often emerging brush-like at the surface (Fucales, Phaeophyceae)

Cuneate—a leaf shape

Cuneated—wedge-shaped; tapering, as some leaves

Cymose—relating to or resembling a cyme; determinate

Cystocarp—a reproductive body in red algae (Rhodophyta), developed after fertilization and consisting of filaments bearing carpospores

Cystostatic—inhibiting or suppressing cellular growth and multiplication

Dalo—a herb of the Pacific Is (*Colocasia esculenta*) grown throughout the tropics for its edible root and in temperate areas as an ornamental for its large glossy leaves

Decumbent—reclining on the substrate

Dichotomous branching—branches that regularly divides in two

Digitiform—shaped like a finger

Dioecious—having the male and female reproductive organs in separate plants

Disciform—flat and rounded in shape; discoid

Discoidal—having a flat, circular form; disk-shaped

Distichous—arranged in two rows along opposite sides of the axis

Distromatic—formed by two layers of cells

Divaricate—widely spreading branching

Dorsiventral—extending from a dorsal to a ventral surface

Ecorticate—being without a cortex

Emarginate—having a notched tip or edge

Endemic—a native organism restricted to a particular area or geographical location

Endophytic—an alga or other organism, which lives within a plant

Epifauna—aquatic animals, such as starfish, flounder, or barnacles, that live on the surface of a sea or lake bottom or on the surface of a submerged substrate, such as rocks or aquatic plants and animals, but that do not burrow into or beneath the surface

Epiflora—benthic flora fixed on the substrate surface

Epilithic—growing on the surface of rocks

Epiphyte—growing on another plant but not parasitic

Erect—forming a frond; not prostrate

Essential amino acids—amino acids that cannot be synthesized by human or animal organisms

Eulittoral—a subdivision of the benthic division of the littoral zone of the marine environment, extending from high-tide level to about 60 m, the lower limit for abundant growth of attached plants

Eutrophication—rich in mineral and organic nutrients that promote a proliferation of algae and aquatic plants, resulting in a reduction of dissolved oxygen, in a lake or pond

Fastigiate—having erect, clustered, almost parallel branches

Filament—a plant or branch composed of a linear group of cells joined at their walls, also a chain of cells forming a hair-like strand

Filiform—filamentous

Fistulose—formed like a fistula; hollow; reedlike

Flagelliform—long, thin, and tapering; whip-shaped

Foliaceous—having the appearance of the leaf of a plant

Foveolate—having small pits or depressions, as the receptacle in some composite flowers

Frond—upright part of a macroalgae

Fucoidan—sulfated polysaccharide found mainly in various species of brown algae

Furcate—to divide into two parts; fork

Fusiform—elongated and tapering at both ends; spindle-shaped

Galeate—shaped like a hood or helmet

Gametagium (pl., Gametangia)—a structure of plants (and algae) that forms gametes

Gametes—the sexual male and female cells, each gamete is haploid (n), has only single set of chromosomes

Gametophyte—which produces gametes; the haploid phase of algae that undergo alternation of generations, with each of its cells containing only a single set of chromosomes

Ganglioid—having or referring to a ganglion-like structure

Germling (pl., Germlings)—the organism produced by germination of an algal or fungal spore, but especially an immature gametophyte produced by the germination of a tetraspore

Germplasm—the genetic material of a species or other related group of organisms, collected for use in study, conservation, and breeding

Glabrous—without hair or a similar growth; smooth

Habitat—the local environment or the place in which a plant or animal lives; for marine environments it is defined according to geographical location, physiographic features and the physical and chemical environment

Hair—a colorless, typically elongate, unicellular or multicellular structure; also a unicellular or multicellular filament growing from the surface of a thallus, often deciduous

Haplostichous—present pseudo-parenchymatous constructions

Haptera—usually cylindrical, multi-branched anchoring structures (usually in Laminariales – Phaeophyceae), more massive than rhizoids

Hapterous—made or provided with haptera

Hemagglutinating—to cause agglutination of red blood cells

Herbivory—an animal that feeds on algae and other plants

Heteromorphous—Heteromorphic

Heteromorphic—having different forms at different periods of the life cycle

Holdfast—basal attachment organ of seaweeds

Inrolled—incurved or rolled inwards

Intercellularly—between cells

Intergenicula—small calcified units of articulated coralline fronds

Internodal—a section or part between two nodes, as of a nerve or stem

Involute—having the margins rolled inward

Isodiametric—having equal diameters or axes

Karyologically—by means of, or with reference to, karyology

Lanceolate—narrow and tapering toward the apex

Laciniate—slashed into narrow pointed lobes

Lectins—carbohydrate-binding proteins (i.e., glycoproteins) found in seaweeds and a wide variety of other life-forms; they are of potential economic importance as topical drug delivery systems

Life cycle, or life history—a period involving all different generations of a species succeeding each other through means of reproduction, whether through asexual reproduction or sexual reproduction

Ligulate—having the shape of a strap

Linear—long and narrow, with parallel margins; also slightly broader than filiform

Littoral—the area of the shore (intertidal zone) that is occupied by marine organisms which are adapted to or need alternating exposure to air and wetting by submersion, splash or spray; it is divided into subzones

Macroalgae—multicellular algae, in general marine and sometimes with considerable size (reaching 50 m in length), who's thalli exhibit a high degree of morphological complexity

Mammiform—resembling a breast, breast-shaped

Mariculture—cultivation of marine organisms in their natural habitats, usually for commercial purposes

Medullary—the central portion of a thallus in certain red or brown algae

Meiosporagium (pl., Meiosporagia)—a sporangium in which meiosis occurs

Membranous—the consistency of a membrane; laminar thalli, thin, sometimes transparent or semitransparent

Membranaceous—thin and rather soft or pliable, as the leaves of the rose, peach tree, and aspen poplar

Meristem—region of cells capable of division and growth in plants (and algae)

Meristoderm—the outer layer of the stipe, which resembles meristem in that its cells divide continually to replace tissue damaged by abrasion against rocks

Moniliform—resembling a string of beads, as the roots of certain plants

Monoecious—having unisexual reproductive organs

Monopodial—a main axis of a plant, that maintains a single line of growth, giving off lateral branches

Monospore—a simple or undivided nonmotile asexual spore; produced by the diploid generation of some algae

Monostromatic—single cell layered

Mucilaginous—containing mucilage; which is consistent

Mucoid—resembling mucus

Mucosy—suggesting mucus; slimy

Mucronate—terminating in a sharp point

Multiaxial—having more than one axis; developing in more than a single line or plain; - opposed to monoaxial

Nemathecium—a peculiar kind of fructification on certain red algæ, consisting of an external mass of filaments at length separating into tetraspores

Nociceptive—causing or reacting to pain

Node—the site on an axis from which blades and/or branches arise, with rings being formed at the junction of each successive axial cell

Nodose (= nodous)—full of nodes; knotty

Noduled—having little knots or lumps

Oblanceolate—lance-shaped, with the thin end at the base

Obovate—inversely ovate

Obovoid—egg-shaped and solid, with the narrow end at the base

Obpyramidal—with the tip of the pyramid turned down

Obtuse—having a blunt or rounded tip

Ocellate—having an ocellus or ocelli

Ocellus—an enlarged discoloured cell in a leaf

Olivaceous—of a deep shade of green

Oogamous—Characterized by or having small motile male gametes and large nonmotile female gametes

Oogonial—concerning oogonium

Oogonium (pl., oogonia)—a single celled female gametangium; also a swollen cell containing one or more ova or eggs

Opposite—referring to a leaf or/and branch arrangement where two leaves or branches arise at the same position at a single node but on reverse sides of the stem; also referring to locations in the same transverse line but removed by 180° on an axis

Orbiculate—circular or nearly circular

Osteoblastogenesis—the production of osteoblasts

Osteoclastogenesis—the development of osteoclasts

Ostiole—any small pore

Ovate—twice or less as long as broad

Palmately—having three or more veins, leaflets, or lobes radiating from one point; digitate

Paniculate—arranged in panicles

Papyraceous—of, relating to, made of, or resembling paper

Paraphyses—erect sterile filaments often occurring among the reproductive organs of certain algae

Parenchymatous—tissue-like structure of uniform, thin-walled, tightly knit cells forming a tissue resembling the parenchyma of vascular plants

Paronychia—is a skin infection that occurs around the nails

Pedately—deeply divided into several lobes with the lateral lobes divided into smaller lobes

Pedicillate—having a pedicel

Pedunculate—having a peduncle

Peltate—having the stalk attached to the center of the lower surface

Percurrent—extending through the entire length of a structure (e.g., from base to apex of a thallus)

Pericarp—a layer of tissue around the reproductive bodies of some algae

Phaeophycean—related to Phaeophyceae class (brown algae)

Phaeoplast—plastid of brown seaweed (Phaeophyceae)

Photic—penetrated by or receiving light

Percurrent—extending throughout the entire length

Perennial—a plant living for three or more years

Pericentral—surrounding the center

Periclinal—used to describe a type of cell division in a layer of cells that occurs parallel to an adjacent layer of cells

Phycocolloid—colloid (gel) extracted from algae; the phycocolloids are molecules of great size, consisting of simple sugars, which form part of the cell walls and intercellular spaces of a large number of algae, mainly brown and red

Phylloid—ribbon-shaped blade

Phytohormones—any of various hormones produced by plants that control or regulate germination, growth, metabolism, or other physiological activities

Pinna (pl., Pinnae)—any leaflet of a pinnate compound leaf

Pinnate—having branchlets set closely together on opposite sides of the main axis, arranged like the plumes of a feather

Pinnated—featherlike; having leaflets on each side of a common axis

Pinnately—resembling a feather in having parts or branches arranged on each side of a common axis

Pinnulate—having each pinna subdivided; - said of a leaf, or of its pinnæ

Pinnule—any of the lobes of a leaflet of a pinnate compound leaf, pinnately divided

Plurilocular—having several cells or loculi; many-celled sporangia, each cell containing a single spore, as in many algae

Polysiphonous—resembling, belonging to, or characteristic of the genus *Polysiphonia* (Rhodophyta)

Polystichous—arranged in two or more series or rows

Pneumatocyst—gas-filled bladder

Proliferous—producing many side branches or offshoots and normally reproducing vegetatively by buds

Prostrate—lying flat on the ground

Pseudodichotomous—forming two unequal branches, or having two equal branches at branch points but with one being derived from a lateral branch

Pulvinate—shaped like a cushion

Pyrenoid—a proteinaceous structure found within the chloroplast of certain algae

Pyriform—pear-shaped

Racemose—resembling or borne in a raceme

Racemosely—resembling or borne in a raceme

Radially—having or characterized by parts so arranged or so radiating

Ramuli (sing., Ramulus)—determinate branchlets

Receptacle—structure with conceptacles, present in some brown algae

Recurved—curved backward or inward

Reniform—bean- or kidney-shaped

Retuse—having a rounded apex and a central depression

Rhizoid—cell or filament responsible for fixing the thalli to the substrate

Rhizoidal—filamentous

Rhodoplast—Plastid of red algae (Rhodophyta)

Rostrate—having a beak

Ruffly—having many ruffles

Ruga—a fold, wrinkle, or crease

Rugose—having a rough, ridged, or wrinkled surface

Saccate—in the form of a sac; pouched

Sagittal—the anteroposterior plane or the section parallel to the median plane of the body

Scalpelliform—having the shape of a scalpel blade

Serrate—having flat small teeth that are projected forward

Seaweed—a macroscopic marine alga (non-vascular plant)

Septate—divided by a septum or septa

Sorus (pl., Sori)—is a cluster of sporangia (structures producing and containing spores)

Spatulate—having a narrow base and a broad rounded apex

Spermatangia—an organ or a cell in which gametes are produced

Spinose—bearing many spines

Spinous—covered with or having spines; thorny

Spinule—a small spine or thorn

Sporocyst—reproductive structure (single cell) of a sporophyte

Sporophyll—a modified leaf-like structure (especially in Laminariales, Phaeophyceae), bearing the spore-producing sporocysts

Sporophyte—spore-producing phase (usually diploid) in the life cycle of a plant (or algae) with alternation of generations

Stellate—resembling a star in shape; radiating from the center

Stichidium (pl., Stichidia)—specialized side branches of a red algal thallus, enclosing reproductive cells, such as tetraspores and spermatia

Stipe—portion located in the base of the algae, situated between the rhizoids and the blade

Stipitate—supported on or having a stipe

Stolons—axes lying or crawling on the substrate and giving rise to upright branches at intervals

Stupose—Composed of, or having, tufted or matted filaments like tow

Subquadrate—nearly or approximately square; almost square

Sublittoral—below the lowest low tide mark; = subtidal

Sub-rotate—having the parts flat and spreading or radiating like the spokes of a wheel

Substratum—surface to which something is attached

Subtidal—below the lowest low tide mark; = sublittoral

Subulate—tapering to a point; awl-shaped

Sympodium—an apparent axis that develops when growth occurs by means of lateral branches rather than continuing along the principal stem, often having a zigzag or irregular form

Tempeh—is made from cooked and slightly fermented soybeans and formed into a patty, similar to a very firm veggie burger

Terete—cylindrical in cross section

Tetrasporangia—a sporangium containing four asexual spores

Tetraspore—one of four spores produced from a tetrasporangium

Thalli (thallus, sing.)—the plant body of seaweeds

Tomentose—covered with short, dense, matted hairs

Tophule—reserve structures, typically ovoid, located in base of the branches of some species of the genus *Cystoseira*

Trichothallic—having a filamentous thallus or one ending in hairs or hairlike branches; used especially of an alga that grows by the action of an intercalary meristem

Trichotomous—having three angles or corners

Trichotomously—division into three parts or elements

Tristichously—having leaves growing in three ranks

Umbilicate—having a central mark or depression resembling a navel

Ultrastructure—The detailed structure of a biological specimen, such as a cell, tissue, or organ, which can be observed only by electron microscopy; also called fine structure

Uncalcified—not calcified

Unilocular—having or consisting of a single chamber or cavity

Unfused—not fused

Uronic (acid)—any of a group of organic acids, such as glucuronic acid, derived from oxidation of aldose sugars and occurring in urine

Verticil—a circular arrangement, as of flowers, leaves, or hairs, growing about a central point; a whorl

Verticillate—arranged in whorls

Vesiculate—to make (an organ or part) vesicular or (of an organ or part) to become vesicular

Vinegared—dressed with vinegar

Zonate—marked with, divided into, or arranged in zones

Abbreviations and Acronyms

μ	:	Micron (sing.)/Micra (pl.)
BAS(s)	:	Biological Active Substance(s); substance(s) having pronounced physiological activity and producing either stimulatory or inhibitory impact on *in vivo* or *in vitro* biological processes
CFS	:	Chronic fatigue syndrome
DHA	:	Decosahexaenoic acid
diam	:	Diameter
FAO	:	Food and Agriculture Organization of the United Nations
g	:	Gram
HIV	:	*Human immunodeficiency* virus
HSV	:	*Herpes simplex* virus
Is	:	Island/Islands
JIRCAS	:	Japan International Research Center for Agricultural Sciences
Kg	:	Kilogram
m	:	Meter
min	:	Minute(s)
N	:	North/Northern
NE	:	Northeast/Northeastern
NW	:	Northwest/Northwestern
pl	:	Plural
PUFA(s)	:	Polyunsaturated fatty acids; remarkable for highest bioactivity, they are the precursors of biosynthesis of some hormone-like substances (prostaglandins)
S	:	South/Southern
SE	:	Southeast/Southeastern
SW	:	Southwest/Southwestern
Tb	:	Tablespoon
Tsp	:	Teaspoon
US	:	United States
USD	:	United States dollar
PP (vitamin)	:	Niacin

Annex I

Recipes with *Alaria*

Vegetable and Atlantic Wakame Seaweed Salad

Ingredients

> 2 cucumbers, chopped
> 2 red peppers, chopped
> 1 red onion, chopped
> 2 tomatoes, chopped
> 3 cups of Atlantic Wakame seaweed (*Alaria esculenta*) (Fig. 5)

For the dressing you will need;
> 2 Tb brown rice vinegar
> 2 Tb Nama shoyu (unpasteurized soy sauce) or wheat-free tamari (gluten-free soy sauce)
> 2 Tb raw honey
> 2 Tb toasted sesame oil

Atlantic Wakame: soaked in hot water to hydrate, then drained well and sliced.

Chop the cucumbers, red peppers, red onion and tomatoes.

Dressing: mix together brown rice vinegar, honey, and sesame oil.

Presentation

Place all vegetables together in a large bowl. Pour the dressing over the vegetables and seaweed. Enjoy! (After Seaweed Iceland—All natural seaweed from pure Icelandic waters, http://www.seaweed.is/).

***Alaria* Pea Soup**

Ingredients

28 g *Alaria* (about ¾ cup tightly packed)
6 cups water
1½ cups split peas, rinsed
1 bay leaf

1 Tsp cumin
1 Tsp coriander
2 to 3 cloves of garlic
1 large onion, diced
2 to 3 cups sliced carrots
1 to 2 stalks celery, chopped
2 Tsp oregano
1 Tb soy sauce, or to taste

Preparation

Snip the *Alaria* into 2.5 cm pieces with scissors.
Add to a large soup pot with the water, split peas, bay leaf, cumin and coriander.
Simmer for 50 to 60 min until the split peas are soft and almost ready to dissolve.
Add the garlic, onion, carrots, and celery, and simmer until the vegetables are soft and brightly colored, about 20 min.
Add the oregano and season to taste with soy sauce.

Optional Preparation Method

Pressure-cook all the ingredients except the oregano and soy sauce for 20 min. Wait until the pressure comes down, and season to taste with oregano and soy sauce.

You can sauté the vegetables before adding them to the pot.

For a lighter soup, use only ¾ cup split peas.

Add 4 to 8 ounces cubed tempeh when you add the vegetables (After Main Cost Sea Vegetables—www.seaveg.com).

Recipes with Agar

How to Make Your Own Agar

To make agar, rinse the *Gracilaria* in fresh water five or six times until it has lost its pink color. Then take one part washed seaweed to 50 parts water and boil gently with one or two teaspoons of white vinegar until the liquid is thick and opaque. Strain it and let it set. When cool, cut into cubes. Some like to add sugar and or fruit juice to the liquid before it jells. Another option is after it jells you can cut it into small pieces and mix with soy sauce, vinegar, garlic juice, salted vegetables, sesame oil, coriander and other spices of your choice. It is served like a salad. A third method is to stir fry the seaweed for a minute and then simmer at least for 30 min or more with salt, soy sauce, sugar, wine, green onion and ginger. This is then cooled and once set, cut into pieces and served cold. Others way is to cook it is with pork, fish or vegetables, stewing it all together. That can be eaten hot or jelled. *Gracilaria* can also be fried in tempura batter (Deane 2014b).

Cold Mousse of Lemon-Lime with Agar

Ingredients

2 Tb of agar flakes
4 Tb of water
4 Eggs
100 g Fructose or sucrose
2 Tb of lemon zest
2 Tb of lime zest
4 Tb of lemon juice
4 Tb of lime juice
175 mL milk cream slightly beaten

Preparation

In a bowl, beat the egg yolks together with the fructose, lime and lemon zest. Cook this mixture in a water bath until they are thick, always stirring to avoid burning. Remove from water bath and continue stirring until cool.

In a separate bowl, cook the agar in water (4 Tb). Once dissolved and warm add the egg yolks. Leave the mixture for 20 min. When the mixture starts the gelation process, add the milk, cream and the egg whites, mix with a wooden spoon in a smooth motion until the mixture is homogeneous.

Put the mixture in a large dessert bowl and let cool in refrigerator for 4–6 hours; serve cool (Pereira 2010a).

Recipes with *Chondrus crispus* (Fig. 32)

Hana Tsunomata™ (Fig. 33) Couscous with Lemon and Wild Rice

1 cup boiling chicken stock
1 cup instant couscous
3 Tb olive oil
1/2 small onion, chopped
1/2 cup cooked wild rice
Juice of 1/2 lemon
1/4 cup dried Hana Tsunomata™, hydrated

Combine boiling chicken stock with couscous—let sit for 5 min and fluff with a fork. Heat olive oil in a small sauté pan and cook onion until translucent. Add wild rice, lemon juice to onion—heat through. Add cooked couscous mix and season with salt and pepper. Add Hana Tsunomata™ just before serving, mix well and enjoy (after www.acadianseaplants.com).

Pudding made with "Irish moss" (*Chondrus crispus*)

Ingredients

30 g of dry Irish moss (*Chondrus crispus*)—previously washed and soaked in water (2 hours)
500 mL milk
100 g cooking chocolate
Sugar to taste

Preparation

Place the seaweed in a saucepan with the milk and the chocolate chips; boil on low heat for 10 min.

Mix the sugar and then strain in a Chinese strainer, and pour into a shape and cool (Pereira and Correia 2015).

Sea Moss (*Chondrus crispus*) Drink

Preparing Sea Moss Gel

Soak the dried Sea Moss in cold water for 15 min. Wash algae thoroughly, but do not soak it overnight, as the algae will lose most of its nutritional value and potency. Next, fill a pot with two quarts of fresh water. Add the clean Sea Moss and two cinnamon sticks. Bring pot to boil. Once boiling, cover and simmer on low for 20–25 min. The algae will soften and the water will thicken. Remove from heat and discard cinnamon sticks. Allow the solution to cool and place it in the refrigerator overnight, allowing it to gel.

Preparing Sea Moss Drink

Combine all ingredients in a blender and mix on low speed until pieces of algae are finely chopped. That's it!

Serve chilled or over ice. Also, for an additional flavor boost, add a shot of rum (After Unicorn Caribbean—http://www.uncommoncaribbean.com).

Recipes with *Codium*

Fuzzy Finger Caviar on Potato Latkes

Ingredients

3 Russet potatoes, scrubbed
1/2 Yellow onion
1/2 Tb salt
1/4 Tb pepper
1/4 cup Canola oil
1 Tb Olive oil
2 cloves garlic, minced
1 shallot, minced

3 bunches *Codium fragile*, minced
Regular or tofu sour cream

Procedure

1. On a box grater, shred the potatoes. Rinse in fresh water and squeeze out any extra liquid with your hands.
2. Also on the box grater, shred the onions. Toss with the potatoes, salt and pepper.
3. In a skillet, heat canola oil until very hot. Form the potato mixture into balls and plop into the hot oil, spreading it out into a thin patty. The mixture will be loose at first but will seal together as it cooks.
4. Fry potato patties for about 2 min per side, until golden brown. Set aside on paper towels.
5. In a skillet, heat the olive oil. Add garlic and shallot and sauté until soft, about 5 min. Add the seaweed and cook until the seaweed turns bright green and most of the goo has simmered away. Chill.
6. To serve, warm up the latkes in the oven; spoon about a tablespoon of sour cream on top of each latke and about a teaspoon of the caviar on top of that. Enjoy! (After Diane De La Veaux—http://www.diannedelaveaux.com).

Recipes with *Durvillaea antarctica* (Cochayuyo)

Cochayuyo Pâté with Reduction of Balsamic Vinegar

Ingredients

40 g Cochayuyo, dry weight
60 g Walnuts
2 generous Tb olive oil
A squeeze of lemon juice
2 heaped Tb of brewer's yeast [if you have it]
1 clove garlic
2 Tb onion
2 Tb parsley
A pinch of cayenne
A Tb of balsamic vinegar reduction

Preparation

Soak the Cochayuyo over night. Drain and blend with the other ingredients. Add water if too dry. Top with additional balsamic vinegar reduction (After Eating Chilean—http://eatingchile.blogspot.pt).

Recipes with *Fucus*

Fish Chowder with Rockweed

Ingredients

2 large potatoes, diced
2 quarts water
2 cups of Rockweed, chopped
500 g Rockfish fillet, chunked
1 cup chicken broth
1 Tb celery salt
½ Tsp sage
½ Tsp thyme
1/8 Tsp cayenne
½ Tsp salt
2 Tb olive oil
1 Tb flour

Preparation

Place potatoes in soup pot with water. Put chopped rockweed in cheesecloth and suspend the bag in the soup pot. Boil until potatoes are almost cooked. Remove the spent Rockweed and discard. Add all other ingredients except flour. Lightly boil for 15 min to cook the Rockfish. Dissolve flour in a few Tsp. of cold water, and stir in into chowder to thicken. Simmer five more min, and then serve (After Garza 2005).

Recipes with *Gracilaria*

Gracilaria Spaghetti and Tomato Sauce

Ingredients

1 Tb of oil (olive)
250 g of ground beef
1 medium onion, chopped
1 clove minced garlic
1/8 cayenne
1 can tomato sauce (8 ounce)
1 cup carrot
2 cups *Gracilaria*
2 quarts boiling water
½ cup cheese, grated (optional)

Preparation

Brown meat in oil with onion and garlic; add tomato sauce, cayenne, carrot and herbs and simmer 15 min. Drop *Gracilaria* into the boiling water for just 15 seconds. Drain, serve with sauce and top with grated cheese (McConnaughey 2012).

Gracilaria Salad

Ingredients

2 cups of *Gracilaria*, raw or briefly steamed and cooled
250 g cottage cheese
4 radishes, grated
2 tomatoes, sliced
1/2 cup scallions, chopped

Preparation

Arrange *Gracilaria* on the bottom of a serving plate. Mix cottage cheese and grated radishes together and place in middle. Sprinkle with chopped scallions. Arrange tomato slices around the outside. Serve with yogurt dressing (Deane 2014b).

Red Ogo Seaweed Salad

One of the newer vendors at the Hollywood and Santa Monica farmers' markets is Carlsbad Aquafarm. Hailing from the Agua Hedionda Lagoon outside San Diego, the family-run business provides farm-raised, sustainably-harvested oysters, mussels, clams, and abalone. (The business was interviewed on Good Food earlier this year.) Being a vegetarian, I must admit I had been passing the stand by, until I discovered that they also carry fresh seaweed.

Carlsbad Aquafarm's Red ogo seaweed is crisp and lightly salty. It's a refreshing change from the dried and reconstituted seaweed I usually get at the grocery store. Their stand gladly provides samples, and I have friends who liked its crunch and were inspired to buy a carton but weren't sure what to do with it once they got home. Ogo, which is considered a "superfood" high in potassium, iron, and calcium, can be used in salads, soups, and Hawaiian pokes. Here's a quick, basic salad recipe. While I highly recommend trying red ogo if you have access to it, the recipe could be adapted for other types of seaweed such as Wakame (Han 2015).

Ingredients

1 cup Red ogo, torn or chopped if seaweed is in large pieces
2 small Japanese or Persian cucumbers, sliced (about 1 cup)
6 ounces firm tofu, cubed

Dressing

2 Tb Rice vinegar
1 Tb Sesame oil

1 1/2 Teaspoons honey or agave nectar
2.5 cm Piece of ginger, grated
Soy sauce, tamari, or liquid amino acids (optional)

Garnish

Chives, chopped
Black sesame seeds

Preparation

In a medium bowl, combine sesame oil, rice vinegar, honey, and ginger. Add seaweed, cucumbers, and tofu and toss to coat. As an option, add soy sauce to taste. The seaweed may be salty enough on its own. Plate and garnish with chives and sesame seeds. Serve immediately (Han 2015).

Recipes with *Himanthalia elongata* (Fig. 17)

Seaweed Rice, Raisins and Pine Nuts

Ingredients for Six People

30 g Sea spaghetti (*Himanthalia elongata*)
600 g Rice
Small onion, finely chopped
100 g raisins
1 handful of pine nuts
Saffron
Parsley and red pepper to decorate
Olive oil
Marine salt
4 Tb soy sauce

Preparation

Soak the seaweed in a bowl with warm water, while fry the chopped onion with oil in a frying pan. Then add the raisins until they swell. Later add the pine nuts without letting them burn. Add the saffron powder and mix everything well. In a separate pan cook the rice with seaweed, with plenty of water (no salt) for 30 min on low heat. At the end add the salt (as needed) and let it cook another 5 min. Drain the excess water from rice and mix with the sautéed, raisins and pine nuts. Finally add the soy sauce and serve warm (Pereira 210a).

Recipes with *Nereocystis* (Bull Whip Kelp)

Bull Whip Kelp, Avocado and Basil Spread

Ingredients

2 ripe avocados, halved, pitted, scooped out of skins, and cubed
Large handful of fresh basil leaves
2 Tb extra-virgin olive oil
2 Tb fresh lemon or lime juice
1 garlic clove, peeled and minced
1 Tb or more powdered bull whip kelp to taste
A pinch of cayenne to taste (optional)

Directions

1. Combine basil, garlic, and a pinch of powdered bull whip kelp and cayenne in a small food processor and blend until ingredients form a paste.
2. Add avocados and process until smooth. Blend in the oil and lemon or lime juice, and then season with the rest of the bull whip kelp.

Use this delicious bull whip, avocado and basil spread in place of mayonnaise—it's particularly tasty on fresh raw veggies and makes a wonderful dip for crackers (After Ocean Sea Vegetable Company—http://www.ohsv.net/).

Recipes with *Palmaria palmata* (Dulse)

Dulse Quiche

Ingredients

17 g Dulse
1 small packet of ready made short-crust pastry (about 250 g)
300 mL milk
3 or 4 eggs
50 g cheese (grated)
Salt and Pepper

Preparation

Roll out the pastry to fit a greased and lined 20 cm pie/flan dish and line the dish with the pastry
Leave at room temperature for 30 min (resting)
After 30 min trim excess pastry off pie/flan dish
Preheat oven to 200°C

Finely chop Dulse—preferably in a blender
Place into a sieve and put into a bowl of water for 10 min
Remove sieve from bowl and pat Dulse dry
Sprinkle Dulse and grated Dulse into pie dish
Put eggs into a bowl and beat
Add milk to eggs and beat
Pour mixture into pie dish
Season with salt and pepper
Bake for 20 min or until egg mixture is firm and check with a skewer—it should come out clean (After Irish Seaweeds—www.irishseaweeds.com).

Recipes with *Postelsia palmaeformis* (Sea Palm)

Sea Palm with Peanuts

Ingredients

1 cup dry Sea palm
2 cups cool water
1/4 cup raw Spanish peanuts, washed and drained
1 medium onion, thin crescents
1 Tb sesame oil
1 to 2 Tb soy sauce

Preparation

Soak sea palm in water until softened, 5 to 10 min. Drain and retain soaking water. Rinse sea palm in fresh water and drain. Cut into 2.5 cm pieces; heat oil in a heavy pan over medium heat. Add onions and sauté until transparent, 1 to 2 min. Add peanuts and sauté 1 min. Add sea palm and sauté 1 min. Add 1 cup of the retained soaking water and the soy sauce. Bring to a boil over medium heat with the cover ajar. Simmer 15 to 20 min until tender. Remove cover and cook away any remaining liquid (After Ocean Harvest Sea Vegetable Company—http://www.ohsv.net/).

Recipes with *Porphyra*

Toasted Nori Sheets for Award Winning Sushi!

6 cups water
3 cups short-grain brown or white rice

Any Combination of the Following Vegetables

Cucumber
Zucchini
Green bell pepper
Red bell pepper
Carrots

Daikon radish
Asparagus
Scallions
Mushrooms
Sesame seeds
Avocado fresh spinach

2/3 cup rice wine vinegar
6 Tb brown sugar
10–20 toasted Nori sheets
Prepared wasabi paste

Preparation

Bring the water to a boil. Add the rice, lower the heat, and simmer for 40 min, stirring occasionally; seed and julienne the cucumber, and julienne the vegetables. Steam these vegetables, except avocado which should be raw, over boiling water for 5 to 7 min. Let it cool to room temperature.

Mix together the vinegar and brown sugar, and stir until the sugar is dissolved. When the rice is cooked, stir in the vinegar and brown sugar mixture, and cool to room temperature.

When the vegetables and rice are cool enough to handle, lay out the first Nori sheet. Place a handful of rice in the center of the sheet, moisten your hands with water, and gently but firmly press the rice to the edges of the sheet so that there is a thin layer of rice in a line on the sheet. Spread a bit of wasabi paste on top of the rice, approximately 4 cm from one edge of the Nori sheet. Lay vegetable strips parallel to the wasabi in a width of approximately 2.5 cm along the wasabi line.

Carefully wrap the closest edge over the vegetables, and then roll the Nori delicately but tightly. Seal by moistening the edge of the Nori. Once the Nori sheet is completely rolled, slice the roll into six pieces and arrange on a platter. Repeat with the remaining Nori sheets.

Tip—If your Nori rolls won't stay rolled, try "sealing" the seam with a little brown rice syrup. To make rolling easier and prevent the Nori sheets from tearing, use an inexpensive bamboo sushi mat (After Florida Herb House—www.floridaherbhouse.com).

Shrimp Broth with Seaweed (Nori - *Porphyra*)

Ingredients

30 g Nori flakes
500 g shrimp
1.7 L water
Salt
2 chillies

3 Tb of fish sauce
½ lime
50 g margarine for cooking
100 g spaghetti noodles

Preparation

Soak the algae in warm water and then cook them for 10 min in boiling water. Cook the shrimp for 15 min in boiling water seasoned with salt and chillies. After cooling remove the heads and shells of shrimps. Add the lime juice and fish sauce, mix well and let stand for 5 min.

In an appropriate container, melt margarine with olive oil; add drained shrimp and sauté them for 5 min; add the shrimp broth and when boiling add the spaghetti noodles and the seaweed. Boil for 5 min and serve hot (Pereira 210a).

Seaweed Pie (Nori - *Porphyra*), Carrot and Coconut

Ingredients

2 Tb of seaweed flakes (Nori)
200 g of grated carrot
200 g coconut
5 eggs
75 g of maize flour
75 g plain flour
150 mL milk
1 Tsp baking soda
Sugar to taste

Preparation

In a pastry bowl beat up the egg whites; then add the remaining ingredients, including gems, until a homogeneous mass; at the end add in the beaten egg whites. Once prepared, the dough is poured in a form of 25 cm in diam. with a separable bottom and leave it in a preheated oven at 180°C; after cooking cool to room temperature before unmolding.

Laverbread Recipe

To make laverbread, you need wash your laver (the algae *Porphyra laciniata*—Fig. 61) and, without any additional water, simmer it until it becomes dark green gelatinous pulp—about 4 hours. Drain the leaves and chop them, adding salt to taste; and there you have it, laverbread, or bara lawr as the Welsh call it. Laverbread is traditionally fried in small balls or patties in bacon fat. It doesn't take long because the laverbread is already cooked.

Take a pound of prepared laverbread and mix in enough fine oatmeal to make soft, pliable dough. Roll into balls and flatten slightly. Fry in bacon fat for a few minutes each side or until nice and golden brown.

Serve with bacon in a mixed grill or a fried breakfast. I did something a little healthier and used the bacon I fried to flavor vegetable soup, and used the laverbread patties as dumplings (After Neil Cooks Grigson—http:// neilcooksgrigson.blogspot.pt).

Recipes with *Saccharina*

Royal Kombu (*Saccharina latissima*) with Horseradich Dip

Ingredients

4 stripes Royal Kombu
1/3 cup (5 Tb) sesame oil
3 carrots cut into 10 cm stripes
1/2 head broccoli, cut into 7 cm florets
3 cups unsalted corn chip

Dip Ingredients

1 cup (200 mL) whole milk yogurt
1 1/2 Tb prepared horseradish
1 Tb rice syrup
2 Tb lemon juice

Preparation

Using scissors cut the Royal Kombu stripes in half lengthwise and again crosswise.

Heat the oil and add the Royal Kombu stripes to the hot oil and fry for 3–4 min on each side until they change to a deep golden-brown color. Remove and drain on paper towels.

Boil 2 quarts of water in saucepan and blanch the carrots for 2 min, remove and set aside. Boil the broccoli for 3 min until bright green and set aside.

Dip: whisk the dip ingredients together in a small serving bowl.

Presentation

Serve as a snack, set the dip bowl in a center of a platter and alternate the Royal Kombu, corn chips, carrots and broccoli around it (After Seaweed Iceland—http://www.seaweed.is).

Kombu Cured Tuna

Ingredients

Kombu leaves (enough to wrap tuna)
500 g sushi-grade tuna steak cut into 5 mm thick pieces
Zest of 1 lime
2 Tb fresh lime juice

50 mL honey
1 Tb soy
6 Tb cooking sake Smoked salt to taste
Watercress leaves to garnish

Lime and Sake Yoghurt Dressing

3 Tp plain yoghurt
1 Tb sake
1/2 lime—zest and juice

Method

In a bowl, cover the Kombu leaves with warm sake and let stand until softened (5–10 min). Drain. Set the tuna on the Kombu and top with the other ingredients. Wrap the Kombu leaves around the tuna and moisten with 2 Tb of the sake juice. Cover with a plastic wrap and refrigerate for 24 hours.

To Make the Lime and Sake Yoghurt Dressing

Combine the yoghurt, 1 Tb sake and zest and juice of half a lime in a bowl. Mix well and serve as a dressing over the cured tuna pieces (After Sea Vegetable Recipes—www.seaweedrecipes.co.nz).

Recipes with *Ulva*

Sea Lettuce Flakes

Roast dried blades of *Ulva* in an oven at 180–200° for about 5–15 min or until the blades are just crisp. They will easily crush to a powder or small flakes in your hand or with a mortar. Store in a jar and use as a seasoning in soups or on salads. Alternatively delicate algae as *Ulva* can be dried in the microwave with a very low power (300 watts) for 3 min (Garza 2005, Pereira 2010c).

***Ulva intestinalis* "Tortilla"**

Ingredients

100 g of fresh *Ulva intestinalis*
Chives
2 eggs
Spicy chili
Garlic cloves

Preparation

Cutting the seaweed with scissors, add the chives, eggs, 2 Tb flour, chopped garlic and ground chili. Fry as a "tortilla" in olive oil (After http://algasquesecomem.weebly.com/tortas-de-erva-do-calhau.html).

Credits of Images

Images on Public Domain

Ernst Haeckel

Figs. 68, 69, 87.

Robert K. Greville

Figs. 5, 80.

William H. Harvey

Figs. 1, 2, 3, 5, 8, 9, 10, 11, 16, 22, 26, 28, 29, 30, 31, 32, 35, 36, 37, 39, 40, 43, 44, 45, 46, 48, 49, 50, 51, 52, 53, 57, 58, 60, 61, 62, 65, 66, 67, 70, 71, 72, 73, 74, 75, 78, 79, 81, 82, 83, 84, 85, 86, 88, 89, 90, 91, 92, 93, 94, 95, 96, 97.

Originals

Acadian Seaplants Limited

Fig. 33.

Leonel Pereira

Figs. 4, 6, 7, 12, 13, 14, 15, 17, 18, 19, 20, 21, 23, 24, 25, 27, 34, 38, 41, 42, 47, 54, 55, 56, 59, 63, 64, 76, 77, 98, 99.

Species Index